# Bacterial Membrane Vesicles

Maria Kaparakis-Liaskos • Thomas A. Kufer
Editors

# Bacterial Membrane Vesicles

## Biogenesis, Functions and Applications

Springer

*Editors*
Maria Kaparakis-Liaskos
Department of Physiology, Anatomy and Microbiology
La Trobe University
Melbourne, VIC, Australia

Research Centre for Extracellular Vesicles, La Trobe Institute for Molecular Science
La Trobe University
Melbourne, VIC, Australia

Thomas A. Kufer
Department of Immunology, Institute of Nutritional Medicine
University of Hohenheim
Stuttgart, Baden-Württemberg, Germany

ISBN 978-3-030-36330-7 ISBN 978-3-030-36331-4 (eBook)
https://doi.org/10.1007/978-3-030-36331-4



This Springer imprint is published by the registered company Springer Nature Switzerland AG.
The registered company address is: Gewerbestrasse 11, 6330 Cham, Switzerland

# Preface

Since the discovery of bacteria as causal agents of infectious disease, scientists have been fascinated by the multitude of interactions in which these small organisms are engaged with their environment. In particular, the communication of bacteria with eukaryotic host cells rapidly became a major focus during the past few decades, coining the term cellular microbiology. In the field of bacterial pathogenesis, this resulted in the generation of an era that mainly focused on understanding the mechanisms whereby bacteria manipulate the host cellular machinery by effector proteins, known as virulence factors.

Although it was well recognized that conserved structures of bacteria and their cell membranes induce strong immune responses in the host, it was only until very recently that a long-known property of bacteria, their capability to produce membrane vesicles, gained increasing interest.

The potential of bacteria to produce membrane vesicles was initially thought to be an artifact of culturing and isolation, and their ability to contribute to biological and pathological functions was often overlooked or dismissed. Now a wealth of data confirms that Gram-negative bacteria, in addition to Gram-positive species that lack an outer membrane, can form outer-membrane vesicles (OMVs) and membrane vesicles (MVs), respectively. Moreover, recent findings support the notion that the production of OMVs and MVs is controlled, and not a random process as was once suggested, further emphasizing that bacteria have the potential to modulate OMV biogenesis.

Due to the plethora of research conducted within the past decade as a result of reinvigorated interest in bacterial membrane vesicles, OMVs and MVs have now emerged as a *bona fide* delivery mechanism of bacteria. They can be used by bacteria for the targeted transport of proteins and nucleic acids, showing that membrane vesicles are versatile tools that can transport a range of bacterial cargo over great distances. In addition, the incorporation of membrane proteins and lipids derived from the parent bacterial membrane into OMVs can be used for receptor-targeted delivery, activation of membrane receptors, and inducing fusion with target recipient cell membranes. Moreover, due to their nanostructure, OMVs and MVs can be used by bacteria to communicate with each other and regulate microbial functions, or conversely, they can be used to attack and kill competing bacteria. The potential applications of OMVs and MVs for bacteria appear to be endless, and it is evident that we are just beginning to understand the vast complexity of their functions.

Currently, we witness an explosion in the field of bacterial membrane vesicle research and the broad acceptance that these once suggested bacterial artifacts have defined biological functions. This has resulted in the identification of exciting novel properties of MVs, including their function in inter- and intra-bacterial communication, in induction and regulation of host immune defense, and as decoys for viral and antibiotic attack. In addition, OMVs may also contribute to bacterial evolution via their potential to facilitate horizontal gene transfer, a field which is currently in its infancy. We also have a greater understanding that bacterial vesicles are produced by pathogenic and commensal bacteria and that they can be found within a diverse range of environments, including the ocean where they can serve as an energy source to other bacteria. Furthermore, our ability to increase their production and modify their composition now sets the stage for testing their potential in targeted interventions such as their use as drug delivery systems or as vaccine candidates.

In this book, leading experts in the field summarize current knowledge regarding the biology, biogenesis, and functions of bacterial OMVs and MVs. Furthermore, interactions of OMV and MV from pathogenic bacterial species with host cells and the immune system in addition to pathogenesis in plants are discussed in detail, as is their presence and functions in the environment and their emerging potential as novel vaccine candidates.

While providing molecular details for the informed reader, this book is also well suited for the non-expert reader who wants to gain insights and a detailed overview of the fascinating world of bacterial membrane vesicles. We hope that this book will provide a useful resource for information about bacterial membrane vesicles and increase the interest and awareness of this intriguing and novel topic in microbiology, cellular microbiology, and immunology.

Finally, we would like to express our gratitude to all of our colleagues and friends who contributed to the generation of this book. We thank each of them for sharing their knowledge in their respective areas of expertise, their time, and dedication to producing this book with us.

Melbourne, Australia — Maria Kaparakis-Liaskos
Stuttgart, Germany — Thomas A. Kufer

# Contents

# Editors and Contributors

## About the Editors

**Maria Kaparakis-Liaskos** Associate Professor Maria Kaparakis-Liaskos is Head of the Host-Pathogen Interactions and Bacterial Membrane Vesicles Laboratory at La Trobe University, Australia. She is also the Deputy Director of the La Trobe Research Centre for Extracellular Vesicles. Her recent research activities focus on understanding the cellular and molecular mechanisms of host–pathogen interactions with particular focus on bacterial membrane vesicles.

Associate Professor Kaparakis-Liaskos obtained her PhD from the Department of Microbiology and Immunology at the University of Melbourne in Australia in 2005. She then undertook postdoctoral studies at Monash University, working on innate immune responses to *Helicobacter pylori* and bacterial outer-membrane vesicles. She then headed a research group at the Hudson Institute of Medical Research, Australia, during which time she identified the mechanisms whereby OMVs were detected by NOD1 and were degraded by autophagy. In 2016, she joined La Trobe University in Melbourne, Australia, where her research is supported by the Australian Research Council (ARC), the National Health and Medical Research Council (NHMRC), and the Victorian Endowment for Science Knowledge and Innovation (veski).

**Thomas A. Kufer** Professor Thomas Kufer is head of the Department of Immunology at the University of Hohenheim, Stuttgart, Germany. His primary research interests are focused on understanding the functions of NLR proteins in host–pathogen interactions, immunity, and inflammation.

Professor Kufer obtained his PhD (Dr. rer. nat.) in Cell Biology from the Ludwig-Maximilians University of Munich in 2004. He then undertook postdoctoral studies in the laboratories of Dana Philpott and Philippe Sansonetti in the Pasteur Institute in Paris working on the intracellular pattern recognition receptors NOD1 and NOD2 in the context of *Shigella flexneri* infection. Subsequently, he headed a junior research group funded by the German Research Foundation (DFG) at the Institute of Medical

Microbiology, Immunology and Hygiene at the University of Cologne in Germany working on innate immune responses mediated by NLR proteins. In 2014, he was recruited at the University of Hohenheim as professor for immunology.

## Contributors

**Josefa Badia** Secció de Bioquímica i Biología Molecular, Departament de Bioquímica i Fisiologia, Facultat de Farmàcia i Ciències de l'Alimentació, Universitat de Barcelona, Barcelona, Spain
Spaina nd Institut de Biomedicina de la Universitat de Barcelona (IBUB), Institut de Recerca Sant Joan de Déu (IRSJD), Barcelona, Spain

**Ofir Bahar** Department of Plant Pathology and Weed Research, Agricultural Research Organization, Volcani Center, Rishon LeZion, Israel

**Laura Baldomà** Secció de Bioquímica i Biología Molecular, Departament de Bioquímica i Fisiologia, Facultat de Farmàcia i Ciències de l'Alimentació, Universitat de Barcelona, Barcelona, Spain
Spaina nd Institut de Biomedicina de la Universitat de Barcelona (IBUB), Institut de Recerca Sant Joan de Déu (IRSJD), Barcelona, Spain

**Steven J. Biller** Department of Biological Sciences, Wellesley College, Wellesley, MA, USA

**Natalie J. Bitto** Department of Physiology, Anatomy and Microbiology, La Trobe University, Melbourne, VIC, Australia
Research Centre for Extracellular Vesicles, La Trobe Institute for Molecular Science, La Trobe University, Melbourne, VIC, Australia

**Julie C. Caruana** American Society for Engineering Education Postdoctoral Research Fellow, Washington, DC, USA

**Arturo Casadevall** Department of Molecular Microbiology and Immunology, Johns Hopkins Bloomberg School of Public Health, Baltimore, MD, USA

**Carolina Coelho** Department of Molecular Microbiology and Immunology, Johns Hopkins Bloomberg School of Public Health, Baltimore, MD, USA
Medical Research Council Centre for Medical Mycology, Institute of Medical Sciences, University of Aberdeen, Aberdeen, UK
Department of Biosciences, College of Life and Environmental Sciences, University of Exeter, Exeter, UK

**Leo Eberl** Department of Plant and Microbial Biology, University of Zürich, Zürich, Switzerland

**Ella L. Johnston** Department of Physiology, Anatomy and Microbiology, La Trobe University, Melbourne, VIC, Australia

Research Centre for Extracellular Vesicles, La Trobe Institute for Molecular Science, La Trobe University, Melbourne, VIC, Australia

**Maria Kaparakis-Liaskos** Department of Physiology, Anatomy and Microbiology, La Trobe University, Melbourne, VIC, Australia
Research Centre for Extracellular Vesicles, La Trobe Institute for Molecular Science, La Trobe University, Melbourne, VIC, Australia

**Thomas A. Kufer** Department of Immunology, Institute of Nutritional Medicine, University of Hohenheim, Stuttgart, Baden-Württemberg, Germany

**Deborah R. Leitner** Institute of Molecular Biosciences, University of Graz, Graz, Austria

**Jose L. Luque-García** Faculty of Chemistry, Department of Analytical Chemistry, Complutense University of Madrid, Madrid, Spain

**Maria Maryam** Department of Molecular Microbiology and Immunology, Johns Hopkins Bloomberg School of Public Health, Baltimore, MD, USA

**Aftab Nadeem** Department of Molecular Biology, Umeå University, Umeå, Sweden
The Umeå Centre for Microbial Research (UCMR), Umeå University, Umeå, Sweden

**Nobuhiko Nomura** Faculty of Life and Environmental Sciences, University of Tsukuba, Ibaraki, Japan

**Jan Oscarsson** Oral Microbiology, Department of Odontology, Umeå University, Umeå, Sweden

**Ainhoa Palacios** CIC bioGUNE, Technological Park Bizkaia, Derio, Spain

**Rafael Prados-Rosales** CIC bioGUNE, Technological Park Bizkaia, Derio, Spain
Department of Preventive Medicine and Public Health, and Microbiology, Autonoma University of Madrid, Madrid, Spain

**Stefan Schild** Institute of Molecular Biosciences, University of Graz, Graz, Austria
BioTechMed-Graz, Graz, Austria

**Yosuke Tashiro** Applied Chemistry and Biochemical Engineering Course, Department of Engineering, Graduate School of Integrated Science and Technology, Shizuoka University, Hamamatsu, Japan

**Masanori Toyofuku** Faculty of Life and Environmental Sciences, University of Tsukuba, Tsukuba, Ibaraki, Japan

**Sun Nyunt Wai** Department of Molecular Biology, Umeå University, Umeå, Sweden
The Umeå Centre for Microbial Research (UCMR), Umeå University, Umeå, Sweden

**Scott A. Walper** U.S. Naval Research Laboratory, Center for Bio/Molecular Science and Engineering, Washington, DC, USA

**Lauren Zavan** Department of Physiology, Anatomy and Microbiology, La Trobe University, Melbourne, VIC, Australia
Research Centre for Extracellular Vesicles, La Trobe Institute for Molecular Science, La Trobe University, Melbourne, VIC, Australia

**Franz G. Zingl** Institute of Molecular Biosciences, University of Graz, Graz, Austria

# Chapter 1
# Introduction, History, and Discovery of Bacterial Membrane Vesicles

**Lauren Zavan, Natalie J. Bitto, and Maria Kaparakis-Liaskos**

**Abstract** The production of extracellular vesicles is a conserved process that is common to all living cells. Both Gram-negative and Gram-positive bacteria produce extracellular vesicles, known as outer membrane vesicles (OMVs) and membrane vesicles (MVs), respectively. Once disregarded as artifacts of bacterial growth, research over the last 50 years has shown that OMVs contribute to numerous bacterial functions. It is now understood that OMVs are purposely secreted by Gram-negative bacteria to aid in bacterial communication and pathogenesis. The OMV field has focused on understanding the mechanisms of OMV biogenesis, the content of OMVs and how OMVs interact with the host immune system and their environment. While there is a wealth of knowledge regarding OMVs, it was only in the last decade that Gram-positive bacteria were found to release MVs. Due to the late discovery of MVs there is little known about MVs in comparison to our knowledge regarding OMVs. However, there is emerging evidence that MVs contain bacterial cargo and may aid in bacterial functions. Research in the field of bacterial vesicles has expanded rapidly within the past decade and continues to be a growing field of interest. Future work aims to manipulate bacterial membrane vesicles as novel therapeutics and nanoparticle technology.

## 1.1 Introduction to Gram-Negative OMVs

All forms of life, prokaryotic and eukaryotic, naturally release extracellular vesicles as part of their normal growth (Brown et al. 2015; Deatherage and Cookson 2012). Vesicles produced by Gram-negative bacteria are called outer membrane vesicles (OMVs) as they are derived from the outer membrane of the Gram-negative bacterial

L. Zavan · N. J. Bitto · M. Kaparakis-Liaskos (✉)
Department of Physiology, Anatomy and Microbiology, La Trobe University, Melbourne, VIC, Australia

Research Centre for Extracellular Vesicles, La Trobe Institute for Molecular Science, La Trobe University, Melbourne, VIC, Australia
e-mail: m.liaskos@latrobe.edu.au



M. Kaparakis-Liaskos, T. A. Kufer (eds.), *Bacterial Membrane Vesicles*,
https://doi.org/10.1007/978-3-030-36331-4_1

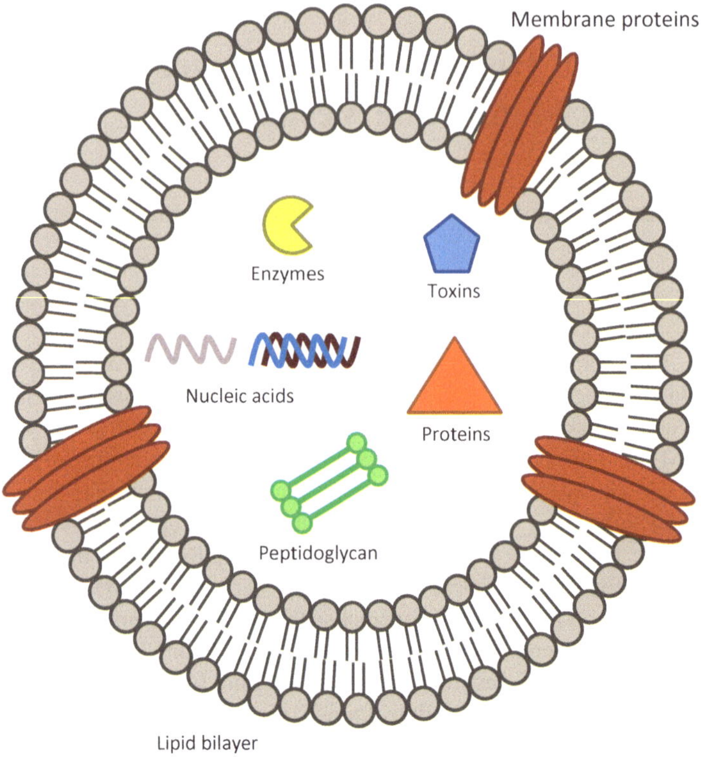


**Fig. 1.1** Schematic overview of a Gram-negative outer membrane vesicle. Outer membrane vesicles are composed of a lipid bilayer and contain membrane and cytoplasmic proteins, nucleic acids, enzymes, toxins, and peptidoglycans that are derived from their parent bacterium

cell (Hoekstra et al. 1976). First dismissed as bacterial artifacts, early studies visualized OMVs being released from the outer membrane of a range of Gram-negative pathogens by electron microscopy (Knox et al. 1966; Chatterjee and Das 1967). However, it was not until OMVs were identified in the spinal fluid of meningococcal patients (DeVoe and Gilchrist 1975), that interest developed in understanding OMV production, their functions in the host and how they benefit bacteria.

It is now accepted that OMVs are purposely secreted by Gram-negative bacteria to aid in an array of bacterial functions. OMVs range from approximately 20–400 nm in size and contain materials derived from their parent bacterium, including nucleic acids, proteins and enzymes (Fig. 1.1) (Kadurugamuwa and Beveridge 1995; Dorward et al. 1989; Dorward and Garon 1989; Haurat et al. 2011; reviewed in Schwechheimer and Kuehn 2015). It was originally thought that OMV cargo was derived from the bacterial outer membrane and periplasm only, as cytoplasmic components were thought to be unable to cross the inner membrane (Hoekstra et al. 1976; Gankema et al. 1980). However, it is now known that OMVs can also contain components derived from the bacterial cytoplasm, such as nucleic acids and proteins (Lee et al. 2007; Perez-Cruz et al. 2015; Bitto et al. 2017; Renelli et al. 2004; Sjostrom et al. 2015).

OMVs are involved in a range of bacterial functions. Research over the last two decades has highlighted the importance of OMVs in cell-to-cell communication (Mashburn and Whiteley 2005), the transfer of genetic material (Yaron et al. 2000; Dorward et al. 1989), biofilm formation (Yonezawa et al. 2009), inflammation and disease progression (Ismail et al. 2003; Kaparakis et al. 2010; reviewed in Bitto and Kaparakis-Liaskos 2017).

### 1.1.1 The First Observations of OMVs

OMVs produced by *Escherichia coli* were the first OMVs to be observed using electron microscopy, appearing as small, spherical "particles" that surrounded the bacterial cell (Knox et al. 1966). These particles were thought to be responsible for the secretion of lipopolysaccharide (LPS) and lipoproteins from bacteria (Knox et al. 1966). Subsequently, OMVs were isolated from the oral bacterium *Veillonella parvula* by phenol–water extraction (Mergenhagen et al. 1966). These isolated OMVs were heterogeneous in size and the outer leaflet was similar in morphology to the outer membrane of *V. parvula* cells (Mergenhagen et al. 1966). The release of OMVs from *Vibrio cholerae* was subsequently observed and it was noted that OMV production occurred only during the log phase of bacterial growth (Chatterjee and Das 1967). Researchers postulated that *V. cholerae* regulated the release of OMVs from bacterial cells during active growth as a mechanism to secrete bacterial toxins into the extracellular environment (Chatterjee and Das 1967). Despite these first observations, the wider field did not consider OMVs as important or necessary products released by bacteria but merely viewed OMVs as artifacts of bacterial growth.

As research progressed, the release of OMVs from the outer membrane of bacterial cells was proposed to be a continuous and essential process, and not the result of cell lysis (Rothfield and Pearlman-Kothencz 1969; Loeb 1974; Gankema et al. 1980). Importantly, for the first time it was discovered that *E. coli* and *Salmonella enterica* serovar Typhimurium (*S.* Typhimurium) OMVs contained bacterial proteins, lipids and LPS derived from their parent bacteria (Rothfield and Pearlman-Kothencz 1969). Additionally, OMV production was seen to increase when bacterial protein synthesis was inhibited, which was speculated to have been due to stress of the outer membrane (Rothfield and Pearlman-Kothencz 1969). OMVs were next observed to be released by *Neisseria meningitidis* and were thought to be associated with the release of *N. meningitidis* toxins into the environment (Devoe and Gilchrist 1973). Similar to *V. cholera*, *N. meningitidis* OMVs were not detected as bacteria progressed to stationary phase of growth (Devoe and Gilchrist 1973) supporting the theory that OMVs were only produced during the log phase of bacterial growth. In addition to being released during the exponential phase of bacterial growth, OMVs were also observed to be released in response to treatment with detergents (Leive et al. 1968) and exposure to stress from bacteriophages (Loeb 1974) suggesting they are produced in response to bacterial stress.

### *1.1.2 Advances in OMV Research*

Until 1975, OMV production had only been observed in vitro. OMVs were identified in primary cultures of spinal fluid taken from patients with meningococcal disease (DeVoe and Gilchrist 1975) indicating that the release of OMVs was a normal part of bacterial infection. This was the first study to identify OMVs released from pathogenic bacteria in a physiologically relevant setting.

A decade after their discovery, OMVs were given the name outer membrane vesicles as they closely resembled the outer membrane of their parent bacterium in composition, and were thought to be lacking in cytoplasmic material (Hoekstra et al. 1976). Subsequently, the release of OMVs from *E. coli* was shown to preferentially occur in locations of the outer membrane that contained newly synthesized proteins (Mug-Opstelten and Witholt 1978). One of the first hypotheses of OMV biogenesis suggested that the incorporation of new proteins into the outer membrane enabled a portion of the outer membrane to bulge from the cell and once large enough, to be released from the bacterial cell (Mug-Opstelten and Witholt 1978). Numerous subsequent studies further elucidated the composition of OMVs from Gram-negative bacterial species including *E. coli* (Gankema et al. 1980; Wensink and Witholt 1981), *Aeromonas* spp. (MacIntyre et al. 1980), *Brucella melitensis* (Gamazo and Moriyon 1987), and *Haemophilus influenzae* (Deich and Hoyer 1982).

As OMVs were known to contain LPS, they were suggested to be able to interact with host cells (MacIntyre et al. 1980). Functional studies of OMVs produced by *Porphyromonas gingivalis* suggested that OMVs may contribute to the progression of periodontal disease, as *P. gingivalis* OMVs contained bacterial toxins and enzymes and promoted bacterial adhesion (Grenier and Mayrand 1987). Importantly, it was found that immunization of mice with OMVs from *H. influenzae* type B resulted in an increase in the permeability of the blood–brain barrier, a similar response to that observed when mice were treated with *H. influenzae* LPS (Wispelwey et al. 1989). These studies highlighted the importance of OMVs in an infection setting and demonstrated the role of OMVs as vehicles for bacterial cargo. Furthermore, these studies were some of the first to describe how OMVs may contribute to bacterial pathogenesis.

Due to their pathogenic cargo, OMVs were also thought to be ideal vaccine candidates. The first OMV vaccine was trialed in 1991 against the pathogen *N. meningitidis*, which causes group B meningococcal disease (Bjune et al. 1991). Subsequent studies determined that three doses of the OMV-based vaccine increased the vaccine efficiency and therefore conferred protection, leading to the production of the MenB vaccine (Rosenqvist et al. 1995) that is now licensed for human use (Arnold et al. 2011; Vernikos and Medini 2014). Currently, there are ongoing efforts to develop new vaccines for other diseases caused by Gram-negative pathogens (Chen et al. 2010; Nieves et al. 2011).

### 1.1.3 Outer Membrane Vesicles Research in the Last Decade

OMVs from numerous pathogenic bacteria have been investigated for their ability to induce an immune response in host cells. OMVs isolated from *Helicobacter pylori* and *Pseudomonas aeruginosa* can induce an interleukin-8 (IL-8) response in host epithelial cells, resulting in inflammation (Ismail et al. 2003; Bauman and Kuehn 2006). Additionally, OMVs isolated from *Treponema denticola* can disrupt the epithelial cell layer and cross to the basolateral side of epithelial cells (Chi et al. 2003). Furthermore *H. pylori* OMVs can enter host epithelial cells via lipid rafts and interact with intracellular nucleotide-binding oligomerization domain-containing protein 1 (NOD1) causing an inflammatory response in host cells (Kaparakis et al. 2010; Allison et al. 2009). We now have a greater understanding of how OMVs can interact with and sometimes cross the host epithelial cell layer to elicit a pro-inflammatory immune response in the host. These studies highlight the important role of OMVs in contributing to the immunogenicity of bacteria.

While there was increasing interest in understanding the inflammatory nature of OMVs, there was still relatively little known about their production. Although early studies hypothesized the mechanisms of OMV biogenesis, the last decade of research has emphasized that OMV biogenesis is a complex and varied process that is still not well understood. It is now known that OMV release can be mediated by membrane proteins, LPS, O-polysaccharides, and phospholipids (Murphy et al. 2014; Roier et al. 2016; Elhenawy et al. 2016). Future studies should aim to identify other novel mechanisms of OMV biogenesis that may be either conserved to specific bacterial species or common to all Gram-negative bacteria.

Due to the nature of their biogenesis, there has been debate as to whether cargo is selectively packaged into OMVs. Selective packaging has since been identified in *P. gingivalis*, where OMVs were enriched in virulence proteins such as gingipains, while excluding numerous outer membrane proteins (Haurat et al. 2011). There is now greater interest in providing in-depth proteomic analyses of OMVs to further elucidate OMV content. For example, the proteomes of *E. coli*, *P. aeruginosa*, and *H. pylori* have been examined to provide insights into how bacterial growth stage, biofilms, and infection settings can determine the protein content of OMVs (Zavan et al. 2019; Ayalew et al. 2013; Pierson et al. 2011; Turner et al. 2018; Park et al. 2015) highlighting that there are a number of conditions that can determine OMV composition.

These works highlight only some areas of OMV research that has been the focus in recent years. Interest in OMVs has vastly increased over the last decade and continued research shows there is still much that remains unknown. Future efforts may focus on understanding what regulates OMV production and composition, how OMV content can be used by recipient bacteria, and further elucidating the role of

OMVs in inflammation and disease. Collectively, these studies will broaden our knowledge regarding OMVs and will facilitate their development as novel therapeutics.

### *1.1.4 Biogenesis of OMVs*

The Gram-negative cell membrane is composed of the outer membrane, inner membrane, and the periplasmic space, which contains a thick peptidoglycan layer (reviewed in Costerton et al. 1974). Embedded within the membranes and connecting them together are proteins which allow the bacterial cell to maintain its shape (Schnaitman 1970). Additionally, the bacterial outer membrane contains lipids, lipoproteins, and LPS that dictate membrane fluidity, curvature, and integrity (reviewed in Schwechheimer and Kuehn 2015). Disruption to these fundamental building blocks of the bacterial outer membrane can result in changes to OMV biogenesis. Here we summarize some of the mechanisms of OMV biogenesis, and a detailed discussion of this topic can be found in Chap. 2.

One mechanism of OMV biogenesis observed in numerous Gram-negative species is a process known as budding or blebbing. Blebbing of OMVs occurs when a portion of the outer membrane bulges at the cell surface and is liberated from the membrane to create a vesicle (reviewed in Schwechheimer and Kuehn 2015). Blebbing of OMVs from the outer membrane has been studied in numerous bacterial species and can be the result of a disruption at the cell membrane during protein modification or lipid remodeling (Bernadac et al. 1998; Elhenawy et al. 2016).

Protein modifications that occur in the outer membrane and surrounding regions are known to affect the production of OMVs in differing ways. For example, mutations in *tolB* or *tolC* of the Tol-Pal complex spanning the inner and outer membrane of *E. coli* cause a decrease in the release of OMVs, while in *H. pylori* Tol-Pal mutants cause an increase in OMV production (Turner et al. 2015; Bernadac et al. 1998). Additionally, the overexpression or misfolding of membrane proteins can cause vesiculation to increase up to 100-fold, which may be a response to the increasing pressure at the outer membrane (McBroom and Kuehn 2007; reviewed in Vasilyeva et al. 2009).

Furthermore, other key components of the bacterial outer membrane such as phospholipids and LPS have been implicated in OMV biogenesis. For example, the accumulation of phospholipids in the outer membrane of *H. influenzae* and *V. cholerae* can regulate OMV biogenesis (Roier et al. 2016). Additionally, the remodeling of Lipid A in phospholipids of *S.* Typhimurium was found to be required for the formation of OMVs (Elhenawy et al. 2016).

Moreover, research has shown that explosive cell lysis events in *P. aeruginosa*, caused by the production of prophage endolysins, can also result in OMV production (Turnbull et al. 2016). This mechanism has not been observed for other bacterial species; however, research continues to investigate the numerous mechanisms of OMV biogenesis.

While there have been numerous studies detailing specific changes in the outer membrane that result in OMV release, there are other circumstances that lead to variation in OMV production. Cellular stresses such as growth conditions (Park et al. 2015), bacterial growth stage (Zavan et al. 2019), changes in temperature (McMahon et al. 2012), or the presence of antibiotics (Kadurugamuwa and Beveridge 1997, 1995) can all alter OMV biogenesis.

The mechanisms described are widely varied but highlight the numerous factors which influence the release of OMVs. However, future research is needed to understand what other factors may impact or regulate OMV production. Expanding our knowledge regarding the mechanisms of Gram-negative OMV biogenesis will provide further understanding of the regulation of OMV composition and subsequent functions by bacteria.

### *1.1.5 Outer Membrane Vesicles in Bacterial Communication*

OMVs contain parental proteins, enzymes, and nucleic acids that can be delivered to surrounding bacterial cells (Kadurugamuwa and Beveridge 1995; Mashburn and Whiteley 2005; Dorward et al. 1989; Bomberger et al. 2009). Importantly, it has been shown that *P. aeruginosa* OMVs carrying proteins and toxins from their parent bacteria could kill neighboring bacterial cells (Kadurugamuwa and Beveridge 1996). This research has since sparked interest into examining the ability of OMVs to confer a selective advantage to their parent bacterium.

OMVs from *Neisseria gonorrhoeae*, *Acinetobacter baylyi* and *P. aeruginosa* can carry both chromosomal and plasmid DNA derived from their parent bacteria (Dorward et al. 1989; Fulsundar et al. 2014; Renelli et al. 2004). DNA contained in some OMVs can be transferred to neighboring cells, including DNA that encodes for antibiotic resistance, highlighting the potential role of OMVs in horizontal gene transfer. Furthermore, it has recently been discovered that OMVs can contain RNA including messenger RNA (mRNA), ribosomal RNA (rRNA), and small RNA (sRNA) (Blenkiron et al. 2016; Koeppen et al. 2016; Sjostrom et al. 2015; Choi et al. 2017). Additionally, it was recently shown that OMVs can deliver sRNA to host cells, where the sRNA modulates the innate immune molecules of the host (Koeppen et al. 2016; Choi et al. 2017). However, it is still not well understood how or why RNA is packaged into OMVs.

OMVs can package other bacterial molecules such as the Pseudomonas quinolone signal (PQS) molecule, which contributes to cell-to-cell communication and coordination of the formation of biofilms (Mashburn and Whiteley 2005; Pesci et al. 1999). Removal of OMVs from *P. aeruginosa* cultures stops bacterial communication and group behaviors that are mediated by PQS, highlighting OMVs as a key contributor to cell-to-cell communication of *P. aeruginosa* (Mashburn and Whiteley 2005). Collectively, these works highlight the variety of materials that can be packaged into OMVs and how bacteria can use OMVs to communicate and interact with neighboring cells. A more detailed discussion about the ability of OMVs to function in inter-bacterial communication can be found in Chap. 5.

### 1.1.6 Outer Membrane Vesicles in Host–Pathogen Interactions

Due to the nature of their contents, OMVs are immunostimulatory to eukaryotic hosts. OMVs contain a number of microbe-associated molecular patterns (MAMPs) such as LPS, DNA, and peptidoglycan, all of which are capable of inducing a host pro-inflammatory response (reviewed in Ellis and Kuehn 2010). Toll-like receptors (TLRs) are pattern recognition receptors (PRRs) located on the membranes of host cells that can detect bacterial MAMPs (reviewed in Kawai and Akira 2010). TLR4 has been shown to detect LPS contained in *E. coli* and *N. meningitidis* OMVs, with the detection of *E. coli* OMVs leading to the production of pro-inflammatory cytokines (Mirlashari and Lyberg 2003; Soderblom et al. 2005). Alternatively, when OMVs interact with epithelial cells they can be internalized by lipid rafts on the cell surface and their peptidoglycan cargo can be detected by cytosolic NOD1 (Kaparakis et al. 2010). The detection of OMVs by PRRs activates a signaling cascade, resulting in the initiation of an innate immune response, the production of pro-inflammatory cytokines and the recruitment of immune cells (Kaparakis et al. 2010; Ismail et al. 2003; Bielig et al. 2011). These reports represent a small portion of the research that has been undertaken to determine how bacterial OMVs interact with innate immune receptors of the host to mediate inflammation. They also highlight the ability of OMVs to function as an important secretory system for immunostimulatory molecules and demonstrates their role in bacterial infection and inflammation. A more detailed discussion of the pathogenic and immunostimulatory functions of OMVs can be found in Chaps. 7 and 8.

Commensal bacteria have adapted mechanisms to enable their persistence in the host (reviewed in Hooper and Gordon 2001; Hooper 2004). Commensal bacteria that reside in the gut cannot cross the mucus layer to interact directly with epithelial cells (Johansson et al. 2008). However, it was recently identified that commensal and probiotic bacteria such as *Bacteroides fragilis* (Shen et al. 2012) and *E. coli* (Cañas et al. 2016) are capable of producing OMVs. These OMVs can be used as a bacterial delivery system, as they can cross the mucus layer and enter host epithelial cells via endocytic pathways (Cañas et al. 2016). It is now known that OMVs from commensal bacteria are able to modulate the host immune system to prevent inflammation and protect against diseases such as colitis (Shen et al. 2012; Kang et al. 2013). Additionally, OMVs isolated from commensal and probiotic *E. coli* prime the host immune system via NOD1 activation which may aid in the elimination of pathogenic bacteria (Cañas et al. 2018). These recent works highlight that OMVs from commensal bacteria may interact with the host immune system to maintain host–microbe homeostasis and may aid in the prevention of infections, and these topics are discussed in further detail in Chap. 9.

### 1.1.7 Distribution of OMVs in the Environment

Although OMVs have been thoroughly studied in the context of human pathogenic bacteria, more recently we have begun to understand their presence and functions in the environment. Here we give a brief overview of the distribution of bacterial OMVs in the environment, and this topic is discussed in further detail in Chap. 4. Two marine bacterial species, *Prochlorococcus* sp. and *Synechoccocus* sp., were found to be able to produce OMVs in their ecosystem (Biller et al. 2014). OMVs produced by *Prochlorococcus* sp. contained carbon and were able to aid in the growth of other marine bacterial species as the sole carbon source provided (Biller et al. 2014). Along with carbon, *Prochlorococcus* sp. OMVs contain DNA, RNA, and a range of proteins. In addition to the presence of OMVs in marine ecosystems, early research had identified OMVs were produced by freshwater bacterial biofilms (Beveridge 1999). However, it was not until recently that freshwater OMVs were further explored. Electron microscopy images of autotrophic freshwater bacterial species showed the release of OMVs from the outer membrane of bacterial cells into the environment (Silva et al. 2014). However, the importance of OMVs in freshwater aquatic environments is currently unknown and requires further investigation.

Environmental bacteria predominantly reside in the form of biofilms (Costerton et al. 1978). Biofilms are composed of a mucus layer containing bacteria in a scaffold-like structure known as the extracellular matrix. The extracellular matrix of biofilms contains exopolysaccharides, proteins, and extracellular DNA (eDNA) (Danese et al. 2000; Allesen-Holm et al. 2006; Jurcisek and Bakaletz 2007; Whitchurch et al. 2002). It was shown that *P. aeruginosa* OMVs make up an important and necessary component of the biofilm matrix (Schooling and Beveridge 2006; Whitchurch et al. 2002). It is now known that OMV size and content can differ between biofilm and planktonic cultures, as demonstrated for *P. aeruginosa* and *H. pylori*, suggesting that the role of OMVs in bacterial biofilms determines their composition (Park et al. 2015; Grande et al. 2015). These works highlight biofilm OMVs as an important component of the extracellular matrix and suggests that changes in OMV composition are in response to the role and necessity of OMVs in biofilms.

Finally, OMVs have recently been found within household environments. Gram-negative OMVs have been identified in household dust in the air and in mattresses (Kim et al. 2013). It was speculated that dust OMVs may be inhaled by residents and internalized by epithelial cells of the airway to cause disease. Mouse models have shown that internalization of dust OMVs leads to an inflammatory response that can be blocked by Polymyxin B (Kim et al. 2013). This suggests that like pathogenic OMVs, LPS from dust OMVs is detected by PRRs and can cause an inflammatory response (Kim et al. 2013).

Collectively, these studies indicate that Gram-negative OMVs can be identified in a number of environments suggesting that they are an essential part of bacterial growth and survival. As research continues, the extent to which OMVs can be found in the environment and the roles that OMVs play in these settings will become apparent.

## 1.2 Introduction to Gram-Positive MVs

The last decade of research has uncovered that Gram-positive bacteria can also produce vesicles, known as membrane vesicles (MVs). The discovery of Gram-positive MVs occurred much later than the discovery of Gram-negative OMVs, as researchers thought that the thick cell wall that surrounds Gram-positive bacteria would prevent the release of MVs. Despite this, MVs were reported to be produced by Gram-positive bacteria as early as 1976 (Bisschop and Konings 1976), as well as in a number of other early reports, however, these findings were dismissed by the wider bacterial vesicle field (Dorward and Garon 1990; Ruhr and Sahl 1985). Here we provide a brief discussion of the discovery, biogenesis, and functions of Gram-positive MVs, and an extensive review of this topic can be found in Chap. 3.

In 2009, electron microscopy showed for the first time the release of MVs from the surface of the Gram-positive organism, *Staphylococcus aureus* (Lee et al. 2009). This renewed interest in the existence of Gram-positive MVs, and soon reports emerged of MVs being produced by other Gram-positive species, including *Bacillus anthracis* (Rivera et al. 2010), *Listeria monocytogenes* (Lee et al. 2013b), *Clostridium perfringens* (Jiang et al. 2014), and *Streptococcus* sp. (Liao et al. 2014; Resch et al. 2016).

Due to the years between the discovery of Gram-negative OMVs and Gram-positive MVs, MVs remain poorly understood in comparison to their Gram-negative counterparts. While interest in MVs is increasing, there is still much to be uncovered surrounding their roles in inter-bacterial communication and host–pathogen interactions.

### *1.2.1 Production and Biogenesis of Gram-Positive MVs*

Gram-positive MVs are similar in size to Gram-negative OMVs, ranging from 20 to 400 nm (Jiang et al. 2014; Brown et al. 2014; Haas and Grenier 2015; Tartaglia et al. 2018). However, the mechanism of Gram-positive MV biogenesis and release through their thick peptidoglycan layer is unclear. Studies have suggested that surfactant-like enzymes may be involved in disrupting the cytoplasmic membrane (Wang et al. 2018; Schlatterer et al. 2018), as well as endolysins that may alter the permeability of the peptidoglycan-rich cell wall thereby enabling the release of MVs (Toyofuku et al. 2017). It has also been suggested that cytoskeletal changes may contribute to the formation of MVs (Mayer and Gottschalk 2003). Furthermore, MV production is increased during stress conditions such as antibiotic exposure (He et al. 2017; Andreoni et al. 2019) suggesting that environmental factors may regulate their biogenesis. However, since MV biogenesis is still in the early stages of exploration, more studies are needed to reach a consensus on their mechanisms of biogenesis and the factors that influence it.

### 1.2.2 Contents of Gram-Positive MVs

The first characterization of Gram-positive MVs was a proteomic study of *S. aureus* MVs (Lee et al. 2009). These findings revealed that *S. aureus* MVs contain a variety of proteins that may serve biological roles in inter-bacterial communication, antibiotic resistance, virulence, and regulation of MV biogenesis (Lee et al. 2009). Moreover, this study suggested an enrichment of specific proteins in MVs compared to their parent bacteria indicating a selective packaging of protein cargo (Lee et al. 2009). Further studies confirmed that *S. aureus* MVs carry a range of pathogenic proteins including beta-lactamase (Lee et al. 2013a), alpha-toxin (Thay et al. 2013), and other virulence-related proteins (Lee et al. 2013b; Tartaglia et al. 2018). Similar findings have since been reported for MVs isolated from other Gram-positive species, including *B. anthracis* (Rivera et al. 2010), *Enterococcus faecium* (Wagner et al. 2018), *C. perfringens* (Jiang et al. 2014), *L. monocytogenes* (Coelho et al. 2019), and *Streptococcus* sp. (Haas and Grenier 2015; Resch et al. 2016). These studies suggest that MVs may serve as a Gram-positive secretion system for the delivery of biologically active proteins.

Early reports suggested that Gram-positive MVs do not carry nucleic acids (Dorward and Garon 1990). However, more recent findings have demonstrated that MVs from a variety of Gram-positive species contain DNA and RNA, including *C. perfringens* (Jiang et al. 2014), *Streptococcus* sp. (Liao et al. 2014; Resch et al. 2016), and *Lactobacillus reuteri* (Grande et al. 2017). While it is unclear how this DNA is packaged, the amount of DNA contained in *Streptococcus* MVs changes at different growth stages, suggesting that this process may be regulated by their parent bacterium during bacterial growth (Liao et al. 2014). Moreover, there are currently only a few reports describing the detection of RNA associated with Gram-positive MVs. RNA species detected in MVs include ribosomal RNA (rRNA) (Resch et al. 2016), transfer RNA (tRNA) (Resch et al. 2016), and small RNA (sRNA) (Choi et al. 2018). Differences in the abundance of RNA species carried by MVs when compared to their parent bacteria suggests that RNA may be selectively packaged into MVs (Resch et al. 2016).

These studies highlight that Gram-positive MVs can contain a range of molecules from their parent bacterium including proteins and nucleic acids. While it is still not well understood as to how and why these molecules are packaged into MVs, researchers are beginning to understand how the contents of MVs may aid in bacterial functions.

### 1.2.3 Role of MVs in Inter-Bacterial Communication

While there are limited studies describing the role of MVs in aiding bacterial functions compared to OMVs, their contents suggest that MVs are involved in inter-bacterial communication and the delivery of molecules between bacteria (reviewed in Brown et al. 2015). Proteins with bacteriolytic function and proteins that may facilitate transfer of molecules in an inter-bacterial manner have been identified in *S. aureus* MVs (Lee et al. 2009). Additionally, MVs have been implicated in biofilm production and formation. For example, DNA contained in *Streptococcus mutans* MVs are thought to be a component of *S. mutans* biofilms (Liao et al. 2014), while *S. aureus* MV production is upregulated during biofilm formation (He et al. 2017). Additionally, recent work determined that *E. faecium* MVs carry proteins that facilitate the production of bacterial biofilms (Wagner et al. 2018). While horizontal gene transfer via MVs is yet to be demonstrated, transfer of functional beta-lactamase protein via MVs has been shown, whereby MVs from an ampicillin resistant *S. aureus* strain transferred resistance to ampicillin-sensitive strains of *E. coli*, *Salmonella enterica* ser. Enteritidis, and *Staphylococcus* sp. via the transfer of the BlaZ protein (Lee et al. 2013a). These studies indicate that MVs may play an important role in communicating with other bacterial cells in the environment to promote bacterial survival.

### *1.2.4 Role of MVs in Host–Pathogen Interactions*

Like OMVs, MVs are able to interact with eukaryotic host cells, including both epithelial cells (Gurung et al. 2011; Kim et al. 2012) and immune cells (Haas and Grenier 2015; Rivera et al. 2010; Jiang et al. 2014). Although there are few studies investigating the mechanisms of MV entry into target cells, there is evidence that *S. aureus* MVs enter host cells via cholesterol-dependent fusion (Thay et al. 2013), and that they are likely to enter host cells via a number of other mechanisms. Entry into host cells enables MVs to deliver their immunogenic cargo to mediate pathogenesis, similar to Gram-negative OMVs.

MVs carry a range of MAMPs, however, there is limited knowledge surrounding the innate and adaptive immune responses that they induce. Reports have shown that *S. aureus* MVs induce inflammation and cell death in host cells (Gurung et al. 2011; Jeon et al. 2016; Hong et al. 2011; Jun et al. 2017). The first reports of MVs activating innate immune pathways showed that *S. aureus* MVs activate TLR2 and NOD2, leading to the production of pro-inflammatory cytokines (Hong et al. 2011; Jun et al. 2017; Kim et al. 2012). Furthermore, MVs isolated from feces were shown to cause sepsis through the activation of TLR2 (Park et al. 2018).

The ability of Gram-positive MVs to induce adaptive immune responses has also been reported. *C. perfringens* MVs were shown to produce high-titer immunoglobulin G1 (IgG1) responses in mice (Jiang et al. 2014), while *B. anthracis* MVs produce a robust IgM response in mice when they encounter toxins carried by the MVs (Rivera et al. 2010). Due to the ability of MVs to activate the adaptive immune response, researchers have investigated their efficacy as a vaccine platform. Studies have shown that MVs from *Streptococcus pneumoniae* induce a protective response in mice when exposed to bacterial challenge (Olaya-Abril et al. 2014). Similarly, administration of *S. aureus* MVs to mice has been shown to be protective against *S. aureus* lung infection (Choi et al. 2015). These studies indicate that MVs warrant further investigation into their potential as alternative vaccine candidates.

Compared to OMV research, these few studies highlight that there is still little knowledge regarding how MVs interact with the innate and adaptive immune system of their host. Future work focused on how MVs from a variety of Gram-positive bacteria can enter host cells and modulate the host immune system will provide better understanding of the role of MVs in the context of Gram-positive bacterial infections.

## 1.3 Conclusions

It has become apparent that OMVs are important biological products that contribute to numerous bacterial functions including cell-to-cell communication and bacterial pathogenesis (Mashburn and Whiteley 2005; reviewed in Kaparakis-Liaskos and Ferrero 2015). OMVs contain a range of materials such as proteins and nucleic acids that aid in bacterial functions (Haurat et al. 2011; Dorward et al. 1989; Ciofu et al. 2000); however, the mechanisms of selective packaging of materials into OMVs remains elusive. Research is now focusing on understanding the mechanisms of OMV biogenesis and how OMVs modulate the innate and adaptive immune system of their host in order to develop their use as novel therapeutics. In the last decade it was shown that Gram-positive bacteria can also produce vesicles as part of their natural growth (Lee et al. 2009). It has become apparent that like OMVs, MVs can carry a range of cargo from their parent bacterium that may be able to aid in bacterial communication and pathogenesis (Resch et al. 2016; Lee et al. 2009). Nevertheless, the roles of MVs in bacterial functions are still not well understood. Research is continuously progressing in both OMV and MV fields to further understand the fundamental production of vesicles and the packaging of materials into them. Importantly, elucidating how OMVs and MVs interact with host epithelial and immune cells is necessary to determine the role of membrane vesicles in contributing to bacterial survival and disease progression. Understanding the production of bacterial membrane vesicles and manipulating vesicles for therapeutic use will have broad implications in how we consider host–pathogen interactions and bacterial diseases. Overall, these works demonstrate the multifaceted, but not exhaustive, roles bacterial membrane vesicles play in contributing to bacterial survival, communication, and pathogenesis.

## References


Allesen-Holm M, Barken KB, Yang L, Klausen M, Webb JS, Kjelleberg S, Molin S, Givskov M, Tolker-Nielsen T (2006) A characterization of DNA release in *Pseudomonas aeruginosa* cultures and biofilms. Mol Microbiol 59(4):1114–1128. https://doi.org/10.1111/j.1365-2958.2005.05008.x

Allison CC, Kufer TA, Kremmer E, Kaparakis M, Ferrero RL (2009) *Helicobacter pylori* induces MAPK phosphorylation and AP-1 activation via a NOD1-dependent mechanism. J Immunol 183(12):8099–8109. https://doi.org/10.4049/jimmunol.0900664

Andreoni F, Toyofuku M, Menzi C, Kalawong R, Mairpady Shambat S, Francois P, Zinkernagel AS, Eberl L (2019) Antibiotics stimulate formation of vesicles in *Staphylococcus aureus* in both phage-dependent and -independent fashions and via different routes. Antimicrob Agents Chemother 63(2):e01439. https://doi.org/10.1128/AAC.01439-18

Arnold R, Galloway Y, McNicholas A, O'Hallahan J (2011) Effectiveness of a vaccination programme for an epidemic of meningococcal B in New Zealand. Vaccine 29 (40):7100–7106. https://doi.org/10.1016/j.vaccine.2011.06.120

Ayalew S, Confer AW, Shrestha B, Wilson AE, Montelongo M (2013) Proteomic analysis and immunogenicity of *Mannheimia haemolytica* vesicles. Clin Vaccine Immunol 20(2):191–196. https://doi.org/10.1128/CVI.00622-12

Bauman SJ, Kuehn MJ (2006) Purification of outer membrane vesicles from *Pseudomonas aeruginosa* and their activation of an IL-8 response. Microbes Infect 8(9–10):2400–2408. https://doi.org/10.1016/j.micinf.2006.05.001

Bernadac A, Gavioli M, Lazzaroni JC, Raina S, Lloubes R (1998) *Escherichia coli* tol-pal mutants form outer membrane vesicles. J Bacteriol 180(18):4872–4878

Beveridge TJ (1999) Structures of Gram-negative cell walls and their derived membrane vesicles. J Bacteriol 181(16):4725–4733

Bielig H, Rompikuntal PK, Dongre M, Zurek B, Lindmark B, Ramstedt M, Wai SN, Kufer TA (2011) NOD-like receptor activation by outer membrane vesicles from *Vibrio cholerae* non-O1 non-O139 strains is modulated by the quorum-sensing regulator HapR. Infect Immun 79 (4):1418–1427. https://doi.org/10.1128/IAI.00754-10

Biller SJ, Schubotz F, Roggensack SE, Thompson AW, Summons RE, Chisholm SW (2014) Bacterial vesicles in marine ecosystems. Science 343(6167):183–186. https://doi.org/10.1126/science.1243457

Bisschop A, Konings WN (1976) Reconstitution of reduced nicotinamide adenine dinucleotide oxidase activity with menadione in membrane vesicles from the menaquinone-deficient *Bacillus subtilis* aro D. relation between electron transfer and active transport. Eur J Biochem 67 (2):357–365

Bitto NJ, Kaparakis-Liaskos M (2017) The therapeutic benefit of bacterial membrane vesicles. Int J Mol Sci 18(6):E1287. https://doi.org/10.3390/ijms18061287

Bitto NJ, Chapman R, Pidot S, Costin A, Lo C, Choi J, D'Cruze T, Reynolds EC, Dashper SG, Turnbull L, Whitchurch CB, Stinear TP, Stacey KJ, Ferrero RL (2017) Bacterial membrane vesicles transport their DNA cargo into host cells. Sci Rep 7(1):7072. https://doi.org/10.1038/s41598-017-07288-4

Bjune G, Hoiby EA, Gronnesby JK, Arnesen O, Fredriksen JH, Halstensen A, Holten E, Lindbak AK, Nokleby H, Rosenqvist E et al (1991) Effect of outer membrane vesicle vaccine against group B meningococcal disease in Norway. Lancet 338(8775):1093–1096

Blenkiron C, Simonov D, Muthukaruppan A, Tsai P, Dauros P, Green S, Hong J, Print CG, Swift S, Phillips AR (2016) Uropathogenic *Escherichia coli* releases extracellular vesicles that are associated with RNA. PLoS One 11(8):e0160440. https://doi.org/10.1371/journal.pone.0160440

Bomberger JM, Maceachran DP, Coutermarsh BA, Ye S, O'Toole GA, Stanton BA (2009) Long-distance delivery of bacterial virulence factors by *Pseudomonas aeruginosa* outer membrane vesicles. PLoS Pathog 5(4):e1000382. https://doi.org/10.1371/journal.ppat.1000382

Brown L, Kessler A, Cabezas-Sanchez P, Luque-Garcia JL, Casadevall A (2014) Extracellular vesicles produced by the Gram-positive bacterium *Bacillus subtilis* are disrupted by the lipopeptide surfactin. Mol Microbiol 93(1):183–198. https://doi.org/10.1111/mmi.12650

Brown L, Wolf JM, Prados-Rosales R, Casadevall A (2015) Through the wall: extracellular vesicles in Gram-positive bacteria, mycobacteria and fungi. Nat Rev Microbiol 13(10):620–630. https://doi.org/10.1038/nrmicro3480

Cañas MA, Gimenez R, Fabrega MJ, Toloza L, Baldoma L, Badia J (2016) Outer membrane vesicles from the probiotic *Escherichia coli* Nissle 1917 and the commensal ECOR12 enter intestinal epithelial cells via clathrin-dependent endocytosis and elicit differential effects on DNA damage. PLoS One 11(8):e0160374. https://doi.org/10.1371/journal.pone.0160374

Cañas M-A, Fábrega M-J, Giménez R, Badia J, Baldomà L (2018) Outer membrane vesicles from probiotic and commensal *Escherichia coli* activate NOD1-mediated immune responses in intestinal epithelial cells. Front Microbiol 9:498. https://doi.org/10.3389/fmicb.2018.00498

Chatterjee SN, Das J (1967) Electron microscopic observations on the excretion of cell-wall material by *Vibrio cholerae*. J Gen Microbiol 49(1):1–11. https://doi.org/10.1099/00221287-49-1-1

Chen DJ, Osterrieder N, Metzger SM, Buckles E, Doody AM, DeLisa MP, Putnam D (2010) Delivery of foreign antigens by engineered outer membrane vesicle vaccines. Proc Natl Acad Sci U S A 107(7):3099–3104. https://doi.org/10.1073/pnas.0805532107

Chi B, Qi M, Kuramitsu HK (2003) Role of dentilisin in *Treponema denticola* epithelial cell layer penetration. Res Microbiol 154(9):637–643. https://doi.org/10.1016/j.resmic.2003.08.001

Choi SJ, Kim MH, Jeon J, Kim OY, Choi Y, Seo J, Hong SW, Lee WH, Jeon SG, Gho YS, Jee YK, Kim YK (2015) Active immunization with extracellular vesicles derived from *Staphylococcus aureus* effectively protects against staphylococcal lung infections, mainly via Th1 cell-mediated immunity. PLoS One 10(9):e0136021. https://doi.org/10.1371/journal.pone.0136021

Choi JW, Kim SC, Hong SH, Lee HJ (2017) Secretable small RNAs via outer membrane vesicles in periodontal pathogens. J Dent Res 96(4):458–466. https://doi.org/10.1177/0022034516685071

Choi J-W, Kwon T-Y, Hong S-H, Lee H-J (2018) Isolation and characterization of a microRNA-size secretable small RNA in *Streptococcus sanguinis*. Cell Biochem Biophys 76(1):293–301. https://doi.org/10.1007/s12013-016-0770-5

Ciofu O, Beveridge TJ, Kadurugamuwa J, Walther-Rasmussen J, Hoiby N (2000) Chromosomal beta-lactamase is packaged into membrane vesicles and secreted from *Pseudomonas aeruginosa*. J Antimicrob Chemother 45(1):9–13

Coelho C, Brown L, Maryam M, Vij R, Smith DFQ, Burnet MC, Kyle JE, Heyman HM, Ramirez J, Prados-Rosales R, Lauvau G, Nakayasu ES, Brady NR, Hamacher-Brady A, Coppens I, Casadevall A (2019) *Listeria monocytogenes* virulence factors, including listeriolysin O, are secreted in biologically active extracellular vesicles. J Biol Chem 294(4):1202–1217. https://doi.org/10.1074/jbc.RA118.006472

Costerton JW, Ingram JM, Cheng KJ (1974) Structure and function of the cell envelope of Gram-negative bacteria. Bacteriol Rev 38(1):87–110

Costerton JW, Geesey GG, Cheng KJ (1978) How bacterial stick. Sci Am 238(1):86–95

Danese PN, Pratt LA, Kolter R (2000) Exopolysaccharide production is required for development of *Escherichia coli* K-12 biofilm architecture. J Bacteriol 180(12):3593–3596

Deatherage BL, Cookson BT (2012) Membrane vesicle release in bacteria, eukaryotes, and archaea: a conserved yet underappreciated aspect of microbial life. Infect Immun 80(6):1948–1957. https://doi.org/10.1128/IAI.06014-11

Deich RA, Hoyer LC (1982) Generation and release of DNA-binding vesicles by *Haemophilus influenzae* during induction and loss of competence. J Bacteriol 152(2):855–864

Devoe IW, Gilchrist JE (1973) Release of endotoxin in the form of cell wall blebs during in vitro growth of *Neisseria meningitidis*. J Exp Med 138(5):1156–1167

DeVoe IW, Gilchrist JE (1975) Pili on meningococci from primary cultures of nasopharyngeal carriers and cerebrospinal fluid of patients with acute disease. J Exp Med 141:297. https://doi.org/10.1084/jem.141.2.297

Dorward DW, Garon CF (1989) DNA-binding proteins in cells and membrane blebs of *Neisseria gonorrhoeae*. J Bacteriol 171(8):4196–4201

Dorward DW, Garon CF (1990) DNA is packaged within membrane-derived vesicles of Gram-negative but not Gram-positive bacteria. Appl Environ Microbiol 56(6):1960–1962

Dorward DW, Garon CF, Judd RC (1989) Export and intercellular transfer of DNA via membrane blebs of *Neisseria gonorrhoeae*. J Bacteriol 171(5):2499–2505

Elhenawy W, Bording-Jorgensen M, Valguarnera E, Haurat MF, Wine E, Feldman MF (2016) LPS remodeling triggers formation of outer membrane vesicles in *Salmonella*. MBio 7(4):e00940–e00916. https://doi.org/10.1128/mBio.00940-16

Ellis TN, Kuehn MJ (2010) Virulence and immunomodulatory roles of bacterial outer membrane vesicles. Microbiol Mol Biol Rev 74(1):81–94. https://doi.org/10.1128/MMBR.00031-09

Fulsundar S, Harms K, Flaten GE, Johnsen PJ, Chopade BA, Nielsen KM (2014) Gene transfer potential of outer membrane vesicles of *Acinetobacter baylyi* and effects of stress on vesiculation. Appl Environ Microbiol 80(11):3469–3483. https://doi.org/10.1128/AEM.04248-13

Gamazo C, Moriyon I (1987) Release of outer membrane fragments by exponentially growing *Brucella melitensis* cells. Infect Immun 55(3):609–615

Gankema H, Wensink J, Guinee PA, Jansen WH, Witholt B (1980) Some characteristics of the outer membrane material released by growing enterotoxigenic *Escherichia coli*. Infect Immun 29 (2):704–713

Grande R, Marcantonio MC, Robuffo I, Pompilio A, Celia C, Marzio L, Paolino D, Codagnone M, Muraro R, Stoodley P, Hall-Stoodley L, Mincione G (2015) *Helicobacter pylori* ATCC 43629/NCTC 11639 outer membrane vesicles (OMVs) from biofilm and planktonic phase associated with extracellular DNA (eDNA). Front Microbiol 6:1369. https://doi.org/10.3389/fmicb.2015.01369

Grande R, Celia C, Mincione G, Stringaro A, Di Marzio L, Colone M, Di Marcantonio MC, Savino L, Puca V, Santoliquido R, Locatelli M, Muraro R, Hall-Stoodley L, Stoodley P (2017) Detection and physicochemical characterization of membrane vesicles (MVs) of *Lactobacillus reuteri* DSM 17938. Front Microbiol 8:1040. https://doi.org/10.3389/fmicb.2017.01040

Grenier D, Mayrand D (1987) Functional characterization of extracellular vesicles produced by *Bacteroides gingivalis*. Infect Immun 55(1):111–117

Gurung M, Moon DC, Choi CW, Lee JH (2011) *Staphylococcus aureus* produces membrane-derived vesicles that induce host cell death. PLoS One 6:e27958. https://doi.org/10.1371/journal.pone.0027958

Haas B, Grenier D (2015) Isolation, characterization and biological properties of membrane vesicles produced by the swine pathogen *Streptococcus suis*. PLoS One 10(6):e0130528. https://doi.org/10.1371/journal.pone.0130528

Haurat MF, Aduse-Opoku J, Rangarajan M, Dorobantu L, Gray MR, Curtis MA, Feldman MF (2011) Selective sorting of cargo proteins into bacterial membrane vesicles. J Biol Chem 286 (2):1269–1276. https://doi.org/10.1074/jbc.M110.185744

He X, Yuan F, Lu F, Yin Y, Cao J (2017) Vancomycin-induced biofilm formation by methicillin-resistant *Staphylococcus aureus* is associated with the secretion of membrane vesicles. Microb Pathog 110:225–231. https://doi.org/10.1016/j.micpath.2017.07.004

Hoekstra D, van der Laan J, de Leij L, Witholt B (1976) Release of outer membrane fragments from normally growing *Escherichia coli*. Biochim Biophys Acta Biomembr 455(3):889–899. https://doi.org/10.1016/0005-2736(76)90058-4

Hong SW, Kim MR, Lee EY, Kim JH, Kim YS, Jeon SG, Yang JM, Lee BJ, Pyun BY, Gho YS, Kim YK (2011) Extracellular vesicles derived from *Staphylococcus aureus* induce atopic dermatitis-like skin inflammation. Allergy 66(3):351–359. https://doi.org/10.1111/j.1398-9995.2010.02483.x

Hooper LV (2004) Bacterial contributions to mammalian gut development. Trends Microbiol 12 (3):129–134. https://doi.org/10.1016/j.tim.2004.01.001

Hooper LV, Gordon JI (2001) Commensal host-bacterial relationships in the gut. Science 292 (5519):1115–1118

Ismail S, Hampton MB, Keenan JI (2003) *Helicobacter pylori* outer membrane vesicles modulate proliferation and interleukin-8 production by gastric epithelial cells. Infect Immun 71 (10):5670–5675

Jeon H, Oh MH, Jun SH, Kim SI, Choi CW, Kwon HI, Na SH, Kim YJ, Nicholas A, Selasi GN, Lee JC (2016) Variation among *Staphylococcus aureus* membrane vesicle proteomes affects cytotoxicity of host cells. Microb Pathog 93:185–193. https://doi.org/10.1016/j.micpath.2016.02.014

Jiang Y, Kong Q, Roland KL, Curtiss R (2014) Membrane vesicles of *Clostridium perfringens* type a strains induce innate and adaptive immunity. Int J Med Microbiol 304(3–4):431–443. https://doi.org/10.1016/j.ijmm.2014.02.006

Johansson ME, Phillipson M, Petersson J, Velcich A, Holm L, Hansson GC (2008) The inner of the two Muc2 mucin-dependent mucus layers in colon is devoid of bacteria. Proc Natl Acad Sci U S A 105(39):15064–15069. https://doi.org/10.1073/pnas.0803124105

Jun SH, Lee JH, Kim SI, Choi CW, Park TI, Jung HR, Cho JW, Kim SH, Lee JC (2017) *Staphylococcus aureus*-derived membrane vesicles exacerbate skin inflammation in atopic dermatitis. Clin Exp Allergy 47(1):85–96. https://doi.org/10.1111/cea.12851

Jurcisek JA, Bakaletz LO (2007) Biofilms formed by nontypeable *Haemophilus influenzae* in vivo contain both double-stranded DNA and type IV pilin protein. J Bacteriol 189(10):3868–3875. https://doi.org/10.1128/JB.01935-06

Kadurugamuwa JL, Beveridge TJ (1995) Virulence factors are released from *Pseudomonas aeruginosa* in association with membrane vesicles during normal growth and exposure to gentamicin: a novel mechanism of enzyme secretion. J Bacteriol 177(14):3998–4008

Kadurugamuwa JL, Beveridge TJ (1996) Bacteriolytic effect of membrane vesicles from *Pseudomonas aeruginosa* on other bacteria including pathogens: conceptually new antibiotics. J Bacteriol 178(10):2767–2774

Kadurugamuwa JL, Beveridge TJ (1997) Natural release of virulence factors in membrane vesicles by *Pseudomonas aeruginosa* and the effect of aminoglycoside antibiotics on their release. J Antimicrob Chemother 40:615. https://doi.org/10.1093/jac/40.5.615

Kang CS, Ban M, Choi EJ, Moon HG, Jeon JS, Kim DK, Park SK, Jeon SG, Roh TY, Myung SJ, Gho YS, Kim JG, Kim YK (2013) Extracellular vesicles derived from gut microbiota, especially *Akkermansia muciniphila*, protect the progression of dextran sulfate sodium-induced colitis. PLoS One 8(10):e76520. https://doi.org/10.1371/journal.pone.0076520

Kaparakis M, Turnbull L, Carneiro L, Firth S, Coleman HA, Parkington HC, Le Bourhis L, Karrar A, Viala J, Mak J, Hutton ML, Davies JK, Crack PJ, Hertzog PJ, Philpott DJ, Girardin SE, Whitchurch CB, Ferrero RL (2010) Bacterial membrane vesicles deliver peptidoglycan to NOD1 in epithelial cells. Cell Microbiol 12(3):372–385. https://doi.org/10.1111/j.1462-5822.2009.01404.x

Kaparakis-Liaskos M, Ferrero RL (2015) Immune modulation by bacterial outer membrane vesicles. Nat Rev Immunol 15(6):375–387. https://doi.org/10.1038/nri3837

Kawai T, Akira S (2010) The role of pattern-recognition receptors in innate immunity: update on toll-like receptors. Nat Immunol 11(5):373–384. https://doi.org/10.1038/ni.1863

Kim MR, Hong SW, Choi EB, Lee WH, Kim YS, Jeon SG, Jang MH, Gho YS, Kim YK (2012) *Staphylococcus aureus*-derived extracellular vesicles induce neutrophilic pulmonary inflammation via both Th1 and Th17 cell responses. Allergy 67(10):1271–1281. https://doi.org/10.1111/all.12001

Kim YS, Choi EJ, Lee WH, Choi SJ (2013) Extracellular vesicles, especially derived from Gram-negative bacteria, in indoor dust induce neutrophilic pulmonary inflammation associated with both Th1 and Th17 cell responses. Clin Exp Allergy 43:443–454

Knox KW, Vesk M, Work E (1966) Relation between excreted lipopolysaccharide complexes and surface structures of a lysine-limited culture of *Escherichia coli*. J Bacteriol 92(4):1206–1217

Koeppen K, Hampton TH, Jarek M, Scharfe M, Gerber SA, Mielcarz DW, Demers EG, Dolben EL, Hammond JH, Hogan DA, Stanton BA (2016) A novel mechanism of host-pathogen interaction through sRNA in bacterial outer membrane vesicles. PLoS Pathog 12(6):e1005672. https://doi.org/10.1371/journal.ppat.1005672

Lee EY, Bang JY, Park GW, Choi DS, Kang JS, Kim HJ, Park KS, Lee JO, Kim YK, Kwon KH, Kim KP, Gho YS (2007) Global proteomic profiling of native outer membrane vesicles derived from *Escherichia coli*. Proteomics 7(17):3143–3153. https://doi.org/10.1002/pmic.200700196

Lee EY, Choi DY, Kim DK, Kim JW, Park JO, Kim S, Kim SH, Desiderio DM, Kim YK, Kim KP, Gho YS (2009) Gram-positive bacteria produce membrane vesicles: proteomics-based characterization of *Staphylococcus aureus*-derived membrane vesicles. Proteomics 9(24):5425–5436. https://doi.org/10.1002/pmic.200900338

Lee J, Lee E-Y, Kim S-H, Kim D-K, Park K-S, Kim KP, Kim Y-K, Roh T-Y, Gho YS (2013a) *Staphylococcus aureus* extracellular vesicles carry biologically active β-lactamase. Antimicrob Agents Chemother 57(6):2589–2595. https://doi.org/10.1128/AAC.00522-12

Lee JH, Choi C-W, Lee T, Kim S (2013b) Transcription factor σB plays an important role in the production of extracellular membrane-derived vesicles in *Listeria monocytogenes*. PLoS One 8 (8):e73196. https://doi.org/10.1371/journal.pone.0073196

Leive L, Shovlin VK, Mergenhagen SE (1968) Physical chemical and immunological properties of lipopolysaccharide released from *Escherichia coli* by Ethylenediaminetetraacetate. J Biol Chem 243(24):6384+

Liao S, Klein MI, Heim KP, Fan Y, Bitoun JP, Ahn SJ, Burne RA, Koo H, Brady LJ, Wen ZT (2014) *Streptococcus mutans* extracellular DNA is upregulated during growth in biofilms, actively released via membrane vesicles, and influenced by components of the protein secretion machinery. J Bacteriol 196(13):2355–2366. https://doi.org/10.1128/jb.01493-14

Loeb MR (1974) Bacteriophage T4-mediated release of envelope components from *Escherichia coli*. J Virol 13:631–641

MacIntyre S, Trust TJ, Buckley JT (1980) Identification and characterization of outer membrane fragments released by *Aeromonas sp*. Can J Biochem 58(10):1018–1025

Mashburn LM, Whiteley M (2005) Membrane vesicles traffic signals and facilitate group activities in a prokaryote. Nature 437(7057):422–425. https://doi.org/10.1038/nature03925

Mayer F, Gottschalk G (2003) The bacterial cytoskeleton and its putative role in membrane vesicle formation observed in a Gram-positive bacterium producing starch-degrading enzymes. J Mol Microbiol Biotechnol 6(3–4):127–132. https://doi.org/10.1159/000077243

McBroom AJ, Kuehn MJ (2007) Release of outer membrane vesicles by Gram-negative bacteria is a novel envelope stress response. Mol Microbiol 63(2):545–558. https://doi.org/10.1111/j.1365-2958.2006.05522.x

McMahon KJ, Castelli ME, Garcia Vescovi E, Feldman MF (2012) Biogenesis of outer membrane vesicles in *Serratia marcescens* is thermoregulated and can be induced by activation of the Rcs phosphorelay system. J Bacteriol 194(12):3241–3249. https://doi.org/10.1128/JB.00016-12

Mergenhagen SE, Bladen HA, Hsu KC (1966) Electron microscopic localization of endotoxic lipopolysaccharide in Gram-negative organisms. Ann N Y Acad Sci 133(2):279–291

Mirlashari MR, Lyberg T (2003) Expression and involvement of toll-like receptors (TLR)2, TLR4, and CD14 in monocyte TNF-alpha production induced by lipopolysaccharides from *Neisseria meningitidis*. Med Sci Monit 9(8):BR316–BR324

Mug-Opstelten D, Witholt B (1978) Preferential release of new outer membrane fragments by exponentially growing *Escherichia coli*. Biochim Biophys Acta 508(2):287–295

Murphy K, Park AJ, Hao Y, Brewer D, Lam JS, Khursigara CM (2014) Influence of O polysaccharides on biofilm development and outer membrane vesicle biogenesis in *Pseudomonas aeruginosa* PAO1. J Bacteriol 196(7):1306–1317

Nieves W, Asakrah S, Qazi O, Brown KA, Kurtz J, Aucoin DP, McLachlan JB, Roy CJ, Morici LA (2011) A naturally derived outer-membrane vesicle vaccine protects against lethal pulmonary *Burkholderia pseudomallei* infection. Vaccine 29(46):8381–8389. https://doi.org/10.1016/j.vaccine.2011.08.058

Olaya-Abril A, Prados-Rosales R, McConnell MJ, Martín-Peña R, González-Reyes JA, Jiménez-Munguía I, Gómez-Gascón L, Fernández J, Luque-García JL, García-Lidón C, Estévez H, Pachón J, Obando I, Casadevall A, L-a P, Rodríguez-Ortega MJ (2014) Characterization of protective extracellular membrane-derived vesicles produced by *Streptococcus pneumoniae*. J Proteome 106:46–60

Park AJ, Murphy K, Surette MD, Bandoro C, Krieger JR, Taylor P, Khursigara CM (2015) Tracking the dynamic relationship between cellular systems and extracellular subproteomes in *Pseudomonas aeruginosa* biofilms. J Proteome Res 14(11):4524–4537. https://doi.org/10.1021/acs.jproteome.5b00262

Park K-S, Lee J, Lee C, Park HT, Kim J-W, Kim OY, Kim SR, Rådinger M, Jung H-Y, Park J, Lötvall J, Gho YS (2018) Sepsis-like systemic inflammation induced by nano-sized extracellular vesicles from feces. Front Microbiol 9:1735. https://doi.org/10.3389/fmicb.2018.01735

Perez-Cruz C, Delgado L, Lopez-Iglesias C, Mercade E (2015) Outer-inner membrane vesicles naturally secreted by Gram-negative pathogenic bacteria. PLoS One 10(1):e0116896. https://doi.org/10.1371/journal.pone.0116896

Pesci EC, Milbank JB, Pearson JP, McKnight S, Kende AS, Greenberg EP, Iglewski BH (1999) Quinolone signaling in the cell-to-cell communication system of *Pseudomonas aeruginosa*. Proc Natl Acad Sci U S A 96(20):11229–11234

Pierson T, Matrakas D, Taylor YU, Manyam G, Morozov VN, Zhou W, van Hoek ML (2011) Proteomic characterization and functional analysis of outer membrane vesicles of *Francisella novicida* suggests possible role in virulence and use as a vaccine. J Proteome Res 10(3):954–967. https://doi.org/10.1021/pr1009756

Renelli M, Matias V, Lo RY, Beveridge TJ (2004) DNA-containing membrane vesicles of *Pseudomonas aeruginosa* PAO1 and their genetic transformation potential. Microbiology 150(Pt 7):2161–2169. https://doi.org/10.1099/mic.0.26841-0

Resch U, Tsatsaronis JA, Le Rhun A, Stubiger G, Rohde M, Kasvandik S, Holzmeister S, Tinnefeld P, Wai SN, Charpentier E (2016) A two-component regulatory system impacts extracellular membrane-derived vesicle production in group a *Streptococcus*. MBio 7(6):e00207–e00216. https://doi.org/10.1128/mBio.00207-16

Rivera J, Cordero RJB, Nakouzi AS, Frases S (2010) *Bacillus anthracis* produces membrane-derived vesicles containing biologically active toxins. Proc Natl Acad Sci U S A 107(44):19002–19007. https://doi.org/10.1073/pnas.1008843107

Roier S, Zingl FG, Cakar F, Durakovic S, Kohl P, Eichmann TO, Klug L, Gadermaier B, Weinzerl K, Prassl R, Lass A, Daum G, Reidl J, Feldman MF, Schild S (2016) A novel mechanism for the biogenesis of outer membrane vesicles in Gram-negative bacteria. Nat Commun 7:10515. https://doi.org/10.1038/ncomms10515

Rosenqvist E, Hoiby EA, Wedege E, Bryn K, Kolberg J, Klem A, Ronnild E, Bjune G, Nokleby H (1995) Human antibody responses to meningococcal outer membrane antigens after three doses of the Norwegian group B meningococcal vaccine. Infect Immun 63(12):4642–4652

Rothfield L, Pearlman-Kothencz M (1969) Synthesis and assembly of bacterial membrane components: a lipopolysaccharide-phospholipid-protein complex excreted by living bacteria. J Mol Biol 44:477

Ruhr E, Sahl HG (1985) Mode of action of the peptide antibiotic nisin and influence on the membrane potential of whole cells and on cytoplasmic and artificial membrane vesicles. Antimicrob Agents Chemother 27(5):841–845

Schlatterer K, Beck C, Hanzelmann D, Lebtig M, Fehrenbacher B, Schaller M, Ebner P, Nega M, Otto M, Kretschmer D, Peschel A (2018) The mechanism behind bacterial lipoprotein release: phenol-soluble modulins mediate toll-like receptor 2 activation via extracellular vesicle release from *Staphylococcus aureus*. MBio 9(6):e01851–e01818. https://doi.org/10.1128/mBio.01851-18

Schnaitman CA (1970) Protein composition of the cell wall and cytoplasmic membrane of *Escherichia coli*. J Bacteriol 104(2):890–901

Schooling SR, Beveridge TJ (2006) Membrane vesicles: an overlooked component of the matrices of biofilms. J Bacteriol 188(16):5945–5957. https://doi.org/10.1128/JB.00257-06

Schwechheimer C, Kuehn MJ (2015) Outer-membrane vesicles from Gram-negative bacteria: biogenesis and functions. Nat Rev Microbiol 13(10):605–619. https://doi.org/10.1038/nrmicro3525

Shen Y, Giardino Torchia ML, Lawson GW, Karp CL, Ashwell JD, Mazmanian SK (2012) Outer membrane vesicles of a human commensal mediate immune regulation and disease protection. Cell Host Microbe 12(4):509–520. https://doi.org/10.1016/j.chom.2012.08.004

Silva TP, Noyma NP, Duque TLA, Gamalier JP, Vidal LO, Lobao LM, Chiarini-Garcia H, Roland F, Melo RCN (2014) Visualizing aquatic bacteria by light and transmission electron microscopy. Antonie Van Leeuwenhoek 105:1–14

Sjostrom AE, Sandblad L, Uhlin BE, Wai SN (2015) Membrane vesicle-mediated release of bacterial RNA. Sci Rep 5:15329. https://doi.org/10.1038/srep15329

Soderblom T, Oxhamre C, Wai SN, Uhlen P, Aperia A, Uhlin BE, Richter-Dahlfors A (2005) Effects of the *Escherichia coli* toxin cytolysin a on mucosal immunostimulation via epithelial $Ca^{2+}$ signalling and toll-like receptor 4. Cell Microbiol 7(6):779–788. https://doi.org/10.1111/j.1462-5822.2005.00510.x

Tartaglia NR, Breyne K, Meyer E, Cauty C, Jardin J, Chrétien D, Dupont A, Demeyere K, Berkova N, Azevedo V, Guédon E, Le Loir Y (2018) *Staphylococcus aureus* extracellular vesicles elicit an immunostimulatory response in vivo on the murine mammary gland. Front Cell Infect Microbiol 8:277. https://doi.org/10.3389/fcimb.2018.00277

Thay B, Wai SN, Oscarsson J (2013) *Staphylococcus aureus* alpha-toxin-dependent induction of host cell death by membrane-derived vesicles. PLoS One 8(1):e54661. https://doi.org/10.1371/journal.pone.0054661

Toyofuku M, Cárcamo-Oyarce G, Yamamoto T, Eisenstein F, Hsiao C-C, Kurosawa M, Gademann K, Pilhofer M, Nomura N, Eberl L (2017) Prophage-triggered membrane vesicle formation through peptidoglycan damage in *Bacillus subtilis*. Nat Commun 8(1):481. https://doi.org/10.1038/s41467-017-00492-w

Turnbull L, Toyofuku M, Hynen AL, Kurosawa M, Pessi G, Petty NK, Osvath SR, Carcamo-Oyarce G, Gloag ES, Shimoni R, Omasits U, Ito S, Yap X, Monahan LG, Cavaliere R, Ahrens CH, Charles IG, Nomura N, Eberl L, Whitchurch CB (2016) Explosive cell lysis as a mechanism for the biogenesis of bacterial membrane vesicles and biofilms. Nat Commun 7:11220. https://doi.org/10.1038/ncomms11220

Turner L, Praszkier J, Hutton ML, Steer D, Ramm G, Kaparakis-Liaskos M, Ferrero RL (2015) Increased outer membrane vesicle formation in a *Helicobacter pylori* tolB mutant. Helicobacter 20(4):269–283. https://doi.org/10.1111/hel.12196

Turner L, Bitto NJ, Steer DL, Lo C, D'Costa K, Ramm G, Shambrook M, Hill AF, Ferrero RL, Kaparakis-Liaskos M (2018) *Helicobacter pylori* outer membrane vesicle size determines their mechanisms of host cell entry and protein content. Front Immunol 9:1466

Vasilyeva NV, Tsfasman IM, Suzina NE, Stepnaya OA, Kulaev IS (2009) Outer membrane vesicles of *Lysobacter sp*. Dokl Biochem Biophys 426:139–142

Vernikos G, Medini D (2014) Bexsero(R) chronicle. Pathog Glob Health 108(7):305–316. https://doi.org/10.1179/2047773214Y.0000000162

Wagner T, Joshi B, Janice J, Askarian F, Škalko-Basnet N, Hagestad OC, Mekhlif A, Wai SN, Hegstad K, Johannessen M (2018) *Enterococcus faecium* produces membrane vesicles containing virulence factors and antimicrobial resistance related proteins. J Proteome 187:28–38. https://doi.org/10.1016/j.jprot.2018.05.017

Wang X, Thompson CD, Weidenmaier C, Lee JC (2018) Release of *Staphylococcus aureus* extracellular vesicles and their application as a vaccine platform. Nat Commun 9(1):1379. https://doi.org/10.1038/s41467-018-03847-z

Wensink J, Witholt B (1981) Outer-membrane vesicles released by normally growing *Escherichia coli* contain very little lipoprotein. Eur J Biochem 116(2):331–335

Whitchurch CB, Tolker-Nielsen T, Ragas PC, Mattick JS (2002) Extracellular DNA required for bacterial biofilm formation. Science 295(5559):1487. https://doi.org/10.1126/science.295.5559.1487

Wispelwey B, Hansen EJ, Scheld WM (1989) *Haemophilus influenzae* outer membrane vesicle-induced blood-brain barrier permeability during experimental meningitis. Infect Immun 57(8):2559–2562

Yaron S, Kolling GL, Simon L, Matthews KR (2000) Vesicle-mediated transfer of virulence genes from *Escherichia coli* O157:H7 to other enteric bacteria. Appl Environ Microbiol 66 (10):4414–4420

Yonezawa H, Osaki T, Kurata S, Fukuda M, Kawakami H, Ochiai K, Hanawa T, Kamiya S (2009) Outer membrane vesicles of *Helicobacter pylori* TK1402 are involved in biofilm formation. BMC Microbiol 9:197. https://doi.org/10.1186/1471-2180-9-197

Zavan L, Bitto NJ, Johnston EL, Greening DW, Kaparakis-Liaskos M (2019) *Helicobacter pylori* growth stage determines the size, protein composition, and preferential cargo packaging of outer membrane vesicles. Proteomics 19:e1800209. https://doi.org/10.1002/pmic.201800209

# Chapter 2
# Biogenesis of Gram-Negative OMVs

**Franz G. Zingl, Deborah R. Leitner, and Stefan Schild**

**Abstract** Initially dismissed as artifacts during sample preparation or as cell debris in bacterial cultures, outer membrane vesicles (OMVs) are now widely accepted as facsimiles of the outer membrane naturally secreted by Gram-negative bacteria. Within the last decades, several studies focused on OMV biogenesis resulting in different, in part complementary models to explain OMV release. Notably, vesicle formation seems to be an essential process as neither a bacterial species nor a mutant lacking vesicle release has been reported so far. Based on the complex physiological roles discussed for OMVs, it is likely that parallel strategies for their production have evolved to ensure their release in diverse conditions. Although, we are still far from a comprehensive mechanistic understanding of OMV release, several studies shed some insights in the processes driving the liberation of OMVs from the bacterial surface. Within this chapter, we will discuss the observations resulting in the current models for OMV formation including their regulation and relevance for bacterial physiology.

## 2.1 Introduction

Over the last years, the biogenesis of Gram-negative outer membrane vesicles (OMVs) has been investigated by several groups. This has resulted in a range of models for OMV release from bacteria (Fig. 2.1). Indeed, it is likely that bacteria encompass several complementary mechanisms for vesicle biogenesis.

First reports indicated that mutants harboring a loss or reduction of linkages between the outer membrane and the periplasmic peptidoglycan layer exhibit increased vesiculation (Deatherage et al. 2009; Suzuki et al. 1978; Schwechheimer

F. G. Zingl · D. R. Leitner
Institute of Molecular Biosciences, University of Graz, Graz, Austria

S. Schild (✉)
Institute of Molecular Biosciences, University of Graz, Graz, Austria

BioTechMed-Graz, Graz, Austria
e-mail: stefan.schild@uni-graz.at



M. Kaparakis-Liaskos, T. A. Kufer (eds.), *Bacterial Membrane Vesicles*,
https://doi.org/10.1007/978-3-030-36331-4_2

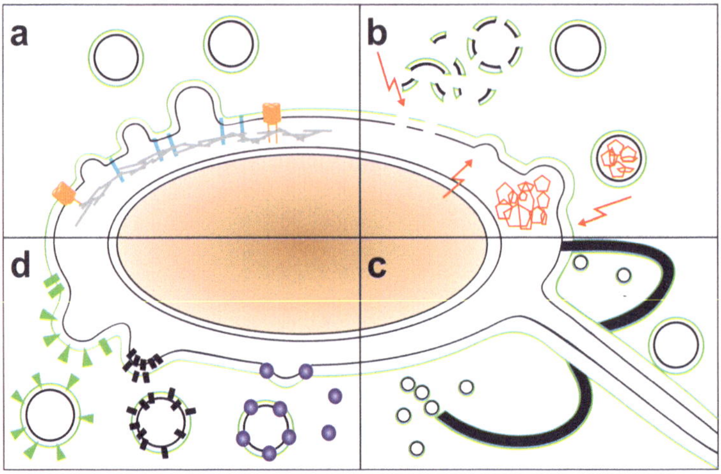


**Fig. 2.1** Schematic overview of OMV biogenesis routes discussed in this chapter. Shown are models of OMV formation based on: modulation of outer membrane–peptidoglycan linkages (**a**), bacterial stress responses (**b**), filamentous and tubular surface structures (**c**), and modulation of outer membrane components or outer membrane composition (**d**). In detail, (**a**) shows the mechanisms of OMV biogenesis due to loss of linkages between the outer membrane [outer leaflet composed of LPS (green) and inner leaflet composed of phospholipids (black)] and the peptidoglycan layer (gray) either by porins (orange) or proteins (light blue). (**b**) shows OMV formation due to stressors (red arrows) ranging from membrane attacking substances, high temperature, DNA damage or bacteriophages and accumulation of osmolytes (red). (**c**) shows OMV formation associated to surface structures like the sheathed flagellum (black) as well as tube-like outer membrane extensions such as nanopods, nanotubes, or nanowires. (**d**) indicates the mechanism of OMV formation via outer membrane modulation ranging from LPS modulation (green rectangles to triangles), intercalation of curvature-inducing molecules (purple circles), and modulation of phospholipid composition (black rectangles)

et al. 2013; Iwami et al. 2007; Moon et al. 2012; Turner et al. 2015; Llamas et al. 2000; Yeh et al. 2010; Mitra et al. 2016; Bernadac et al. 1998). Although these observations were predominantly achieved by loss-of-function mutants, current data suggest that bacteria can rearrange outer membrane–peptidoglycan linkages or regulate their abundance thereby affecting OMV production (Song et al. 2008; Choi et al. 2017; Schwechheimer et al. 2013).

Envelope stress is a general trigger for OMV release, highlighted by induction of vesiculation upon the presence of physical, chemical, and biological membrane stressors (Kadurugamuwa and Beveridge 1997). Moreover, accumulation of misfolded proteins and metabolites (e.g., peptidoglycan breakdown products) in the periplasm can result in increased OMV secretion, most likely via increased turgor pressure and bulging of the outer membrane (Schwechheimer et al. 2014; Hayashi et al. 2002). This could represent a response to harmful environments to increase the bacterial fitness and allow effective removal of undesired factors.

More recent studies revealed OMV secretion processes independent of mutations and presence of stressors, but allow refined regulation by the bacteria (Mashburn-

Warren et al. 2008; Tashiro et al. 2011; Roier et al. 2016; Bonnington and Kuehn 2016; Elhenawy et al. 2016). A first report in this direction investigated small bacterial molecules produced during defined stages of the bacterial life cycle, which induce surface curvature via intercalation into the outer membrane and consequently promote vesiculation (Mashburn-Warren et al. 2008). So far, this strategy seems restricted to certain species as curvature-inducing molecules have only been identified in *Pseudomonas aeruginosa* (Mashburn and Whiteley 2005). However, alternative OMV biogenesis models relying on modulation of the outer membrane composition that are applicable to a wide range of Gram-negative bacteria have been described. This includes several studies demonstrating that OMV release is affected by alterations in the lipopolysaccharide (LPS) composition, LPS modifications or phospholipid accumulation in the outer membrane via silencing of a retrograde lipid trafficking system (Tashiro et al. 2011; Roier et al. 2016; Bonnington and Kuehn 2016; Elhenawy et al. 2016; Sabra et al. 2003; Haurat et al. 2011). As LPS as well as the retrograde lipid trafficking system are quite conserved among Gram-negative species, these models probably reflect general mechanisms broadly applicable to a diverse set of bacteria.

OMV isolation and quantification protocols have improved within the last years. As OMV research spans over decades, different approaches were used to identify and characterize mechanisms of OMV biogenesis such as phenotypical analyses of OMV formation under differential cultivation conditions or random and site-directed mutants. Furthermore, standard operating procedures for isolation and quantification of bacterial membrane vesicles are currently lacking. Methods for OMV analyses range from protein measurement (e.g., Bradford assay, SDS, BCA assay), dry weight measurement, immune-based assays (e.g., ELISA, immunoblot, dotblot), lipid measurements (e.g., FM4-46, phospholipid measurement with ammonium ferrothiocyanate), LPS quantification (e.g., purpald assay, mass spectrometry analyses), or microscopical analyses (e.g., fluorescence microscopy, electron microscopy, nanosight). Thus, comparison of the results from different OMV biogenesis studies may require careful evaluation of the diverse methodologies used to isolate and quantify OMVs. We kindly refer the interested readers to the original articles cited throughout the chapter.

## 2.2 Modulation of Outer Membrane–Peptidoglycan Linkages

The cell envelope of Gram-negative bacteria is composed of two bilayered membranes. The so-called inner and the outer membranes are separated by the periplasm, which contains a thin layer of peptidoglycan. While the inner membrane has phospholipids in both leaflets, the outer membrane is composed of phospholipids in the inner leaflet and predominantly LPS in the outer leaflet. This asymmetric structure functions as a selective barrier against the extracellular environment (Ruiz et al. 2006). The peptidoglycan is covalently linked to the outer membrane and determines the shape of the cell's envelope. Moreover, it prevents Gram-negative cells from lysis due to osmotic changes or mechanical stress (Vollmer and Bertsche 2008). The peptidoglycan is a highly dynamic polymer, which consists of glycan chains that are crosslinked by short peptides (Vollmer and Bertsche 2008). Outer membrane proteins can interact with the peptidoglycan layer, thereby stabilizing the cell envelope by crosslinking the inner and outer membranes with the peptidoglycan. Among these, the main players regarding envelope cross linkage are the Braun's lipoprotein (Lpp), outer membrane porin A (OmpA) and the Tol-Pal complex. Several reports revealed that modulation of these covalent cross-linkages reduces outer membrane integrity and induces vesiculation, which will be discussed within this chapter.

### *2.2.1 Braun's Lipoprotein (Lpp)*

Lpp represents the Braun's lipoprotein in the outer membrane of *Escherichia coli*, with homologs found in a variety of Gram-negative species including all *Enterobacteriaceae* (Chang et al. 2012; Deatherage et al. 2009). It is the most abundant lipoprotein in *E. coli* and was shown to exist in a free or bound form, whereby the latter is covalently linked to the peptidoglycan (Braun 1975). About one third of bacterial Lpp is connected to peptidoglycan via the C-terminal lysine residue, thereby the outer membrane is anchored to the peptidoglycan via Lpp (Braun and Rehn 1969; Braun 1975). While Lpp is quite abundant in the outer membrane of *E. coli*, its OMVs contain relatively low levels of Lpp (Wensink and Witholt 1981; Hoekstra et al. 1976). In fact, OMVs contained almost none of the peptidoglycan-bound Lpp and only 35% of the free form compared to the corresponding bacterial outer membrane (Wensink and Witholt 1981). These results led to early models of OMV formation, wherein membrane blebbing occurs when the outer membrane expands faster than the underlying peptidoglycan or in surface areas with reduced prevalence of covalent linkages between Lpp in the outer membrane and peptidoglycan (Wensink and Witholt 1981; Hoekstra et al. 1976). Consequently, deletion of *lpp* leads to increased OMV formation in *E. coli* and in *Salmonella typhimurium* (Deatherage et al. 2009; Suzuki et al. 1978). Notably, loss of Lpp is associated with an impairment of the structural integrity of the outer membrane and cellular leakage (Suzuki et al. 1978; Deatherage et al. 2009). Thus, elevated OMV production might simply result from elevated cellular disintegration of *lpp* mutants. However, Lpp-dependent OMV formation might be more precisely regulated in wild-type cells. This regulation could be based on the reduction of Lpp crosslinks via the modulation of peptidoglycan or a change in the overall amount of Lpp via the small RNA Reg26 (Schwechheimer et al. 2013). Noteworthy, an uneven distribution of (bound) Lpp resulting in outer membrane areas with reduced Lpp amounts, as suggested for cell division sites (Hoekstra et al. 1976), might be more relevant for OMV formation than the overall Lpp cross-linkages as almost no Lpp is found in *E. coli* OMVs (Schwechheimer et al. 2014; Wensink and Witholt 1981).

### 2.2.2 Outer Membrane Protein A (OmpA)

OmpA is a major porin in the outer membrane of many Gram-negative species (Palva 1983; Wang 2002). Through its ability to non-covalently bind peptidoglycan via the conserved residues D271 and R286 of the C-terminus, it contributes to the structural integrity and stability of the cell envelope as it connects the peptidoglycan with the outer membrane (Samsudin et al. 2016; Iwami et al. 2007). Indeed, loss of OmpA results in an increase of OMV formation in *S. typhimurium*, *Porphyromonas gingivalis*, and *Acinetobacter baumannii* (Iwami et al. 2007; Moon et al. 2012; Deatherage et al. 2009). Both, Lpp and OmpA interactions, support membrane integrity. Consequently, OMV release by the *lpp*/*ompA* double mutant is significantly increased compared to each single mutant in *S. typhimurium* (Deatherage et al. 2009). In a similar manner, Sonntag and coworkers could demonstrate that an *lpp*/*ompA* double mutant in *E. coli* produced more OMVs than a *lpp* single mutant (Sonntag et al. 1978). Notably, Lpp-peptidoglycan linkages seem to be a dominant factor for increased OMV production in most bacteria probably due to the higher abundance of Lpp in the outer membrane (Deatherage et al. 2009). Concordantly, the impact of an *ompA* deletion in *S. typhimurium* is less pronounced as for *lpp* mutants and the loss of OmpA–peptidoglycan linkages was not necessarily associated with reduced membrane integrity (Deatherage et al. 2009). Noteworthy, bacteria can regulate *ompA* expression. For example, the small RNA VrrA of *Vibrio cholerae* interferes with *ompA* translation and its overexpression leads to increased vesiculation (Song et al. 2008). Similar results were obtained in *E. coli* and *S. typhimurium* by MicA, a homologue of VrrA, which downregulates *ompA* expression, thereby inducing OMV production in both organisms (Choi et al. 2017). As highlighted above, a similar regulation has also been observed for Lpp via the small RNA Reg26 (Schwechheimer et al. 2013). Interestingly, these small RNAs (VrrA, MicA, and Reg26) are under positive control of $\sigma^E$, an alternative sigma factor induced upon membrane stress, linking OMV formation to the envelope stress response (See Sect. 2.3.1) (Schwechheimer et al. 2013; Udekwu and Wagner 2007; Song et al. 2008).

### 2.2.3 Tol-Pal Complex

The Tol-Pal complex is widely conserved among Gram-negative bacteria and is crucial for multiple physiological roles including the maintenance of membrane integrity. In *E. coli* the apparatus consists of five proteins (TolQ-R-A-B and Pal) (Lazzaroni et al. 1999). While TolQ, TolR, and TolA form a complex in the cytoplasm, the periplasmic protein TolB interacts with Pal, which is non-covalently linked to the peptidoglycan (Lazzaroni et al. 1999; Parsons et al. 2006). Moreover, TolB and Pal interact with Lpp and OmpA at a protein level (Clavel et al. 1998; Cascales and Lloubes 2004), indicating that the Tol-Pal complex belongs to a protein network, which connects the periplasmic peptidoglycan layer with the outer membrane.

Accordingly, *tol-pal* mutants exhibit elevated OMV formation associated with loss of membrane integrity and leakage of periplasmic proteins into the extracellular milieu in *E. coli* (Bernadac et al. 1998). Concordantly, *tolQ-, tolR-, and tolA*-mutants in *Pseudomonas putida* produce more OMVs, loss of *tolA* resulted in more vesicle protein amount in *Shigella boydii*, deletion of *tolB* in *Helicobacter pylori* resulted in increased vesiculation and in *Caulobacter crescentus* extensive blebbing of OMVs were observed upon mutation of *pal, tolA*, or *tolB* (Turner et al. 2015; Llamas et al. 2000; Yeh et al. 2010; Mitra et al. 2016). In *tol-pal* mutants of *E. coli* and *C. crescentus*, a substantial fraction of OMVs originate from the cell poles and division sites where Tol-Pal usually accumulates in wild-type cells. Notably, *tol-pal* mutants fail to connect the outer membrane and peptidoglycan during cell division, which could explain the increased OMV release (Gerding et al. 2007; Yeh et al. 2010).

## 2.3 Bacterial Stress Responses Affecting Vesiculation

Facing the outside world, the cell envelope of bacterial cells is exposed to varying environmental conditions, which can be detrimental for microbial viability. Variations in temperature, pH, or the exposure to antimicrobial or toxic compounds can cause impaired protein folding and may result in cell death. Thus, the ability to rapidly sense and respond to these adverse conditions is essential for the survival of bacterial cells. Therefore, bacteria have developed regulatory networks to control gene expression according to their specific requirements. Some of them have been linked to increased vesiculation, which can enhance fitness under stressful conditions.

### *2.3.1 Envelope Stress Response*

In *E. coli*, adaptation to harsh environments is mediated by the activation of the signal transduction pathways of the bacterial envelope stress responses (Raivio 2005). One of them is the $\sigma^E$ stress response activated via DegS, a periplasmic sensor protease, which recognizes misfolded outer membrane proteins. Upon its activation, DegS cuts the anti-sigma factor RseA, which is again cleaved by RseP and ClpXP and finally releases the alternative sigma factor $\sigma^E$. Liberated $\sigma^E$ subsequently induces the transcription of genes involved in outer membrane protein and cell wall biogenesis, including DegP, a temperature-dependent periplasmic chaperone and protease (Hayden and Ades 2008). Recently, studies in *E. coli* and *S. typhimurium* linked the $\sigma^E$-mediated stress response to OMV biogenesis demonstrating increased vesiculation for mutants of the $\sigma^E$ pathway such as *degP*-, *degS*-, and *rseA*-mutants for *E. coli* and the *degP* mutant in case of *S. typhimurium* (McBroom and Kuehn 2007; McBroom et al. 2006). The authors concluded that the impairment of the functionality of the $\sigma^E$ pathway leads to an accumulation of misfolded products in the periplasm causing increased turgor pressure, which can be resolved by vesiculation.

The importance of OMV formation as a stress relief mechanism in *E. coli* was confirmed by mutation of *nlpA* (resulting in reduced vesiculation) in combination with *degP* (resulting in accumulation of periplasmic misfolded proteins) (Schwechheimer and Kuehn 2013). The reduced vesiculation of the *nlpA*/*degP*-mutant compared to a *degP* mutant caused a severe growth defect upon higher temperatures as misfolded proteins accumulated in the periplasm and could not be removed via vesiculation (Schwechheimer and Kuehn 2013). By the construction of a fusion protein, which mimicked an unfolded outer membrane protein, McBroom and coworkers could reinforce the model that vesiculation might be an alternative stress relief mechanism as this protein was significantly enriched in OMVs compared to other periplasmic proteins (McBroom and Kuehn 2007). Thus, they introduced OMV formation as a stress response of Gram-negative bacteria, which supports the cell's efforts to reduce the consequences of unfolded proteins.

Furthermore, the model of OMV formation as a stress relief mechanism could be extended from misfolded proteins to other compounds accumulating in the periplasm. For example, vesiculation was also increased in *E. coli* upon deletion of *ampG,* which is an inner membrane permease and *amiD,* an amidase breaking down large peptidoglycan fragments as well as upon deletion of the autolysin homologue *ami* in *P. gingivalis*. These mutations resulted in the accumulation of peptidoglycan fragments in the periplasm and a subsequent increase of turgor pressure (Schwechheimer et al. 2014; Hayashi et al. 2002). In a similar manner, OMV formation was increased in *rfaC* or *rfaG* deletion strains in *E. coli*. Both mutations led to an accumulation of LPS in the periplasm, due to an impaired LPS maturation in the outer membrane (Schwechheimer et al. 2014). Consequently, OMVs of *rfaC* or *rfaG* deletion mutants in *E. coli* exhibited an increase of lipid-to-outer membrane protein ratio (Schwechheimer et al. 2014). The detrimental effect of excessive membrane products was demonstrated by deletion of *yciM* in *E. coli*, which increases LPS production and its periplasmic accumulation. Interestingly, *E. coli* strains lacking *yciM* required specific suppressor mutations for survival, either involved in LPS synthesis (*lpxA*, *lpxC*, or *lpxD*) or enhance vesiculation (*lpp*, *tolA*, *pal, galU*) (Mahalakshmi et al. 2014; Kulp et al. 2015).

The Kuehn lab extended their OMV biogenesis model by showing that the $\sigma^E$-homologue AlgU of *P. aeruginosa* also plays a major role in OMV formation (Macdonald and Kuehn 2013). In concordance with earlier observations in *E. coli*, impairment of the envelope stress response in *P. aeruginosa* also resulted in enhanced OMV formation (Tashiro et al. 2009; Macdonald and Kuehn 2013). Interestingly, hypervesiculating mutants often showed no altered $\sigma^E$ activity in *E. coli* and deletion of *algU* in *P. aeruginosa* did not alter high vesiculation phenotypes of stressed cells, indicating that various mechanisms exist for OMV formation (Macdonald and Kuehn 2013; McBroom and Kuehn 2007). In the case of *P. aeruginosa*, these additional mechanisms might be modulations of surface structures or composition (see Sects. 2.5.1 and 2.5.2.3).

### 2.3.2 Cell Wall-Directed Agents

Various chemical and biological substances targeting the outer membrane have been shown to stimulate OMV formation. Increased OMV formation upon presence of sublethal concentrations of membrane-attacking substances (e.g., polymyxin B, LL-37 and colistin) was shown to be beneficial for bacterial survival as OMVs can act as sink for these antimicrobial compounds (Manning and Kuehn 2011; Duperthuy et al. 2013). The increased vesiculation induced by polymyxin B and colistin is based on the displacement of $Mg^{2+}$ and $Ca^{2+}$ cations in the outer membrane, which destabilize the electrostatic interactions between negatively charged LPS molecules (Moore and Hancock 1986; Storm et al. 1977). Upon treatment with polymyxin B, increased OMV levels were observed for *P. aeruginosa* (Macdonald and Kuehn 2013), *E. coli* and enterotoxigenic *Escherichia coli* (ETEC) (Manning and Kuehn 2011) as well as enterohemorrhagic *Escherichia coli* (EHEC) (Bauwens et al. 2017a; Bauwens et al. 2017b). Exposure to colistin resulted in elevated OMV formation in *E. coli* (Manning and Kuehn 2011). Similarly, the human cathelicidin LL-37, a cationic antimicrobial peptide was shown to induce pore formation in the outer membrane (Brogden 2005) and increases OMV formation in EHEC (Urashima et al. 2017). Presence of bile salts, acting as an emulsifier of biological membranes, also promotes vesicle formation in *Campylobacter jejuni* (Taheri et al. 2018). Moreover, gentamicin was shown to increase OMV formation in *P. aeruginosa* via destabilization of the outer membrane by binding to the LPS (Kadurugamuwa and Beveridge 1997).

Furthermore, highly hydrophobic carbon sources such as hexadecane or phenanthrene were reported to interact with the outer membrane of Gram-negative bacteria and are thought to induce vesiculation. Hexadecane stimulates OMV release in *Acinetobacter calcoaceticus* and phenanthrene induces vesiculation in *Delftia acidovorans* (Shetty and Hickey 2014; Borneleit et al. 1988). Although this is not a comprehensive list of cell wall-directed agents, it still indicates that substances that disturb or disrupt the outer membrane of Gram-negative bacteria are connected to OMV formation.

It is likely that several other cell wall-directed agents destabilize the outer membrane by similar modes of action as described above and thereby increase bacterial vesiculation. Elevated OMV levels caused by cell wall-directed agents could rather be a consequence of physical or chemical membrane damage, than a controlled OMV formation process actively regulated by the bacterium. Indeed, vesicles can reassemble from membrane material originating from lysed cells (Sych et al. 2018; Turnbull et al. 2016). However, cell wall-directed agents might activate the $\sigma^E$ pathway. As a matter of fact, polymyxin B was shown to induce *micA* in a $\sigma^E$-dependent manner in *S. typhimurium* (Papenfort et al. 2006) and $\sigma^E$ in *V. cholerae* (Mathur et al. 2007). Moreover, bile induces the SOS response in *E. coli* and *S. typhimurium* (Prieto et al. 2006). Thus, OMV formation by sublethal concentrations of cell wall-directed agents could be linked to regulatory pathways implicated in OMV biogenesis (see Sects. 2.3.1 and 2.3.3).

### 2.3.3 SOS Response and Bacteriophages

In *E. coli*, one regulatory network facilitating bacterial survival upon DNA damage is the so-called SOS response, representing an inducible DNA repair system (Simmons et al. 2008). Along the SOS response two regulatory proteins, the LexA repressor and the RecA protein, modulate the expression of more than 50 genes comprising, for example, *sulA*, an inhibitor of cell division; *uvr* genes, which are involved in DNA repair or *sbmC*, a DNA gyrase inhibitor (Simmons et al. 2008). While LexA inhibits the expression of the SOS genes during normal cell growth, RecA binds to single-stranded DNA caused by DNA damaging agents (e.g., antibiotics or UV radiation). Binding of RecA to single-stranded DNA activates RecA and stimulates the autocatalytic cleavage of LexA, thereby relieving the repressor from the SOS genes and leading to the expression of LexA repressed genes (Simmons et al. 2008). Recently, some studies have shown that the SOS response is closely linked to OMV biogenesis inducing OMV formation. Such an upregulation of OMV formation has been observed for EHEC and *Shigella dysenteriae* upon treatment with SOS response stimulating substances like ciprofloxacin or mitomycin C (Bauwens et al. 2017a; Dutta et al. 2004). In *P. aeruginosa*, Maredia and coworkers demonstrated that the OMV amount was higher under ciprofloxacin treatment compared to non-treated wild-type cells (Maredia et al. 2012). Since ciprofloxacin is an efficient SOS response stimulator, a noncleavable LexA strain (no induction of the SOS response) was used to test if the enhanced vesiculation can be solely attributed to the SOS response. Under ciprofloxacin treatment, OMV levels were higher in wild-type cells than in the noncleavable LexA strain. However, an induction of vesiculation upon ciprofloxacin treatment was also detected in the noncleavable LexA strain, indicating that vesiculation is linked to the SOS response, but also to alternative pathways (Maredia et al. 2012). Moreover, the study of Toyofuku and coworkers confirmed the association of the SOS response with enhanced OMV formation in *P. aerugionsa* as an increase of the OMV amount was also observed under denitrifying conditions, which induces the SOS response (Toyofuku et al. 2014). Although a concise mechanism is lacking, Maredia and coworkers speculated that the enhanced OMV formation is a consequence of the delay in cell division and cell surface alterations during the SOS response (Maredia et al. 2012).

DNA damage also activates lysogenic bacteriophages, which infect, parasitize, and lyse their host cell (Weinbauer 2004). These phages integrate their genome into the bacterial chromosome and replicate through the cell cycle of their host. Upon stressors such as UV radiation or antibiotics, resulting in bacterial DNA damage and induction of the SOS response, lysogenic phages become activated causing excision from the chromosome, proliferation of new phages, and lysis of the bacterial cell (Weinbauer 2004; Feiner et al. 2015). Recently, a novel OMV formation mechanism based on a phage effector was proposed for *P. aeruginosa* in biofilms (Turnbull et al. 2016). Turnbull and coworkers demonstrated that the SOS-dependent activation of the cryptic phages endolysin, which degrades the bacterial peptidoglycan, results in explosive cell lysis and high amounts of liberated OMVs. These OMVs are formed by shattered membrane fragments of the exploding bacteria, which self-assemble to vesicle structures (Turnbull et al. 2016). Notably, activation of the SOS response in *Stenotrophomonas maltophilia* triggers formation of vesicles containing outer and inner membranes (Devos et al. 2017). Such cell lysis-induced outer–inner membrane vesicles were also reported for *P. aeruginosa*, *Pseudoalteromonas marina*, and *Shewanella vesiculosa* (Perez-Cruz et al. 2013; Hagemann et al. 2014; Kadurugamuwa and Beveridge 1995). A similar mechanism based on cell lysis was shown to induce vesicle formation in the Gram-positive bacterium *Bacillus subtilis* mediated by the SOS-dependent activation of an endolysin encoded by a defective prophage (Toyofuku et al. 2017).

In contrast, lytic phages immediately replicate their genome upon infection until the bacterial cell bursts, thereby transferring the phages to new cells. During this process, increased vesicle formation was observed for *Yersinia pseudotuberculosis* upon infection with a pseudotuberculous diagnostic bacteriophage and for *E. coli* upon infection with the T4 phage, which is thought to occur due to increased osmotic stress and subsequent self-assembly of outer membrane fragments (Byvalov et al. 2018; Tarahovsky et al. 1994). Prophage(-like) elements are frequently found in bacterial genomes and lytic phages are constantly attacking bacterial cells (Casjens 2003; Weinbauer 2004). Thus, the induction of vesicle formation by phages might be a ubiquitous mechanism for the production of membrane vesicles ranging from Gram-negative to Gram-positive bacteria. Similar to cell wall-directed agents (see Sect. 2.3.2), phage-induced vesicle formation likely represents less of an active mechanism than a physical or chemical consequence of the self-assembly properties of biological membranes, which re-organize themselves to vesicles and vesicle-like structures (Sych et al. 2018).

## 2.4 Filamentous or Tubular Surface Structures

Recently, new mechanisms of OMV biogenesis have been addressed by several studies dealing with the outer membrane sheathed flagellum or nanotube-like structures of Gram-negative bacteria. Several bacterial species like *Brucella melitensis*, *Vibrio ssp.*, or *H. pylori* assemble flagella covered with an outer membrane-derived sheath (Fuerst and Perry 1988; Geis et al. 1993; Ferooz and Letesson 2010). Interestingly, for *Vibrio fischeri*, *B. melitensis*, and *H. pylori* vesicle-like structures were found along the sheath, frequently localized at the distal tip of the flagella (Ferooz and Letesson 2010; Millikan and Ruby 2004; Qin et al. 2016). In the case of *V. fischeri*, rotation of the flagellum results in release of more LPS molecules, which were later shown to be shed via OMVs (Brennan et al. 2014; Aschtgen et al. 2016). Interestingly, Aschtgen and coworkers also observed different OMV sizes depending on the presence and functionality of the bacterium's sheathed flagellum. A hyperflagellated mutant produced a higher proportion of smaller OMVs whereas a nonmotile mutant produced fewer OMVs that tend to be slightly larger in electron microscopy than those generated by wild-type cells (Aschtgen et al. 2016). Hampton and coworkers reported similar observations for *Vibrio vulnificus* comparing the amount and size of OMVs via electron microscopy derived from motile and nonmotile strains (Hampton et al. 2017). Notably, recent studies show that the bacterial growth phase determines OMV size and protein composition in *H. pylori*, a pathogen with growth phase-dependent expression of flagellar genes (Niehus et al. 2002; Zavan et al. 2019). Taken together, these results not only demonstrate a novel mechanism of OMV formation by bacteria with sheathed flagella, but also raise the question if at least two OMV populations exits, being either released from the cellular body or from the flagellar sheath. Due to their different site of origin, these vesicles may differ in their size and composition (Aschtgen et al. 2016).

Recently, studies showed that bacteria are also able to produce other outer membrane structures such as nanotubes, nanopods, or nanowires, which stay connected to their originating cell and enable them to interact directly with distant surfaces. Among them, the presence of nanopods and nanowires has just recently been reported (Shetty et al. 2011). Nanopods are described as outer membrane tubes filled with OMVs and may represent a new bacterial organelle, still their biogenesis and regulation are currently not well defined (Shetty et al. 2011). In *D. acidovorans* nanopods were shown to play an important role in the biodegradation of the highly hydrophobic phenanthrene as cells producing almost no nanopods were massively attenuated upon growth on this carbon source (Shetty and Hickey 2014).

*Shewanella oneidensis*, a metal-reducing bacterium was found to produce extensions of the outer membrane and periplasm called nanowires, which were shown to enable the cells to link the respiratory chain to external (metal) electron acceptors (Pirbadian et al. 2014).

Nanotubes have been described as tube-shaped membranous structures and are considered to be a specialized form of OMV production (McCaig et al. 2013). It has been suggested that these structures connect neighboring cells thereby playing important roles in cell–cell communication or the exchange of molecules in *Myxococcus xanthus*, enteropathogenic *Escherichia coli* (EPEC), *V. vulnificus*, and *Francisella novicida* (Remis et al. 2014; Hampton et al. 2017; McCaig et al. 2013; Pal et al. 2019). Though their nature is not completely understood, in EPEC nanotubes are thought to be evolutionary related to the injectosome and the flagellum (Pal et al. 2019). A first step toward a regulatory mechanism is a study showing that nanotube production is induced upon amino acid starvation in *F. novicida* indicating their importance for nutrient acquisition (Sampath et al. 2018).

Taken together, these findings suggest that nanotubes, nanopods and nanowires represent a specialized type of OMVs that stay connected to the cell, thereby facilitating nutrient degradation and acquisition of highly specialized bacteria, aid in energy production cycles or enable cell to cell communication.

## 2.5 Modulation of Outer Membrane Components or Composition

All domains of life share a bilayered membrane mainly composed of phospholipids in both leaflets or at least in the inner leaflet (i.e., outer membrane of Gram-negative bacteria). For eukaryotic cells, the curvature alteration of membranes plays an important role in defining the cell morphology, organelles, and local subdomains and is therefore crucial for compartmentalizing proteins and enzymatic reactions (McMahon and Boucrot 2015). Besides changes in the phospholipid composition, curvature induction in eukaryotes is predominately mediated via proteins, which may be integral components of the membrane or are able to induce curvature via binding from the inside or outside of the membrane (Zimmerberg and Kozlov 2006; McMahon and Boucrot 2015). Notably, mechanisms for the alteration of membrane curvatures have also been identified in bacteria, which range from the insertion of curvature-inducing small molecules (e.g., *Pseudomonas* quinolone signal PQS), modification of membrane components (e.g., LPS), or alteration of the membrane composition (e.g., phospholipid accumulation) (Mashburn-Warren et al. 2008; Tashiro et al. 2011; Roier et al. 2016; Bonnington and Kuehn 2016; Elhenawy et al. 2016). All of these processes have also been demonstrated to affect OMV formation, which will be discussed in this chapter.

### 2.5.1 *Pseudomonas Quinolone Signal (PQS)*

The quorum sensing signal molecule 2-heptyl-3-hydroxy-4-quinolone (*Pseudomonas* quinolone signal; PQS) is involved in communication and growth phase coordination of *P. aeruginosa*. The inter-bacterial delivery of this hydrophobic molecule in aqueous solution was not understood until Mashburn and coworkers demonstrated that PQS is transferred via OMVs (Mashburn and Whiteley 2005). Aside from its role in quorum sensing signaling, PQS also induces OMV formation. In *P. aeruginosa* it seems to be a dominant biogenesis mechanism as a *pqsA* mutant, unable to produce PQS, exhibits massively reduced OMV levels (Mashburn and Whiteley 2005). Due to its chemical properties, it is able to interact with the acyl chains of lipid A, a component of LPS, bringing them into a more ordered gel-like state. Thereby PQS reduces membrane fluidity, which facilitates bulging of the membrane and OMV release. Importantly, the high fluidity of the *P. aeruginosa* outer membrane likely prevents OMV blebbing without PQS (Schertzer and Whiteley 2012; Mashburn and Whiteley 2005; Mashburn-Warren et al. 2008). Consistent with the observation that PQS is strictly produced under aerobic growth, OMV formation under anaerobic conditions is massively reduced (Schertzer and Whiteley 2013). Interestingly, vesicle formation can also be induced upon the addition of PQS to cultures of other Gram-negative or Gram-positive bacteria including *E. coli*, *Burkholderia cepacia*, and *B. subtilis* and is thought to interact with their membranes (Mashburn-Warren et al. 2008; Tashiro et al. 2010).

This fascinating observation implicates a unique strategy allowing interspecies controlled induction of vesicle formation in bacteria.

### 2.5.2 *LPS Modifications*

LPS is a major component of the outer membrane of Gram-negative bacteria and constitutes its outer leaflet. This molecule consists of three components and is highly variable between species. The O-antigen is the outermost part of the LPS with the highest diversity and is mainly recognized by the host immune system. The core region is in parts highly conserved and plays an important role in outer membrane stability. The lipid A, or so-called endotoxin, represents the membrane anchor of the LPS. In many species, it is a target molecule for defined, inducible modifications to increase resistance against antimicrobial factors, which need to penetrate through the outer membrane or directly attack the outer membrane integrity (Hankins et al. 2012; Gunn 2001). Due to its central role in the bacterial surface it is not surprising that modifications in all three parts of the LPS molecule have been reported to affect OMV formation in various species, which will be discussed below.

#### 2.5.2.1 Lipid A Modifications

The biosynthesis of lipid A is a multistep process wherein many modifications can be introduced such as glycinations, changes in acylation patterns or the attachment of phosphoethanolamine or aminoarabinose (Hankins et al. 2012; Raetz et al. 2007; Brozek and Raetz 1990; Gunn et al. 1998). These lipid A modifications have been shown to promote resistance to antimicrobial peptides, reduce host immune system activation, change in membrane stability, or change in vesicle formation (Raetz et al. 2007; Kawasaki et al. 2004b; Gwozdzinski 2018).

A recent study showed that the addition of a phosphoethanolamine group onto the N-acetylglucosamine disaccharide of lipid A via *mcr-1* in *E. coli* increases OMV formation (Gwozdzinski 2018). The current model suggests this modification causes a charge repulse inducing vesiculation (Gwozdzinski 2018). In *S. typhimurium* PagL can deacetylate lipid A to downregulate the host immune response. PagL is activated via the two-component system PhoPQ upon entering the macrophages (Kawasaki et al. 2004a; Trent et al. 2001). The deacylation in the 3-position of the lipid A also increases OMV formation inside of macrophages, where OMVs are thought to interfere with host cellular pathways. The loss of the acyl chain is thought to generate a positive curvature of the outer membrane by changing the shape of the lipid A and thereby induce vesiculation (Kawasaki et al. 2004a; Elhenawy et al. 2016).

#### 2.5.2.2 LPS Core Modifications

Mutations in the core region of the LPS can have a massive impact on the bacterial surface structure, colony morphology, and outer membrane integrity. Several LPS core mutants showed elevated OMV levels in *P. aeruginosa* (Ruhal et al. 2015; Salkinoja-Salonen and Nurmiaho 1978). Similar results were observed in *E. coli* (Nakao et al. 2012; Schwechheimer et al. 2014). These studies often relied on mutants lacking the majority of the LPS core and, as a consequence, the entire O-antigen. As full length LPS has a pivotal impact on outer membrane stability, such core mutants have an impaired outer membrane integrity, which could explain the increased liberation of outer membrane material. Yet, a study by Schwechheimer and coworkers described a mechanism for the increased vesiculation observed in LPS core mutants (Schwechheimer et al. 2014). According to their model, enhanced vesiculation in *E. coli rfaC* or *rfaB* mutants lacking full-length LPS results from accumulation of periplasmic LPS and subsequent increase in turgor pressure (Schwechheimer et al. 2014).

#### 2.5.2.3 O-Antigen Modifications

*P. aeruginosa* was shown to express two distinct types of LPS varying in the O-antigen composition. One is a homopolymer of D-rhamnose and is named common polysaccharide antigen (CPA, formerly termed A-band), the other is a heteropolymer of three to five distinct sugars in its repeat units known as O-specific antigen (OSA, formerly termed B-band). *P. aeruginosa* mutants lacking OSA or CPA were shown to produce OMVs with altered size and protein composition (Murphy et al. 2014). While OSA is more immunogenic, CPA plays an important role in binding to human cells (Lam et al. 2011). Interestingly, *P. aeruginosa* OMVs were shown to be enriched with the negatively charged OSA (Kadurugamuwa and Beveridge 1995). Enrichment of OMVs with OSA is believed to be the consequence of the charge repulsive forces between adjacent OSA and other LPS molecules in the outer membrane. Interestingly, exposure to oxidative stress increased the abundance of negatively charged OSA and OMV formation (Sabra et al. 2003). Thus, increased vesiculation during stress conditions might be induced by the AlgU pathway (see Sect. 2.3.1) and alterations in O-antigen composition to facilitate survival of *P. aeruginosa*.

*P. gingivalis* also produces two classes of LPS, carrying either neutral (O-LPS) or negatively charged (A-LPS) O-antigen chains (Paramonov et al. 2001; Rangarajan et al. 2008; Paramonov et al. 2005). OMVs of this human oral pathogen are enriched in A-LPS, which may implicate a role for this O-antigen species in OMV release (Haurat et al. 2011). Indeed, for species with neutral and negatively charged O-antigens, the latter has been found to be enriched in OMVs. Thus, it could be suggested that an interaction between negatively charged O-antigen species contributes to OMV formation comparable to the destabilization of the OM of *P. aeruginosa* by negatively charged gentamicin (see Sect. 2.3.2; Kadurugamuwa and Beveridge 1997).

### 2.5.3 Modulation of Phospholipid Composition

The outer membrane of Gram-negative bacteria is composed of an asymmetric bilayer with phospholipids in the inner leaflet and LPS on the outside. This asymmetry is maintained by a putative retrograde phospholipid trafficking system transporting phospholipids from the outer to the inner membrane (Malinverni and Silhavy 2009). This system was first described in *E. coli*, but is highly conserved among Gram-negative bacteria (Roier et al. 2016; Malinverni and Silhavy 2009). In *E. coli*, the system is composed of the outer membrane lipoprotein VacJ (MlaA), the periplasmic binding/transport protein YrbC (MlaC), and a permease complex consisting of YrbB (MlaB), YrbD (MlaD), YrbE (MlaE), and YrbF (MlaF) (Malinverni and Silhavy 2009).

Deletion or downregulation of this retrograde phospholipid transport system has been shown to increase OMV formation in multiple species including *E. coli*, *Haemophilus influenzae*, *V. cholerae*, and *Neisseria gonorrhoeae* (Roier et al. 2016; Baarda et al. 2019). In *H. influenzae*, comprehensive lipid analyses revealed a higher phospholipid abundance in the OMVs derived from *yrbE* and *vacJ* deletion mutants compared to wild type (Roier et al. 2016). This is accompanied by distinct changes of the lipid species composition, such as the enrichment of short-chain fatty acids, in the outer membrane and OMVs of *yrbE* and *vacJ* mutants compared to wild type (Roier et al. 2016). Together with the observation that the fatty acid composition of *P. aeruginosa* OMVs differs from the outer membrane, it can be hypothesized that arrangement of certain phospholipid might promote OMV biogenesis (Tashiro et al. 2011). Generalized models based on accumulation of defined lipid species are hampered by the heterogeneous phospholipid compositions found in the outer membrane of Gram-negative bacteria.

Interestingly, the phospholipid transport system was shown to be transcriptionally silenced upon iron limitation in a Fur (ferric uptake regulator)-dependent manner. As iron limitation is a signal for host environment for many bacteria, it is very likely that vesiculation plays a major role in colonization strategies of pathogenic bacteria (Roier et al. 2016). Along this line, an increased in vivo fitness upon deletion of this system has been reported for *P. aeruginosa* and *N. gonorrhoeae* (Shen et al. 2012; Baarda et al. 2019). In contrast, other studies report that mutants in the retrograde lipid transporter of *Shigella flexneri* and *Haemophilus parasuis* are attenuated in vivo (Carpenter et al. 2014; Zhao et al. 2017; Suzuki et al. 1994), indicating a current lack of understanding of the complex physiological processes involved. Another example of increased vesiculation due to phospholipid accumulation was observed in *Neisseria meningitidis*, where higher production of phospholipids triggered by sulfate depletion led to increased OMV production. Moreover, sulfate depletion is associated with oxidative stress, which has already been shown to be a signal for increased vesiculation in *P. aeruginosa* (see Sect. 2.5.1; Gerritzen et al. 2019; Macdonald and Kuehn 2013).

## 2.6 Conclusion

Long neglected within the scientific community, the vital role of membrane vesicles is nowadays increasingly studied and has been proven for all branches of life ranging from Gram-negative and Gram-positive bacteria to archaea and even eukaryotic cells (Deatherage and Cookson 2012). Among them, OMVs of Gram-negative bacteria have been most extensively studied. Pioneer work of the Beveridge and Kuehn labs on OMV physiology and biogenesis provided basic knowledge on the existence and multiple physiological roles of these spherical surface derivatives, which ultimately opened this field to become a striving research topic (Horstman and Kuehn 2000; Kadurugamuwa et al. 1993). Noteworthy, a lack of vesiculation has neither been observed for the investigated bacterial species nor any mutants, implicating that outer membrane vesiculation might be an essential process. This chapter provided a concise overview of the most prominent models of OMV formation of Gram-negative bacteria, which have been reported so far (Fig. 2.1). Importantly, all of these mechanisms could act in concert without excluding each other.

While vesiculation due to the loss of outer membrane and peptidoglycan interaction has mostly been observed in mutant strains lacking important structural proteins, recent studies showed that bacteria may actively regulate the abundance of such integral molecules via small RNAs (Song et al. 2008; Choi et al. 2017; Schwechheimer et al. 2013). In a similar manner, bacteria can modulate their surface composition to alter vesiculation. For example, vesiculation models based on modulation of phospholipid abundance or LPS act on structures ubiquitously present in Gram-negative bacteria (Gerritzen et al. 2019; Roier et al. 2016; Kawasaki et al. 2004a; Elhenawy et al. 2016). In contrast, the only curvature-inducing molecule known to date (PQS) is exclusively produced by *P. aeruginosa* although it induces vesicle formation in other strains including Gram-positive bacteria (Mashburn-Warren et al. 2008; Tashiro et al. 2010). Vesiculation due to external stressors like cell wall-directed agents, UV radiation or high temperature is not only a consequence of bursting cells or increased turgor pressure, but can also be induced by modulation of the $\sigma^E$ pathway or the SOS response (McBroom and Kuehn 2007; McBroom et al. 2006; Maredia et al. 2012). Indeed, the increased release of vesicles upon exposure to hostile conditions can be beneficial for bacterial cells due to their abilities to act as decoys or relief of turgor pressure.

The multitude of OMV biogenesis mechanisms as well as the lack of strains showing no vesiculation emphasizes an important role of these surface facsimiles for bacterial fitness. Interestingly, OMVs play an important role in host–pathogen interactions including toxin delivery, immunomodulation, binding of antimicrobial factors as well as the acquisition of nutrients. Thus, it could be speculated that pathogenic or human-associated bacteria are generally prone to high vesiculation in order to aid in pathogenesis.

Recent work has already revealed quite comprehensive and complementary OMV biogenesis models. It is likely that in the same species, several OMV biogenesis mechanisms are simultaneously active or conditionally induced. Thus, one species might produce a variety of OMV types of different sizes or composition, reflecting diverse OMV biogenesis routes. The fast progression of techniques to study OMVs ranging from their isolation, composition analyses, and microscopic visualization will drive future investigation of OMV biogenesis, morphology, and content.

**Acknowledgements** This work was supported by the BioTechMed-Graz Flagship Project SECRETOME to S.S, the Austrian FWF grants P25691 to S.S., P27654 to S.S., and W901-B12 (DK Molecular Enzymology) to F.G.Z. and S.S.

## References

Aschtgen MS, Lynch JB, Koch E, Schwartzman J, McFall-Ngai M, Ruby E (2016) Rotation of *Vibrio fischeri* flagella produces outer membrane vesicles that induce host development. J Bacteriol 198 (16):2156–2165. https://doi.org/10.1128/JB.00101-16

Baarda BI, Zielke RA, Le Van A, Jerse AE, Sikora AE (2019) *Neisseria gonorrhoeae* MlaA influences gonococcal virulence and membrane vesicle production. PLoS Pathog 15(3): e1007385. https://doi.org/10.1371/journal.ppat.1007385

Bauwens A, Kunsmann L, Karch H, Mellmann A, Bielaszewska M (2017a) Antibiotic-mediated modulations of outer membrane vesicles in Enterohemorrhagic *Escherichia coli* O104:H4 and O157:H7. Antimicrob Agents Chemother 61(9):e00937–e00917. https://doi.org/10.1128/AAC.00937-17

Bauwens A, Kunsmann L, Marejkova M, Zhang W, Karch H, Bielaszewska M, Mellmann A (2017b) Intrahost milieu modulates production of outer membrane vesicles, vesicle-associated Shiga toxin 2a and cytotoxicity in *Escherichia coli* O157:H7 and O104:H4. Environ Microbiol Rep 9(5):626–634. https://doi.org/10.1111/1758-2229.12562

Bernadac A, Gavioli M, Lazzaroni JC, Raina S, Lloubes R (1998) *Escherichia coli* tol-pal mutants form outer membrane vesicles. J Bacteriol 180(18):4872–4878

Bonnington KE, Kuehn MJ (2016) Outer membrane vesicle production facilitates LPS remodeling and outer membrane maintenance in *Salmonella* during environmental transitions. mBio 7(5): e01532-16. https://doi.org/10.1128/mBio.01532-16

Borneleit P, Hermsdorf T, Claus R, Walther P, Kleber HP (1988) Effect of hexadecane-induced vesiculation on the outer membrane of *Acinetobacter calcoaceticus*. J Gen Microbiol 134 (7):1983–1992

Braun V (1975) Covalent lipoprotein from the outer membrane of *Escherichia coli*. Biochim Biophys Acta 415(3):335–377

Braun V, Rehn K (1969) Chemical characterization, spatial distribution and function of a lipoprotein (murein-lipoprotein) of the *E. coli* cell wall. The specific effect of trypsin on the membrane structure. Eur J Biochem 10(3):426–438

Brennan CA, Hunt JR, Kremer N, Krasity BC, Apicella MA, McFall-Ngai MJ, Ruby EG (2014) A model symbiosis reveals a role for sheathed-flagellum rotation in the release of immunogenic lipopolysaccharide. elife 3:e01579. https://doi.org/10.7554/eLife.01579

Brogden KA (2005) Antimicrobial peptides: pore formers or metabolic inhibitors in bacteria? Nat Rev Microbiol 3(3):238–250. https://doi.org/10.1038/nrmicro1098

Brozek KA, Raetz CR (1990) Biosynthesis of lipid A in *Escherichia coli*. Acyl carrier protein-dependent incorporation of laurate and myristate. J Biol Chem 265(26):15410–15417

Byvalov AA, Malkova MA, Chernyad'ev AV, Dudina LG, Litvinets SG, Martinson EA (2018) Influence of specific bacteriophage on the level of vesicle formation and morphology of cells of

*Yersinia pseudotuberculosis*. Bull Exp Biol Med 165(3):403–407. https://doi.org/10.1007/s10517-018-4180-0
Carpenter CD, Cooley BJ, Needham BD, Fisher CR, Trent MS, Gordon V, Payne SM (2014) The Vps/VacJ ABC transporter is required for intercellular spread of *Shigella flexneri*. Infect Immun 82(2):660–669. https://doi.org/10.1128/IAI.01057-13
Cascales E, Lloubes R (2004) Deletion analyses of the peptidoglycan-associated lipoprotein Pal reveals three independent binding sequences including a TolA box. Mol Microbiol 51 (3):873–885
Casjens S (2003) Prophages and bacterial genomics: what have we learned so far? Mol Microbiol 49(2):277–300
Chang TW, Lin YM, Wang CF, Liao YD (2012) Outer membrane lipoprotein Lpp is Gram-negative bacterial cell surface receptor for cationic antimicrobial peptides. J Biol Chem 287(1):418–428. https://doi.org/10.1074/jbc.M111.290361
Choi HI, Kim M, Jeon J, Han JK, Kim KS (2017) Overexpression of MicA induces production of OmpC-enriched outer membrane vesicles that protect against *Salmonella* challenge. Biochem Biophys Res Commun 490(3):991–996. https://doi.org/10.1016/j.bbrc.2017.06.152
Clavel T, Germon P, Vianney A, Portalier R, Lazzaroni JC (1998) TolB protein of *Escherichia coli* K-12 interacts with the outer membrane peptidoglycan-associated proteins Pal, Lpp and OmpA. Mol Microbiol 29(1):359–367
Deatherage BL, Cookson BT (2012) Membrane vesicle release in bacteria, eukaryotes, and archaea: a conserved yet underappreciated aspect of microbial life. Infect Immun 80(6):1948–1957. https://doi.org/10.1128/IAI.06014-11
Deatherage BL, Lara JC, Bergsbaken T, Rassoulian Barrett SL, Lara S, Cookson BT (2009) Biogenesis of bacterial membrane vesicles. Mol Microbiol 72(6):1395–1407
Devos S, Van Putte W, Vitse J, Van Driessche G, Stremersch S, Van Den Broek W, Raemdonck K, Braeckmans K, Stahlberg H, Kudryashev M, Savvides SN, Devreese B (2017) Membrane vesicle secretion and prophage induction in multidrug-resistant *Stenotrophomonas maltophilia* in response to ciprofloxacin stress. Environ Microbiol 19(10):3930–3937. https://doi.org/10.1111/1462-2920.13793
Duperthuy M, Sjostrom AE, Sabharwal D, Damghani F, Uhlin BE, Wai SN (2013) Role of the *Vibrio cholerae* matrix protein Bap1 in cross-resistance to antimicrobial peptides. PLoS Pathog 9(10):e1003620. https://doi.org/10.1371/journal.ppat.1003620
Dutta S, Iida K, Takade A, Meno Y, Nair GB, Yoshida S (2004) Release of Shiga toxin by membrane vesicles in *Shigella dysenteriae* serotype 1 strains and in vitro effects of antimicrobials on toxin production and release. Microbiol Immunol 48(12):965–969
Elhenawy W, Bording-Jorgensen M, Valguarnera E, Haurat MF, Wine E, Feldman MF (2016) LPS remodeling triggers formation of outer membrane vesicles in *Salmonella*. mBio 7(4):e00940-16. https://doi.org/10.1128/mBio.00940-16
Feiner R, Argov T, Rabinovich L, Sigal N, Borovok I, Herskovits AA (2015) A new perspective on lysogeny: prophages as active regulatory switches of bacteria. Nat Rev Microbiol 13 (10):641–650. https://doi.org/10.1038/nrmicro3527
Ferooz J, Letesson JJ (2010) Morphological analysis of the sheathed flagellum of *Brucella melitensis*. BMC Res Notes 3:333. https://doi.org/10.1186/1756-0500-3-333
Fuerst JA, Perry JW (1988) Demonstration of lipopolysaccharide on sheathed flagella of *Vibrio cholerae* O:1 by protein A-gold immunoelectron microscopy. J Bacteriol 170(4):1488–1494
Geis G, Suerbaum S, Forsthoff B, Leying H, Opferkuch W (1993) Ultrastructure and biochemical studies of the flagellar sheath of *Helicobacter pylori*. J Med Microbiol 38(5):371–377. https://doi.org/10.1099/00222615-38-5-371
Gerding MA, Ogata Y, Pecora ND, Niki H, de Boer PA (2007) The trans-envelope Tol-Pal complex is part of the cell division machinery and required for proper outer-membrane invagination during cell constriction in *E. coli*. Mol Microbiol 63(4):1008–1025. https://doi.org/10.1111/j.1365-2958.2006.05571.x

Gerritzen MJH, Martens DE, Uittenbogaard JP, Wijffels RH, Stork M (2019) Sulfate depletion triggers overproduction of phospholipids and the release of outer membrane vesicles by *Neisseria meningitidis*. Sci Rep 9(1):4716. https://doi.org/10.1038/s41598-019-41233-x

Gunn JS (2001) Bacterial modification of LPS and resistance to antimicrobial peptides. J Endotoxin Res 7(1):57–62

Gunn JS, Lim KB, Krueger J, Kim K, Guo L, Hackett M, Miller SI (1998) PmrA-PmrB-regulated genes necessary for 4-aminoarabinose lipid A modification and polymyxin resistance. Mol Microbiol 27:1171–1182

Gwozdzinski K (2018) Macromolecular modification of the cell wall of Gram-negative bacteria leading to antibiotic resistance and formation of outer membrane vesicles. Justus-Liebig-Universität Gießen, Gießen

Hagemann S, Stoger L, Kappelmann M, Hassl I, Ellinger A, Velimirov B (2014) DNA-bearing membrane vesicles produced by *Ahrensia kielensis* and *Pseudoalteromonas marina*. J Basic Microbiol 54(10):1062–1072. https://doi.org/10.1002/jobm.201300376

Hampton CM, Guerrero-Ferreira RC, Storms RE, Taylor JV, Yi H, Gulig PA, Wright ER (2017) The opportunistic pathogen *Vibrio vulnificus* produces outer membrane vesicles in a spatially distinct manner related to capsular polysaccharide. Front Microbiol 8:2177. https://doi.org/10.3389/fmicb.2017.02177

Hankins JV, Madsen JA, Giles DK, Brodbelt JS, Trent MS (2012) Amino acid addition to *Vibrio cholerae* LPS establishes a link between surface remodeling in gram-positive and gram-negative bacteria. Proc Natl Acad Sci U S A 109(22):8722–8727

Haurat MF, Aduse-Opoku J, Rangarajan M, Dorobantu L, Gray MR, Curtis MA, Feldman MF (2011) Selective sorting of cargo proteins into bacterial membrane vesicles. J Biol Chem 286 (2):1269–1276. https://doi.org/10.1074/jbc.M110.185744

Hayashi J, Hamada N, Kuramitsu HK (2002) The autolysin of *Porphyromonas gingivalis* is involved in outer membrane vesicle release. FEMS Microbiol Lett 216(2):217–222

Hayden JD, Ades SE (2008) The extracytoplasmic stress factor, sigmaE, is required to maintain cell envelope integrity in *Escherichia coli*. PLoS One 3(2):e1573. https://doi.org/10.1371/journal.pone.0001573

Hoekstra D, van der Laan JW, de Leij L, Witholt B (1976) Release of outer membrane fragments from normally growing *Escherichia coli*. Biochim Biophys Acta 455(3):889–899

Horstman AL, Kuehn MJ (2000) Enterotoxigenic *Escherichia coli* secretes active heat-labile enterotoxin via outer membrane vesicles. J Biol Chem 275(17):12489–12496

Iwami J, Murakami Y, Nagano K, Nakamura H, Yoshimura F (2007) Further evidence that major outer membrane proteins homologous to OmpA in *Porphyromonas gingivalis* stabilize bacterial cells. Oral Microbiol Immunol 22(5):356–360

Kadurugamuwa JL, Beveridge TJ (1995) Virulence factors are released from *Pseudomonas aeruginosa* in association with membrane vesicles during normal growth and exposure to gentamicin: a novel mechanism of enzyme secretion. J Bacteriol 177(14):3998–4008

Kadurugamuwa JL, Beveridge TJ (1997) Natural release of virulence factors in membrane vesicles by *Pseudomonas aeruginosa* and the effect of aminoglycoside antibiotics on their release. J Antimicrob Chemother 40(5):615–621

Kadurugamuwa JL, Clarke AJ, Beveridge TJ (1993) Surface action of gentamicin on *Pseudomonas aeruginosa*. J Bacteriol 175(18):5798–5805

Kawasaki K, Ernst RK, Miller SI (2004a) 3-O-deacylation of lipid A by PagL, a PhoP/PhoQ-regulated deacylase of *Salmonella typhimurium*, modulates signaling through Toll-like receptor 4. J Biol Chem 279(19):20044–20048. https://doi.org/10.1074/jbc.M401275200

Kawasaki K, Ernst RK, Miller SI (2004b) Deacylation and palmitoylation of lipid A by Salmonellae outer membrane enzymes modulate host signaling through Toll-like receptor 4. J Endotoxin Res 10(6):439–444. https://doi.org/10.1179/096805104225006264

Kulp AJ, Sun B, Ai T, Manning AJ, Orench-Rivera N, Schmid AK, Kuehn MJ (2015) Genome-wide assessment of outer membrane vesicle production in *Escherichia coli*. PLoS One 10(9): e0139200. https://doi.org/10.1371/journal.pone.0139200

Lam JS, Taylor VL, Islam ST, Hao Y, Kocincova D (2011) Genetic and functional diversity of *Pseudomonas aeruginosa* lipopolysaccharide. Front Microbiol 2:118. https://doi.org/10.3389/fmicb.2011.00118
Lazzaroni JC, Germon P, Ray MC, Vianney A (1999) The Tol proteins of *Escherichia coli* and their involvement in the uptake of biomolecules and outer membrane stability. FEMS Microbiol Lett 177(2):191–197
Llamas MA, Ramos JL, Rodriguez-Herva JJ (2000) Mutations in each of the *tol* genes of *Pseudomonas putida* reveal that they are critical for maintenance of outer membrane stability. J Bacteriol 182(17):4764–4772
Macdonald IA, Kuehn MJ (2013) Stress-induced outer membrane vesicle production by *Pseudomonas aeruginosa*. J Bacteriol 195(13):2971–2981. https://doi.org/10.1128/JB.02267-12
Mahalakshmi S, Sunayana MR, SaiSree L, Reddy M (2014) *yciM* is an essential gene required for regulation of lipopolysaccharide synthesis in *Escherichia coli*. Mol Microbiol 91(1):145–157. https://doi.org/10.1111/mmi.12452
Malinverni JC, Silhavy TJ (2009) An ABC transport system that maintains lipid asymmetry in the gram-negative outer membrane. Proc Natl Acad Sci U S A 106(19):8009–8014. https://doi.org/10.1073/pnas.0903229106
Manning AJ, Kuehn MJ (2011) Contribution of bacterial outer membrane vesicles to innate bacterial defense. BMC Microbiol 11:258. https://doi.org/10.1186/1471-2180-11-258
Maredia R, Devineni N, Lentz P, Dallo SF, Yu J, Guentzel N, Chambers J, Arulanandam B, Haskins WE, Weitao T (2012) Vesiculation from *Pseudomonas aeruginosa* under SOS. TheScientificWorldJournal 2012:402919. https://doi.org/10.1100/2012/402919
Mashburn LM, Whiteley M (2005) Membrane vesicles traffic signals and facilitate group activities in a prokaryote. Nature 437(7057):422–425. https://doi.org/10.1038/nature03925
Mashburn-Warren L, Howe J, Garidel P, Richter W, Steiniger F, Roessle M, Brandenburg K, Whiteley M (2008) Interaction of quorum signals with outer membrane lipids: insights into prokaryotic membrane vesicle formation. Mol Microbiol 69(2):491–502
Mathur J, Davis BM, Waldor MK (2007) Antimicrobial peptides activate the *Vibrio cholerae* sigmaE regulon through an OmpU-dependent signalling pathway. Mol Microbiol 63(3):848–858
McBroom AJ, Kuehn MJ (2007) Release of outer membrane vesicles by Gram-negative bacteria is a novel envelope stress response. Mol Microbiol 63(2):545–558
McBroom AJ, Johnson AP, Vemulapalli S, Kuehn MJ (2006) Outer membrane vesicle production by *Escherichia coli* is independent of membrane instability. J Bacteriol 188(15):5385–5392
McCaig WD, Koller A, Thanassi DG (2013) Production of outer membrane vesicles and outer membrane tubes by *Francisella novicida*. J Bacteriol 195(6):1120–1132. https://doi.org/10.1128/JB.02007-12
McMahon HT, Boucrot E (2015) Membrane curvature at a glance. J Cell Sci 128(6):1065–1070. https://doi.org/10.1242/jcs.114454
Millikan DS, Ruby EG (2004) *Vibrio fischeri* flagellin A is essential for normal motility and for symbiotic competence during initial squid light organ colonization. J Bacteriol 186(13):4315–4325. https://doi.org/10.1128/JB.186.13.4315-4325.2004
Mitra S, Sinha R, Mitobe J, Koley H (2016) Development of a cost-effective vaccine candidate with outer membrane vesicles of a *tolA*-disrupted *Shigella boydii* strain. Vaccine 34(15):1839–1846. https://doi.org/10.1016/j.vaccine.2016.02.018
Moon DC, Choi CH, Lee JH, Choi CW, Kim HY, Park JS, Kim SI, Lee JC (2012) *Acinetobacter baumannii* outer membrane protein A modulates the biogenesis of outer membrane vesicles. J Microbiol 50(1):155–160. https://doi.org/10.1007/s12275-012-1589-4
Moore RA, Hancock RE (1986) Involvement of outer membrane of *Pseudomonas cepacia* in aminoglycoside and polymyxin resistance. Antimicrob Agents Chemother 30(6):923–926. https://doi.org/10.1128/aac.30.6.923

Murphy K, Park AJ, Hao Y, Brewer D, Lam JS, Khursigara CM (2014) Influence of O polysaccharides on biofilm development and outer membrane vesicle biogenesis in *Pseudomonas aeruginosa* PAO1. J Bacteriol 196(7):1306–1317. https://doi.org/10.1128/JB.01463-13

Nakao R, Ramstedt M, Wai SN, Uhlin BE (2012) Enhanced biofilm formation by *Escherichia coli* LPS mutants defective in Hep biosynthesis. PLoS One 7(12):e51241. https://doi.org/10.1371/journal.pone.0051241

Niehus E, Ye F, Suerbaum S, Josenhans C (2002) Growth phase-dependent and differential transcriptional control of flagellar genes in *Helicobacter pylori*. Microbiology 148 (Pt 12):3827–3837. https://doi.org/10.1099/00221287-148-12-3827

Pal RR, Baidya AK, Mamou G, Bhattacharya S, Socol Y, Kobi S, Katsowich N, Ben-Yehuda S, Rosenshine I (2019) Pathogenic *E. coli* extracts nutrients from infected host cells utilizing injectisome components. Cell 177(3):683–696 e618. https://doi.org/10.1016/j.cell.2019.02.022

Palva AM (1983) ompA gene in the detection of *Escherichia coli* and other Enterobacteriaceae by nucleic acid sandwich hybridization. J Clin Microbiol 18(1):92–100

Papenfort K, Pfeiffer V, Mika F, Lucchini S, Hinton JC, Vogel J (2006) SigmaE-dependent small RNAs of Salmonella respond to membrane stress by accelerating global omp mRNA decay. Mol Microbiol 62(6):1674–1688

Paramonov N, Bailey D, Rangarajan M, Hashim A, Kelly G, Curtis MA, Hounsell EF (2001) Structural analysis of the polysaccharide from the lipopolysaccharide of *Porphyromonas gingivalis* strain W50. Eur J Biochem 268(17):4698–4707

Paramonov N, Rangarajan M, Hashim A, Gallagher A, Aduse-Opoku J, Slaney JM, Hounsell E, Curtis MA (2005) Structural analysis of a novel anionic polysaccharide from *Porphyromonas gingivalis* strain W50 related to Arg-gingipain glycans. Mol Microbiol 58(3):847–863. https://doi.org/10.1111/j.1365-2958.2005.04871.x

Parsons LM, Lin F, Orban J (2006) Peptidoglycan recognition by Pal, an outer membrane lipoprotein. Biochemistry 45(7):2122–2128. https://doi.org/10.1021/bi052227i

Perez-Cruz C, Carrion O, Delgado L, Martinez G, Lopez-Iglesias C, Mercade E (2013) New type of outer membrane vesicle produced by the Gram-negative bacterium *Shewanella vesiculosa* M7T: implications for DNA content. Appl Environ Microbiol 79(6):1874–1881. https://doi.org/10.1128/AEM.03657-12

Pirbadian S, Barchinger SE, Leung KM, Byun HS, Jangir Y, Bouhenni RA, Reed SB, Romine MF, Saffarini DA, Shi L, Gorby YA, Golbeck JH, El-Naggar MY (2014) *Shewanella oneidensis* MR-1 nanowires are outer membrane and periplasmic extensions of the extracellular electron transport components. Proc Natl Acad Sci U S A 111(35):12883–12888. https://doi.org/10.1073/pnas.1410551111

Prieto AI, Ramos-Morales F, Casadesus J (2006) Repair of DNA damage induced by bile salts in *Salmonella enterica*. Genetics 174(2):575–584. https://doi.org/10.1534/genetics.106.060889

Qin Z, Lin WT, Zhu S, Franco AT, Liu J (2016) Imaging the motility and chemotaxis machineries in *Helicobacter pylori* by cryo-electron tomography. J Bacteriol 199:e00695–e00616. https://doi.org/10.1128/JB.00695-16

Raetz CR, Reynolds CM, Trent MS, Bishop RE (2007) Lipid A modification systems in Gram-negative bacteria. Annu Rev Biochem 76:295–329. https://doi.org/10.1146/annurev.biochem.76.010307.145803

Raivio TL (2005) Envelope stress responses and Gram-negative bacterial pathogenesis. Mol Microbiol 56(5):1119–1128. https://doi.org/10.1111/j.1365-2958.2005.04625.x

Rangarajan M, Aduse-Opoku J, Paramonov N, Hashim A, Bostanci N, Fraser OP, Tarelli E, Curtis MA (2008) Identification of a second lipopolysaccharide in *Porphyromonas gingivalis* W50. J Bacteriol 190(8):2920–2932. https://doi.org/10.1128/JB.01868-07

Remis JP, Wei D, Gorur A, Zemla M, Haraga J, Allen S, Witkowska HE, Costerton JW, Berleman JE, Auer M (2014) Bacterial social networks: structure and composition of *Myxococcus xanthus* outer membrane vesicle chains. Environ Microbiol 16(2):598–610. https://doi.org/10.1111/1462-2920.12187

Roier S, Zingl FG, Cakar F, Durakovic S, Kohl P, Eichmann TO, Klug L, Gadermaier B, Weinzerl K, Prassl R, Lass A, Daum G, Reidl J, Feldman MF, Schild S (2016) A novel mechanism for the biogenesis of outer membrane vesicles in Gram-negative bacteria. Nat Commun 7:10515. https://doi.org/10.1038/ncomms10515

Ruhal R, Antti H, Rzhepishevska O, Boulanger N, Barbero DR, Wai SN, Uhlin BE, Ramstedt M (2015) A multivariate approach to correlate bacterial surface properties to biofilm formation by lipopolysaccharide mutants of *Pseudomonas aeruginosa*. Colloids Surf B Biointerfaces 127:182–191. https://doi.org/10.1016/j.colsurfb.2015.01.030

Ruiz N, Kahne D, Silhavy TJ (2006) Advances in understanding bacterial outer-membrane biogenesis. Nat Rev Microbiol 4(1):57–66. https://doi.org/10.1038/nrmicro1322

Sabra W, Lunsdorf H, Zeng AP (2003) Alterations in the formation of lipopolysaccharide and membrane vesicles on the surface of *Pseudomonas aeruginosa* PAO1 under oxygen stress conditions. Microbiology 149(Pt 10):2789–2795. https://doi.org/10.1099/mic.0.26443-0

Salkinoja-Salonen M, Nurmiaho EL (1978) The effect of lipopolysaccharide composition on the ultrastructure of *Pseudomonas aeruginosa*. J Gen Microbiol 105(1):23–28. https://doi.org/10.1099/00221287-105-1-23

Sampath V, McCaig WD, Thanassi DG (2018) Amino acid deprivation and central carbon metabolism regulate the production of outer membrane vesicles and tubes by Francisella. Mol Microbiol 107(4):523–541. https://doi.org/10.1111/mmi.13897

Samsudin F, Ortiz-Suarez ML, Piggot TJ, Bond PJ, Khalid S (2016) OmpA: a flexible clamp for bacterial cell wall attachment. Structure 24(12):2227–2235. https://doi.org/10.1016/j.str.2016.10.009

Schertzer JW, Whiteley M (2012) A bilayer-couple model of bacterial outer membrane vesicle biogenesis. mBio 3(2):e00297-11. https://doi.org/10.1128/mBio.00297-11

Schertzer JW, Whiteley M (2013) Bacterial outer membrane vesicles in trafficking, communication and the host-pathogen interaction. J Mol Microbiol Biotechnol 23(1–2):118–130. https://doi.org/10.1159/000346770

Schwechheimer C, Kuehn MJ (2013) Synthetic effect between envelope stress and lack of outer membrane vesicle production in *Escherichia coli*. J Bacteriol 195(18):4161–4173. https://doi.org/10.1128/JB.02192-12

Schwechheimer C, Sullivan CJ, Kuehn MJ (2013) Envelope control of outer membrane vesicle production in Gram-negative bacteria. Biochemistry 52(18):3031–3040. https://doi.org/10.1021/bi400164t

Schwechheimer C, Kulp A, Kuehn MJ (2014) Modulation of bacterial outer membrane vesicle production by envelope structure and content. BMC Microbiol 14:324. https://doi.org/10.1186/s12866-014-0324-1

Shen L, Gao X, Wei J, Chen L, Zhao X, Li B, Duan K (2012) PA2800 plays an important role in both antibiotic susceptibility and virulence in *Pseudomonas aeruginosa*. Curr Microbiol 65 (5):601–609. https://doi.org/10.1007/s00284-012-0196-2

Shetty A, Hickey WJ (2014) Effects of outer membrane vesicle formation, surface-layer production and nanopod development on the metabolism of phenanthrene by *Delftia acidovorans* Cs1-4. PLoS One 9(3):e92143. https://doi.org/10.1371/journal.pone.0092143

Shetty A, Chen S, Tocheva EI, Jensen GJ, Hickey WJ (2011) Nanopods: a new bacterial structure and mechanism for deployment of outer membrane vesicles. PLoS One 6(6):e20725. https://doi.org/10.1371/journal.pone.0020725

Simmons LA, Foti JJ, Cohen SE, Walker GC (2008) The SOS Regulatory Network. EcoSal Plus 3 (1). https://doi.org/10.1128/ecosalplus.5.4.3

Song T, Mika F, Lindmark B, Liu Z, Schild S, Bishop A, Zhu J, Camilli A, Johansson J, Vogel J, Wai SN (2008) A new *Vibrio cholerae* sRNA modulates colonization and affects release of outer membrane vesicles. Mol Microbiol 70(1):100–111

Sonntag I, Schwarz H, Hirota Y, Henning U (1978) Cell envelope and shape of *Escherichia coli*: multiple mutants missing the outer membrane lipoprotein and other major outer membrane proteins. J Bacteriol 136(1):280–285

Storm DR, Rosenthal KS, Swanson PE (1977) Polymyxin and related peptide antibiotics. Annu Rev Biochem 46:723–763. https://doi.org/10.1146/annurev.bi.46.070177.003451

Suzuki H, Nishimura Y, Yasuda S, Nishimura A, Yamada M, Hirota Y (1978) Murein-lipoprotein of *Escherichia coli*: a protein involved in the stabilization of bacterial cell envelope. Mol Gen Genet 167(1):1–9

Suzuki T, Murai T, Fukuda I, Tobe T, Yoshikawa M, Sasakawa C (1994) Identification and characterization of a chromosomal virulence gene, *vacJ*, required for intercellular spreading of *Shigella flexneri*. Mol Microbiol 11(1):31–41

Sych T, Mely Y, Romer W (2018) Lipid self-assembly and lectin-induced reorganization of the plasma membrane. Philos Trans R Soc Lond Ser B Biol Sci 373(1747):20170117. https://doi.org/10.1098/rstb.2017.0117

Taheri N, Mahmud A, Sandblad L, Fallman M, Wai SN, Fahlgren A (2018) *Campylobacter jejuni* bile exposure influences outer membrane vesicles protein content and bacterial interaction with epithelial cells. Sci Rep 8(1):16996. https://doi.org/10.1038/s41598-018-35409-0

Tarahovsky YS, Ivanitsky GR, Khusainov AA (1994) Lysis of *Escherichia coli* cells induced by bacteriophage T4. FEMS Microbiol Lett 122(1–2):195–199. https://doi.org/10.1111/j.1574-6968.1994.tb07164.x

Tashiro Y, Sakai R, Toyofuku M, Sawada I, Nakajima-Kambe T, Uchiyama H, Nomura N (2009) Outer membrane machinery and alginate synthesis regulators control membrane vesicle production in *Pseudomonas aeruginosa*. J Bacteriol 191(24):7509–7519

Tashiro Y, Ichikawa S, Nakajima-Kambe T, Uchiyama H, Nomura N (2010) *Pseudomonas* quinolone signal affects membrane vesicle production in not only gram-negative but also gram-positive bacteria. Microbes Environ 25(2):120–125. https://doi.org/10.1264/jsme2.me09182

Tashiro Y, Inagaki A, Shimizu M, Ichikawa S, Takaya N, Nakajima-Kambe T, Uchiyama H, Nomura N (2011) Characterization of phospholipids in membrane vesicles derived from *Pseudomonas aeruginosa*. Biosci Biotechnol Biochem 75(3):605–607

Toyofuku M, Zhou S, Sawada I, Takaya N, Uchiyama H, Nomura N (2014) Membrane vesicle formation is associated with pyocin production under denitrifying conditions in *Pseudomonas aeruginosa* PAO1. Environ Microbiol 16(9):2927–2938. https://doi.org/10.1111/1462-2920.12260

Toyofuku M, Carcamo-Oyarce G, Yamamoto T, Eisenstein F, Hsiao CC, Kurosawa M, Gademann K, Pilhofer M, Nomura N, Eberl L (2017) Prophage-triggered membrane vesicle formation through peptidoglycan damage in *Bacillus subtilis*. Nat Commun 8(1):481. https://doi.org/10.1038/s41467-017-00492-w

Trent MS, Pabich W, Raetz CR, Miller SI (2001) A PhoP/PhoQ-induced lipase (PagL) that catalyzes 3-O-deacylation of lipid A precursors in membranes of *Salmonella typhimurium*. J Biol Chem 276(12):9083–9092. https://doi.org/10.1074/jbc.M010730200

Turnbull L, Toyofuku M, Hynen AL, Kurosawa M, Pessi G, Petty NK, Osvath SR, Carcamo-Oyarce G, Gloag ES, Shimoni R, Omasits U, Ito S, Yap X, Monahan LG, Cavaliere R, Ahrens CH, Charles IG, Nomura N, Eberl L, Whitchurch CB (2016) Explosive cell lysis as a mechanism for the biogenesis of bacterial membrane vesicles and biofilms. Nat Commun 7:11220. https://doi.org/10.1038/ncomms11220

Turner L, Praszkier J, Hutton ML, Steer D, Ramm G, Kaparakis-Liaskos M, Ferrero RL (2015) Increased outer membrane vesicle formation in a *Helicobacter pylori* tolB mutant. Helicobacter 20(4):269–283. https://doi.org/10.1111/hel.12196

Udekwu KI, Wagner EG (2007) Sigma E controls biogenesis of the antisense RNA MicA. Nucleic Acids Res 35(4):1279–1288. https://doi.org/10.1093/nar/gkl1154

Urashima A, Sanou A, Yen H, Tobe T (2017) Enterohaemorrhagic *Escherichia coli* produces outer membrane vesicles as an active defence system against antimicrobial peptide LL-37. Cell Microbiol 19(11). https://doi.org/10.1111/cmi.12758

Vollmer W, Bertsche U (2008) Murein (peptidoglycan) structure, architecture and biosynthesis in *Escherichia coli*. Biochim Biophys Acta 1778(9):1714–1734. https://doi.org/10.1016/j.bbamem.2007.06.007

Wang Y (2002) The function of OmpA in *Escherichia coli*. Biochem Biophys Res Commun 292(2):396–401. https://doi.org/10.1006/bbrc.2002.6657

Weinbauer MG (2004) Ecology of prokaryotic viruses. FEMS Microbiol Rev 28(2):127–181. https://doi.org/10.1016/j.femsre.2003.08.001

Wensink J, Witholt B (1981) Outer-membrane vesicles released by normally growing *Escherichia coli* contain very little lipoprotein. Eur J Biochem 116(2):331–335

Yeh YC, Comolli LR, Downing KH, Shapiro L, McAdams HH (2010) The *Caulobacter* Tol-Pal complex is essential for outer membrane integrity and the positioning of a polar localization factor. J Bacteriol 192(19):4847–4858. https://doi.org/10.1128/JB.00607-10

Zavan L, Bitto NJ, Johnston EL, Greening DW, Kaparakis-Liaskos M (2019) *Helicobacter pylori* growth stage determines the size, protein composition, and preferential cargo packaging of outer membrane vesicles. Proteomics 19(1–2):e1800209. https://doi.org/10.1002/pmic.201800209

Zhao L, Gao X, Liu C, Lv X, Jiang N, Zheng S (2017) Deletion of the *vacJ* gene affects the biology and virulence in *Haemophilus parasuis* serovar 5. Gene 603:42–53. https://doi.org/10.1016/j.gene.2016.12.009

Zimmerberg J, Kozlov MM (2006) How proteins produce cellular membrane curvature. Nat Rev Mol Cell Biol 7(1):9–19. https://doi.org/10.1038/nrm1784

# Chapter 3
# Biogenesis and Function of Extracellular Vesicles in Gram-Positive Bacteria, Mycobacteria, and Fungi

**Ainhoa Palacios, Carolina Coelho, Maria Maryam, Jose L. Luque-García, Arturo Casadevall, and Rafael Prados-Rosales**

**Abstract** The production of membrane vesicles (MVs) has been documented in all domains of life. Justification for the historical lack of interest in the study of vesicle biogenesis in Gram-positive bacteria, mycobacteria, and fungi is based on the difficulty in explaining how MV can traverse the thick cell wall. For this reason, the scientific landscape has been dominated by studies examining vesicle biogenesis in microorganisms that lack cell walls or Gram-negative bacteria, since they possess an outer membrane layer. Evidence of MV production by cell-walled microorganisms is now available from different experimental approaches including, isolation of MVs from culture supernatant, compositional analysis, visualization of vesiculation events, and genetic studies. Strikingly, more recent studies have shown that beside

A. Palacios
CIC bioGUNE, Technological Park Bizkaia, Derio, Spain

C. Coelho
Department of Molecular Microbiology and Immunology, Johns Hopkins Bloomberg School of Public Health, Baltimore, MD, USA

Medical Research Council Centre for Medical Mycology, Institute of Medical Sciences, University of Aberdeen, Aberdeen, UK

Department of Biosciences, College of Life and Environmental Sciences, University of Exeter, Exeter, UK

M. Maryam · A. Casadevall
Department of Molecular Microbiology and Immunology, Johns Hopkins Bloomberg School of Public Health, Baltimore, MD, USA

J. L. Luque-García
Department of Analytical Chemistry, Faculty of Chemistry, Complutense University of Madrid, Madrid, Spain

R. Prados-Rosales (✉)
CIC bioGUNE, Technological Park Bizkaia, Derio, Spain

Faculty of Medicine, Department of Preventive Medicine and Public Health, and Microbiology, Autonoma University of Madrid, Madrid, Spain
e-mail: rafael.prados@uam.es



M. Kaparakis-Liaskos, T. A. Kufer (eds.), *Bacterial Membrane Vesicles*,
https://doi.org/10.1007/978-3-030-36331-4_3

the fundamental differences in the architecture of the cell envelope of Gram-positive and Gram-negative bacteria, a similar mechanism of cell wall remodeling may govern the release of MVs. Here, we describe the current understanding of vesicle biogenesis in cell-walled microorganisms, discussing novel mechanisms of vesicle production, methods to study MVs, cargo, and functions of MVs as well as medical applications of naturally produced MVs.

## 3.1 Introduction

The notion that Gram-negative bacteria produce outer membrane vesicles (OMVs) has been an accepted phenomenon for some time. However, the realization that cell-walled organisms such as fungi, mycobacteria, and Gram-positive bacteria also make MVs has only taken root in the past decade. The major hurdle in considering MV production by cell-walled microbes was the belief that the cell wall was a rigid structure that prevented vesicular transit. In 2007, the fungus *Cryptococcus neoformans* was shown to produce MVs and to use them for export of macromolecular compounds including virulence factors (Rodrigues et al. 2007, 2008b). Subsequently, several other fungal species including *Candida albicans* (Vargas et al. 2015), *Saccharomyces cerevisiae* (Oliveira et al. 2010), *Histoplasma capsulatum* (Albuquerque et al. 2008; Nimrichter et al. 2016), *Sporothrix brasiliensis* (Ikeda et al. 2018), *Paracoccidioides brasiliensis* (Vallejo et al. 2011), *Alternaria infectoria* (Silva et al. 2014), and *Malassezia sympodialis* (Johansson et al. 2018) were shown to be producers of MVs (Table 3.1). The experience with fungal MVs prompted a search for comparable structures in Gram-positive bacteria. The report of the proteomic composition of isolated *Staphylococcus aureus* MVs provided an early example in cell-walled bacteria (Lee et al. 2009). Subsequent studies with the bacterium *Bacillus anthracis* showed that it packaged its toxins into vesicles (Rivera et al. 2010). An initial hint that anthrax toxin components were packaged into vesicles came from immunogold electron microscopy, indicating that these were secreted at discrete sites in the cell wall (Rivera et al. 2010). The production of MVs has now been identified to occur by numerous Gram-positive bacteria (Liu et al. 2018) and mycobacteria (Prados-Rosales et al. 2011) (Table 3.1). The demonstration that vesicle biogenesis is altered in some mutants of cell-walled organisms (Brown et al. 2015) suggest that, as in Gram-negative bacteria, MV production by Gram-positive organisms might be a genetically regulated process. Recent reports have demonstrated a similar mechanism for MV release from *Bacillus subtilis* (Toyofuku et al. 2017) and OMV release from *Pseudomonas aeruginosa* (Turnbull et al. 2016), indicating that beside the fundamental differences in cell envelope architecture, common mechanisms may govern this process in prokaryotes. The clinical relevance of MVs produced by pathogenic species of cell-walled organisms is apparent from several studies. Both beneficial and detrimental effects have been associated with MVs. Moreover, several groups have provided evidence that isolated MVs may represent a good platform for vaccine development in animal

**Table 3.1** List of cell-walled microorganisms where MVs have been demonstrated

| Species | References |
|---|---|
| Gram-positive bacteria | |
| *Bacillus anthracis* | Rivera et al. (2010) |
| *Bacillus subtilis* | Brown et al. (2014) |
| *Bifidobacterium longum* | Kim et al. (2016a) |
| *Clostridium perfringens* | Jiang et al. (2014) |
| *Enterococcus faecalis* | Kim et al. (2016b) |
| *Lactobacillus casei* | Dominguez Rubio et al. (2017) |
| *Lactobacillus plantarum* | Li et al. (2017) |
| *Lactobacillus reuteri* | Grande et al. (2017) |
| *Lactobacillus rhamnosus* | Behzadi et al. (2017) |
| *Listeria monocytogenes* | Lee et al. (2013b) |
| *Propionibacterium acnes* | Jeon et al. (2017) |
| *Streptococcus agalactiae* (group B streptococcus) | Surve et al. (2016) |
| *Streptomyces coelicolor* | Schrempf et al. (2011) |
| *Streptomyces lividans* | Lee et al. (2009) |
| *Staphylococcus aureus* | Lee et al. (2009) |
| *Streptococcus mutans* | Liao et al. (2014) |
| *Streptococcus pneumoniae* | Olaya-Abril et al. (2014) |
| *Streptococcus pyogenes* (group A streptococcus) | Resch et al. (2016) |
| *Streptococcus suis* | Haas and Grenier (2015) |
| Mycobacteria | |
| *Mycobacterium avium* | Prados-Rosales et al. (2011) |
| *Mycobacterium bovis* BCG | Prados-Rosales et al. (2011) |
| *Mycobacterium kansasii* | Prados-Rosales et al. (2011) |
| *Mycobacterium phlei* | Prados-Rosales et al. (2011) |
| *Mycobacterium smegmatis* | Prados-Rosales et al. (2011) |
| *Mycobacterium tuberculosis* | Prados-Rosales et al. (2011) |
| *Mycobacterium ulcerans* | Marsollier et al. (2007) |
| Fungi | |
| *Candida albicans* | Albuquerque et al. (2008) |
| *Cryptococcus gatii* | Bielska et al. (2018) |
| *Candida parapsilosis* | Albuquerque et al. (2008) |
| *Cryptococcus neoformans* | Rodrigues et al. (2007) |
| *Histoplasma capsulatum* | Albuquerque et al. (2008) |
| *Malassezia sympodialis* | Gehrmann et al. (2011) |
| *Paracoccidioides brasiliensis* | Vallejo et al. (2011) |
| *Saccharomyces cerevisiae* | Albuquerque et al. (2008) |
| *Sporothrix schenckii* | Albuquerque et al. (2008) |

models (see Chap. 10). Given the relevance of MVs in clinical settings, the feasibility of MVs as a diagnostic element is starting to be appreciated.

## 3.2 Evidence for MVs in Cell-Walled Organisms

Although the existence of MV transport systems in cell-walled microbes is now generally accepted, acceptance of this notion took a while to acquire traction. The problem with MVs gaining general acceptance was the result of criticisms that ranged from physical concerns to intellectual arguments and these took almost a decade to be resolved. The three major criticisms will be considered separately.

1. *Cell walls are rigid structures that prevent the passage of such large structures as MVs*. The notion that cell walls precluded passage of MVs due to their rigidity and the absence of large pores is known as the physical criticism. This criticism persisted despite the fact that cell walls were known to be flexible structures that could be easily rearranged during budding and, in the case of the fungi, hyphal formation. This criticism was addressed by demonstrating vesicles in the cell wall during what appeared to be a transfer process. For *C. neoformans* there was electron microscopic evidence of vesicles transiting the cell wall (Rodrigues et al. 2007). For *S. aureus*, scanning electron microscopy showed vesicle-like structures with comparable dimensions to MVs protruding from the cell wall surface (Lee et al. 2009). Subsequent work using freeze–fracture electron microscopy revealed vesicles entering or creating pores in the fungal cell wall (Wolf et al. 2014). Recently, liposomes containing amphotericin b were shown to transit the cell wall of *C. neoformans* and *C. albicans* from the outside to the inside, when added exogenously (Walker et al. 2018), making a compelling case for the notion that the living cell wall is a pliable and deformable structure and selectively porous, that is, not a barrier to vesicular transit.
2. *MVs form from self-assembly of lipids released from cells*. The criticism that MVs were artifacts from the self-assembly of lipid molecules released from live or dead cells was difficult to counter because lipids are notoriously able to form micelles. Furthermore, there was data that lysis of *B. subtilis* cells by phages could produce voluminous amounts of membrane lipids that could self-assemble into vesicles. For example, the expression of endolysins by phage in *Bacillus subtilis* led to explosive lysis of the bacterial cells that was associated with vesicle formation (Toyofuku et al. 2017, 2019). The lipid self-assembly criticism was answered by a series of experiments that made a compelling case against this explanation. Several bacteria MVs were shown to have lipid and protein composition different than whole bacterial cells, which argued against their emergence as a consequence of cellular lysis (Coelho et al. 2019; Olaya-Abril et al. 2014). In an experiment designed to rule out the self-assembly explanation *C. neoformans* polysaccharide was added to *B. subtilis* cultures and then localized by immunogold labeling (Brown et al. 2014). If MVs were the result of lipid self-

assembly one would have expected to see the fungal polysaccharide inside the vesicles, but this was not the case providing strong evidence against the lipid self-assembly origin explanation (Brown et al. 2014). In fungi, the finding that MVs were used in two-way communication between fungal cells indicated a new specific function for these structures that made a strong case against an artifactual nature (Bielska et al. 2018; Rodrigues and Casadevall 2018). Finally, the visualization of MVs emerging from *Listeria monocytogenes* inside epithelial cells through live cell imaging provided compelling visual evidence for their physiological existence (Coelho et al. 2019).

3. *The absence of null mutants for MV production* in vesicle-producing organisms. This criticism followed from the viewpoint of genetic reductionism, which posited that if MVs were real then there must be a complex machinery that could allow the isolation of null mutants for MV production. Remarkably, those who espoused such views never doubted the existence of cell membranes despite the fact that no mutant could exist without a cell membrane. The absence of null mutants for MV production could reflect the fact that vesicle formation is an integral part of cell membrane remodeling, and as such, would be produced by any microbe. Nevertheless, this criticism has now been muted by the discovery of genes that regulate MV production in fungi (Oliveira et al. 2010), mycobacteria (Rath et al. 2013), and Gram-positive bacteria (Resch et al. 2016).

Today, fungi are accepted to be producers of MVs and these have been described in *C. neoformans, C. albicans* (Vargas et al. 2015), *S. cerevisiae* (Oliveira et al. 2010), *H. capsulatum* (Nimrichter et al. 2016), *Sporothrix brasiliensis* (Ikeda et al. 2018), *P. brasiliensis* (Vallejo et al. 2011), and *A. infectoria* (Silva et al. 2014), as well as the skin commensal *M. sympodialis* (Johansson et al. 2018). For recent reviews on this topic see also de Toledo Martins et al. (2019); Joffe et al. (2016); Rodrigues et al. (2015).

The first observation that MVs could be produced by a Gram-positive bacterium was reported in 1990 but this was not investigated further (Dorward and Garon 1990). It was not until 2009 when a proteomic study of *S. aureus* MVs set the stage for modern research into the biogenesis of MVs by Gram-positive bacteria (Lee et al. 2009). That finding was followed by studies in *B. anthracis*, *L. monocytogenes*, several species of *Streptococcus* (Haas and Grenier 2015; Olaya-Abril et al. 2014; Surve et al. 2016), and *Lactobacillus* (Li et al. 2017), to name a few.

The first evidence that a bacterial strain from the *Mycobacterium* genus releases MVs was reported in 2007 in *M. ulcerans* (Marsollier et al. 2007). MVs were isolated from the extracellular matrix of *M. ulcerans* and were enriched in mycolactone, the main virulence factor of this pathogen. Subsequent studies validated this study and supported the notion that MV biogenesis is a conserved phenomenon in the *Mycobacterium* genus. Indeed, a 2011 report demonstrated MV production in the two most medically important strains of this genus, *Mycobacterium tuberculosis* (Mtb) and Bacillus Calmette-Guérin (BCG) (Prados-Rosales et al. 2011). Analysis of MV biogenesis was extended to other slow growers and fast growers from the same genus and all of them showed the capacity to release MVs

(Prados-Rosales et al. 2011). Evidence of MV production was not only demonstrated by the possibility of MV isolation from culture supernatant but also by the visualization of MV release events using electron microscopy techniques such as transmission electron microscopy (TEM) and scanning electron microscopy (SEM) during an ongoing infection (Prados-Rosales et al. 2011). Further characterization of Mtb and BCG MVs showed a similar size distribution than that of Gram-negative OMVs (Prados-Rosales et al. 2011).

## 3.3 Methods for the Study of MVs in Organisms with a Thick Cell Wall

Most of the studies on MVs in Gram-positive bacteria, fungi, and mycobacteria involve an initial purification from laboratory culture supernatants followed by the concentration and characterization of these structures. Few studies have attempted MV isolation from host-related compartments. Mass spectrometry has been used to define the MV-associated protein and lipid cargo, and electron microscopy to visualize isolated MVs or vesiculation events from living cells. The methodology to study MVs should be designed carefully attending to the final use of isolated material. If the study involves the interaction of isolated MVs with host cells, the isolation and purification steps should go as far as eliminating most of the unwanted contaminants. This would also apply for cargo identification by mass spectrometry.

### *3.3.1 Methods of Isolation*

As previously mentioned, the isolation process is one of the most challenging approaches during the study of MVs (Mateescu et al. 2017). Since there is not an ideal technique to isolate MVs, most groups combine several approaches with the aim of reducing contaminants, which are typically co-isolated with the MVs pool, and to minimize sample loss. The protocol that has been mostly used to isolate MVs derived from Gram-positive bacteria, mycobacteria, and fungi involve clarification of the supernatant using low-speed centrifugation followed by filtration to remove microorganisms and the collection of an MV-rich faction after subsequent ultracentrifugation (Brown et al. 2015; Prados-Rosales et al. 2014a). Although many studies have performed MVs characterization from the pelleted MV preparation after ultracentrifugation, additional separation steps are desired to increase the purity of the MV pool and avoid the carryover of unwanted cellular material. For instance, an additional step in the form of sucrose cushion density gradient was applied during isolation of MVs from *Streptococcus* or *Streptomyces* (Gurung et al. 2011; Schrempf et al. 2011). Similarly, density gradient ultracentrifugation was used to further purify ultracentrifuge-isolated MVs from Mtb (Prados-Rosales et al. 2011), *S. aureus* (Lee et al. 2009), *S. pneumoniae* (Olaya-Abril et al. 2014), and *C. perfringens* (Jiang et al. 2014) cultures. More recently, size-exclusion chromatography was combined with clarification and ultracentrifugation of culture supernatants to isolate MVs from *C. albicans* biofilms (Zarnowski et al. 2018). The major disadvantage in applying additional purification steps during MV isolation is the decrease in vesicle yield (Chutkan et al. 2013).

MVs have also been isolated from samples different from axenic cultures. In a study reporting a pulmonary allergic reaction to *S. aureus* MVs in mice, vesicles from this organism were isolated from house dust by removing bacteria from dust previously dissolved PBS and subsequent ultracentrifugation (Kim et al. 2012). MVs from *S. aureus* have also been isolated from skin lavage fluids obtained from patients with atopic dermatitis (Hong et al. 2011) and lung tissue from *S. aureus*-infected mice (Gurung et al. 2011) using a similar approach.

One of the major limitations of methods used to isolate MVs, in general, is the difficulty in separating MV subpopulations. The development of new approaches with the sufficient resolution capacity to achieve the separation of such populations will increase the understanding of the intrinsic heterogeneity of MVs.

## 3.3.2 Cargo Identification

The definition of MV composition is critical to understanding their role in different biological processes. MVs contain a diverse variety of cargo including lipids, proteins, nucleic acids, and metabolites. In the case of Gram-positive bacteria, mycobacteria, and fungi derived-MVs, most studies have been focused on the definition of the protein composition of isolated MVs by biological mass spectrometry; or by biochemical assays such as antibody-based assays, which include ELISA or Immunoblot (Lee et al. 2009; Prados-Rosales et al. 2011). Fewer studies report the lipid profile of MVs from cell-walled microorganisms and to a lesser extent the presence of nucleic acids or metabolites. Compared to Gram-negative bacterial OMVs, quantitative data regarding the relative incorporation of cellular material, including lipids and proteins into MVs are not available for cell-walled microorganisms. This information is necessary to better understand the process of Gram-positive MV biogenesis. Moreover, one aspect usually unattended to fully establish the unique association of a specific cellular component to MVs is the lack of additional experiments including the composition of the cell membrane. Elucidation of MV composition will shed light into the enrichment phenomenon of specific MV components.

### 3.3.2.1 Identification of MV-Associated Proteins

The protein content of MVs derived from cell-walled microorganisms can be assessed by antibody-based assays or by high-throughput mass spectrometry approaches. Usually, the selection of the antibody to detect the MV-associated protein is preceded by a mass spectrometry approach where the catalogue of MV proteins has been defined. In the latter, a cleanup step in the form of acetone precipitation is usually performed followed by in-solution trypsin digestion and peptide purification by solid phase extraction (ZipTip) before analysis by mass spectrometry (Prados-Rosales et al. 2014a). As mentioned above, to fully establish the enrichment of a specific protein within the MV it is critical to compare the abundance of the protein hit in the MV relative to the one in the cellular membrane. Most of the approaches used to analyze MV protein composition from cell-walled microorganisms are based on electrospray ionization (ESI)-based nano liquid chromatography coupled to mass spectrometry (nanoLC-MS/MS) (Kim et al. 2015). To increase the number of identified proteins, some studies incorporate a one-dimensional gel electrophoresis separation step before in-gel trypsin digestion (Lee et al. 2009). However, one of the potential issues associated to this step is the low recovery of large and/or highly hydrophobic peptides from the gel. This shortcoming is especially relevant for the study of the protein content of MVs given that these are likely to have a significant content of membrane proteins.

One of the hallmarks of the protein catalogue of MVs from pathogenic cell-walled microorganisms is the fact that virulence-associated proteins are overrepresented. This is true for Mtb (Lee et al. 2015; Prados-Rosales et al. 2011), *S. aureus* (Gurung et al. 2011; Lee et al. 2009), *L. monocytogenes* (Lee et al. 2013b), *B. anthracis* (Rivera et al. 2010), *Streptomyces coelicolor* (Schrempf et al. 2011), *S. pneumoniae* (Olaya-Abril et al. 2014), *C. neoformans* (Rodrigues et al. 2008a), *H. capsulatum* (Albuquerque et al. 2008), *P. brasilensis* (Vallejo et al. 2011), and *C. perfringens* (Jiang et al. 2014). In the particular case of Mtb, an enrichment of lipoproteins in isolated MVs was reported. These classical toll-like receptor 2 (TLR2) ligands that are involved in the interference of the antigen presentation process during their interaction with host cells (Fulton et al. 2004), were not detected in isolated MV from the environmental mycobacterial strain *M. smegmatis* (Prados-Rosales et al. 2011).

#### 3.3.2.2 Identification of MV-Associated Lipids

A few studies have attempted the lipid characterization of isolated MV from cell-walled microorganisms. Lipidomic analysis of isolated MVs from the two Gram-positive bacteria *B. anthracis* and *S. pneumoniae* shows an enrichment in short-chain saturated fatty acids relative to the corresponding cell membranes (Olaya-Abril et al. 2014; Rivera et al. 2010), suggesting that membrane fluidity might be an important factor for MV release. In a different study with Mtb, MV lipid analysis was performed by thin-layer chromatography (TLC) and showed enrichment in polar lipids including phosphatidylinositol mannosides (PIMs) and cardiolipins, suggesting that the origin of mycobacterial MV might be the cellular membrane (Prados-Rosales et al. 2011). Lipidomic analysis of isolated *C. neoformans* MVs showed the presence of the glycosphingolipid (GlcCer) and sterol derivatives

(Rodrigues et al. 2007). Considering that these lipids are enriched in membrane microdomains (Muniz and Riezman 2000), their presence in isolated MVs would suggest that MV release events occur at these discrete regions of the membrane. Similar to *C. neoformans*, lipidomic analysis of isolated MVs from *H. capsulatum* showed the increased abundance of two species of hydroxylated fatty acids, containing 16 or 18 carbons (Albuquerque et al. 2008). As discussed by the authors, the similarity between the lipid composition of mammalian exosomes and fungal MVs in these particular lipids, would suggest a close mechanism of biogenesis. Lipidomic studies of MVs from cell-walled microorganisms are critical to understanding vesicle biogenesis. Consequently, more studies incorporating state-of-the-art methodology are needed to shed light to this conserved biological phenomenon.

#### 3.3.2.3 Identification of MV-Associated Nucleic Acids

Several important recommendations have been proposed by the international society for extracellular vesicles when attempting the analysis of nucleic acids, especially RNA, in extracellular vesicles to exclude the possibility that they are contaminants (Mateescu et al. 2017). Similarly, the study of the presence of DNA in isolated MVs should include enzymatic digestion of isolated material to determine the degree of DNA association with MVs. MVs from *C. neoformans*, *C. albicans*, *P. brasiliensis*, *M. sympodialis* (Rayner et al. 2017), and *S. cerevisiae* contain different types of RNA, including mature tRNAs, mRNAs, and noncoding RNAs (Peres da Silva et al. 2015). The potential role that RNA delivered by MVs might have in the target cells is still a matter of investigation.

### 3.3.3 *Quantification, Labeling, and Visualization of MVs*

Determination of the number and size of MVs is critical to understand how these structures are produced. This information would be essential to assess the feasibility of MVs as novel vaccines. Quantitative measurements of MVs from Gram-positive bacteria, mycobacteria, and fungi have been mostly performed by the determination of either vesicle mass, protein content, or the relative amount of lipids. Alternatively, determination of the amount of MVs has been performed by radiolabeling of growing cells with lipid precursors and by measuring radioactive counts in cells, vesicle pellet, and supernatant after ultracentrifugation (Brown et al. 2014). However, none of these methods provides absolute quantification of MVs. Labeling of MVs has also been used as a method for quantification or visualization of MVs. The first attempts at labeling MVs were performed with lipophilic dyes such as DiI, DiO, and PKH26 (Morales-Kastresana and Jones 2017; Morales-Kastresana et al. 2017). However, this approach has proven problematic given that these lipophilic dyes caused unacceptable alterations in size and morphology, presumably due to aggregation events (Morales-Kastresana et al. 2017). The authors of these works have cautioned against the use of DiI or PKH26 and we join them in urging the discontinuation of these protocols. We have used a fatty-acid-BODIPY conjugate that is added to the culture medium and incorporated into bacterial cells (Coelho et al. 2019), which allowed us to isolate MVs that maintained the same size and density, as observed by DLS characteristics and density gradient centrifugation. This technique has the advantage that it can be tailored to each individual organism, i.e., it is easy to change the lipid species that are tagged, particularly when combined with the knowledge of lipid composition of the MVs under study. Another widely used tool is to detect MVs by immunodetection to one of the surface-associated proteins (Kowal et al. 2016). This technique is only limited by the quality of the antibody used and has allowed affinity purification of MVs subpopulations in mammalian cells (Kowal et al. 2016). Additional tools are needed to perform in vivo tracking and highly sensitive tracking, to allow for more specific labeling and separation. A very successful alternative was genetic coupling of fluorophores (or luciferase) to proteins secreted in MVs (Lai et al. 2015, 2014).

The diversity of microbes entails that for each microbe, MV-specific markers (or at least MV-enriched markers) need to be selected and validated. The task of selecting an appropriate, robust MV marker is further complicated since MVs may be significantly different depending on media or culture conditions used, and therefore one would have to find the one marker that is conserved throughout a wide variety of experimental conditions. However, the rewards to be reaped would be immense: in addition to live-tracking of MVs these tools could allow quick and selective purification, via columns or sorting, and could allow for detection of very low amounts of these MVs.

Technologies have been developed to perform the automatic analysis and quantification of individual MVs, including the tunable resistive pulse sensing (tRPS), nanoparticle tracking analysis (NTA), high-resolution flow cytometry (hFC), or dynamic light scattering (DLS) (Maas et al. 2015). While all of these methods acquire information related to particle size, only hFC has the potential of providing an absolute number of MVs. Advantages and limitations of these technologies have been discussed previously (Szatanek et al. 2017). In the context of Gram-positive bacteria, NTA has been successfully applied to the study of *Sporothrix brasiliensis* MVs (Ikeda et al. 2018) and tRPS was used to determine vesicle size in Mtb (Athman et al. 2015). Also very extended in the bacterial and fungal MVs field is the use of either DLS or transmission electron microscopy (TEM) to assess the size distribution of the vesicle population. Electron microscopy is another complementary approach to characterize MVs populations. TEM was used to visualize MVs and to study size and morphology of these structures in many Gram-positive, mycobacteria, and fungal species (Brown et al. 2015). TEM provides valuable information about the heterogeneity of the MV population. Limitations associated with this technique are the required number of steps for sample processing that usually introduces many morphological artifacts. Scanning electron microscopy (SEM) has also been utilized to visualize vesicle release events in bacterial and fungal cells (Prados-Rosales et al. 2011; Silva et al. 2014). Although this information is valuable to ascertain the capacity of a specific strain to release MVs, the possibility that these structures are confounded with other cell surface-associated molecules requires strict validation of the obtained results. Demonstration of MV production in many species of Gram-positive bacteria usually combines several microscopy approaches such as TEM, SEM, FC, or atomic force microscopy (AFM) (Liu et al. 2018). The recently developed technique of cryoelectron microscopy (cryo-EM) reduces considerably the steps in sample processing and allows the determination of sample morphology in a close-to-native state (Glaeser and Hall 2011). A recent study combined NTA, tRPS, and cryo-EM to determine the size distribution of isolated *S. aureus* MVs. Interestingly, each method provided a different size distribution and showed the monodispersed nature of the MV population, and highlighted the necessity of combining complementary approaches to characterize MVs in cell-walled microorganisms (Tartaglia et al. 2018).

## 3.4 Mechanisms of MV Biogenesis in Cell-Walled Microorganisms

The limited research into MVs from Gram-positive bacteria, mycobacteria, and fungi was influenced by the belief of the potential interference from the thick cell wall for MV transport to outside the cell. This, in turn, makes it more challenging to provide an explanation for the mechanism(s) of vesicle biogenesis in cell-walled microorganisms. It seems intuitive that a cell wall remodeling process that would allow vesicular transit must govern MV biogenesis in these microorganisms (Fig. 3.1). Hypotheses for MV production in cell-walled microorganisms are mostly based on proteomics data from isolated vesicles, assuming that MV-associated proteins are involved in such remodeling processes. The proposal that cell wall-modifying enzymes can alter cell wall thickness and facilitate MV release provides an example for this notion. Isolated MVs from *S. aureus* (Lee et al. 2009) or *M. tuberculosis* (Prados-Rosales et al. 2011) were found to contain peptidoglycan degrading enzymes. Similarly, glycosyl hydrolases with substrate specificity for cell wall polymers have been identified in proteomic screens of EVs from several fungal species (Albuquerque et al. 2008; Oliveira et al. 2010; Rodrigues et al. 2007) (Fig. 3.1a, b). Alternatively, it was proposed that MVs can exert a change in turgor pressure after budding from the cell membrane forcing their pass through the cell wall (Rodrigues et al. 2008a; Vallejo et al. 2012a) (Fig. 3.1b). Derived from ultrastructural studies is the hypothesis that MVs can transit through pores originated within the cell wall (Rodrigues et al. 2008a; Vallejo et al. 2012a). Of note, as previously demonstrated the pore size in *S. cerevisiae* varies from 50 to 300 nm, depending on the growing conditions (de Souza Pereira and Geibel 1999). Modification of cell wall permeability has also been observed during melanization by *C. neoformans*. In this case, melanization reduces the cell wall pore size and induces the accumulation of MVs in the cell wall (Jacobson and Ikeda 2005) (Fig. 3.1b).

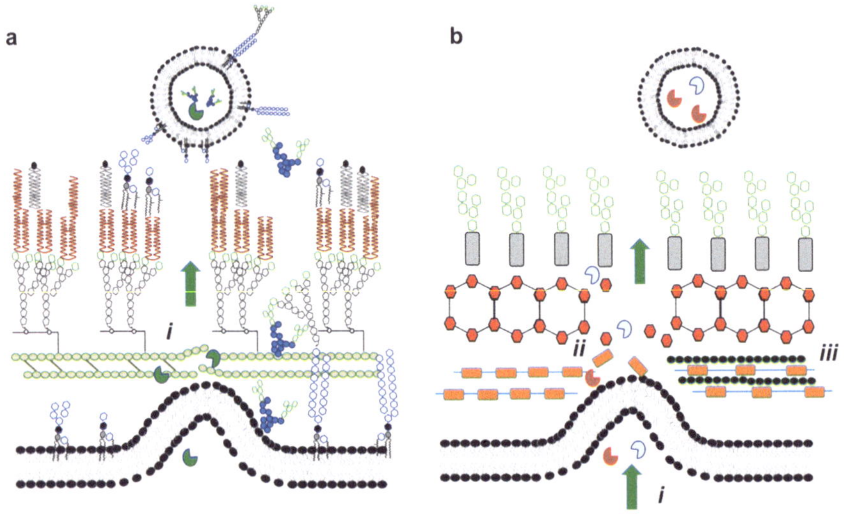


**Fig. 3.1** MV formation in cell-walled microorganisms. Hypothesis on how MVs get through thick cell walls. (**a**) Scheme of the mycobacterial cell envelope. As documented in proteomics studies for either *M. tuberculosis* or Gram-positive bacteria such as *S. aureus*, the presence of cell wall-modifying enzymes (**i**) could loosen the wall skeleton facilitating the release of MVs (green arrow). (**b**) Scheme of the fungal cell wall. Release of MVs (green arrow) by fungal cells may be regulated by the porosity of the cell wall, which can be modified by turgor pressure (**i**), cell wall modifying enzymes (**ii**) or melanization (**iii**)

### 3.4.1 Fungi

The complexity involved in generating vesicles and loading them with such varied cargo implies the existence of complex machinery. The current view is that MVs in fungi are assembled in multivesicular bodies that then release these structures outside of the cell membrane to transit through the cell wall into the extracellular space (Casadevall et al. 2009). Consistent with this notion, numerous genes and signaling pathways have been implicated in MV production. Analysis of MV production in *S. cerevisiae* strains deficient in *Sec4p* and *Snf7*, which are involved in Golgi-derived exocytosis and multivesicular body (MVB) formation, respectively, showed that these produced vesicles with altered composition (Oliveira et al. 2010). Golgi reassembly and stacking protein (GRASP) was implicated in vesicle release and the packaging on RNA in *C. neoformans* vesicles (Kmetzsch et al. 2011; Peres da Silva et al. 2018). In *C. albicans*, the *Sap2* gene and TOR pathway have been implicated in extracellular vesicle production (Gil-Bona et al. 2015).

### 3.4.2 Gram-Positive Bacteria and Mycobacteria

As in Gram-negative bacteria, no gene mutation abolishing MV has been found for either Gram-positive bacteria or mycobacteria. The absence of null mutants has led to suggestions that vesicle biogenesis is either an essential process or that physical changes may give rise to vesicles irrespective as an integral property of membranes. Nevertheless, few reports have established the connection between defects in vesicle biogenesis and genetics in cell-walled microorganisms relative to Gram-negative bacteria, raising the possibility that null mutants may still be found in the future. Deletions in *L. monocytogenes sigB* (RNA polymerase sigma factor $\sigma^B$) (Lee et al. 2013b) or *M. tuberculosis* vesicle biogenesis and immune response regulator (*virR*) (Rath et al. 2013) and transmembrane component of the phosphate-specific transport (Pst) *pstA1* (White et al. 2018) were shown to alter MV production. In the particular case of *L. monocytogenes*, quantification of MVs in the wild type and *sigB* mutant strains was determined by protein concentration, raising the question whether such differences could be attributable to the potential differential capacity of both strains to load proteins on vesicles. Similarly, there is evidence that a two-component system, CovRS (control of virulence regulator-sensor) regulates the production of MVs in *Streptococcus pyogenes* (group A streptococcus (GAS)) (Resch et al. 2016). It was observed that those GAS strains harboring a CovRS genetically disrupted manifested increased MV production compared to those with an intact system. Nevertheless, the exact mechanism by which MV biogenesis is controlled is still unknown. Recently, two independent reports on the Gram-positive bacteria *B. subtilis* and *S. aureus*, have shed some light into mechanisms for MV biogenesis. Mytomicin C-induced genotoxic stress in *B. subtilis* stimulated the expression of endolysins encoded by prophages, which altered cell wall permeability and facilitated MV production (Toyofuku et al. 2017). Of note, a similar mechanism was demonstrated in the Gram-negative bacterium *P. aeruginosa* (Turnbull et al. 2016). However, while in *B. subtilis* MV are released through pores originated in the peptidoglycan, *P. aeruginosa* cells explode and MVs are formed from membrane fragments (Turnbull et al. 2016). In *S. aureus*, the release of MV enriched in lipoproteins was dependent on surfactant-like molecules such a phenol-soluble modulins (PSMs). Apparently, PSMs facilitates MV release by increasing cell membrane fluidity (Schlatterer et al. 2018). In the same bacteria, antibiotics induced MV production by a phage-dependent or phage-independent fashion, depending on the mode of action of the antibiotic. Thus, oxidative stress-inducing antibiotics can induce *S. aureus* MV production through endolysin-triggered cell death (phage-dependent) and β-lactams do it by increasing the permeability of the peptidoglycan layer in phage-independent way (Andreoni et al. 2019). The finding that iron limitation regulates MV release in *M. tuberculosis* (Prados-Rosales et al. 2014c) supports the notion that MV biogenesis is important for the survival strategy of this pathogenic bacterium. In that study, it was shown that in vitro growth of *M. tuberculosis* under iron starvation triggered MV production and those vesicles were enriched in Fe-loaded siderophores. Importantly, growth restoration was achieved when siderophore-deficient *M. tuberculosis* mutants were supplemented with low iron MVs, indicating that they could serve as iron donors. These findings entertain the notion that those MVs can potentially contribute to the bacterial persistence by benefiting the local community.

## 3.5 MV Cargo

### *3.5.1 Fungi*

Fungal MVs cargo includes proteins, lipids, nucleic acids, and carbohydrates. Compositional analysis of MVs from different fungal species revealed proteins from different functional categories such as metabolism of proteins or carbohydrates, response to stress, transport, or signaling (de Toledo Martins et al. 2019). Common to other pathogenic organisms is the fact that MVs from pathogenic fungal species are loaded with proteins associated with virulence. For instance, laccase, urease, and phosphatase activities have been demonstrated in isolated MVs from *C. neoformans* (Rodrigues et al. 2008a). Similarly, phosphatase activity was detected in MVs from *P. brasiliensis* (Vallejo et al. 2012a). Other proteins related to fungal virulence and also found in MVs are those involved in controlling the REDOX balance of the fungal cell, such as catalase B or superoxide dismutase (Albuquerque et al. 2008; Vallejo et al. 2012b). Several hypotheses of MV function have been proposed based on protein composition of fungal MVs. MVs may participate in the fungal cell wall remodeling process due to the fact that cell wall glycosyl hydrolases such as endochitinase and glucanase were identified in isolated MVs (Albuquerque et al. 2008); MVs may participate in cell wall biogenesis as they deliver glycoconjugates to the outermost part of the cell wall (Rodrigues et al. 2007; Vallejo et al. 2011).

Although there are few studies of RNA analysis in fungal MVs, a comparative study of RNA content of fungal MVs from *S. cerevisiae*, *C. neformans*, and *C. albicans* revealed the intraluminal association of RNA with MVs as demonstrated by enzymatic degradation studies, demonstrating the role of vesicles as RNA transporters (Peres da Silva et al. 2015). For a recent review on this topic see also de Toledo Martins et al. (2019).

### *3.5.2 Gram-Positive Bacteria*

Gram-positive MVs harbor a variety of components including proteins, toxins, nucleic acids, lipids and polysaccharides, small metabolites, and antibiotics. These data offer a first insight into their possible functions, as well as mechanisms required for MV biogenesis. While there is more available information in protein and nucleic acid composition, due to the wider availability of technology, information on other types of MV components is starting to emerge.

The production of MV in Gram-positive bacteria was first studied in the case of *S. aureus* that led to the identification of 90 MV-associated proteins using a proteomic approach. The proteins found in MVs were comprised mostly of cytoplasmic proteins in addition to extracellular and membrane-associated proteins (Lee et al. 2009). That study was followed by proteomic analysis of various other Gram-positive bacteria MVs including *B. subtilis* (Brown et al. 2014), *B. anthracis* (Rivera et al. 2010), *L. monocytogenes* (Coelho et al. 2019), *S. coelicolor* (Schrempf et al. 2011), C. *perfringens* (Jiang et al. 2014), and *S. pneumoniae* (Olaya-Abril et al. 2014).

Virulence factors, including toxins, constitute a significant portion of the protein cargo. MVs isolated from *B. anthracis* consisted of biologically active toxins including edema factor, lethal factor, protective antigen, and anthrolysin (Rivera et al. 2010). MVs containing α-hemolysin and various pore-forming toxins have been identified in several other Gram-positive bacteria including *S. aureus*, *S. pneumoniae*, *S. pyogenes*, and *S. agalactiae* (Jeon et al. 2016; Olaya-Abril et al. 2014; Resch et al. 2016; Surve et al. 2016; Thay et al. 2013). The virulence factors Internalin B (InlB) and Listeriolysin O (LLO) were reported in MVs isolated from *L. monocytogenes* (Coelho et al. 2019; Lee et al. 2013b).

MVs can carry components important for bacterial survival, including a role in nutrient scavenging. MVs of *B. subtilis* contain the virulence factor SunI that provide immunity against the bacterial antibiotic sublancin (Dubois et al. 2009). Proteomic analysis of MVs revealed the presence of proteins involved in regulating the levels of drug resistance in Gram-positive bacteria. Specifically, MVs isolated from *S. aureus* ATCC 14458 were loaded with beta-lactamase protein BlaZ, that can bind to beta-lactam antibiotics and protect the neighboring bacteria against ampicillin drug (Lee et al. 2013a). Another study showed the presence of MsrR, a membrane-associated protein, in the MVs of *S. aureus* (Tartaglia et al. 2018). MsrR is known to be involved in providing resistance against methicillin and teicoplanin and also affects the synthesis of various virulence factors like alpha-toxin and protein A (Rossi et al. 2003). *S. coelicolor* and *S. aureus* produce MVs containing iron-binding proteins that help in storage of iron and survival in iron restricted environment (Schrempf et al. 2011; Surve et al. 2016), in agreement with what has been found in other classes of bacteria, as discussed above.

Schrempf et al., studied blue-pigmented exudate droplets, approximately 1 mm in size, obtained on the sporulated lawns of *S. coelicolor* M110. Inside these droplets, they found antibiotics such as actinorhodin. Electron microscopic view showed that these droplets were filled with vesicular structures. The proteomic analysis of the exudates revealed the presence of several other proteins important in survival and defense mechanisms (Schrempf et al. 2011).

MV cargo from commensal bacteria can also mediate beneficial effects on the mammalian hosts. MVs isolated from *Bifidobacterium longum*, *Lactobacillus plantarum*, *Lactobacillus casei*, and *Lactobacillus rhamnosus* contain proteins responsible for the probiotic effects of the bacterium (Behzadi et al. 2017; Dominguez Rubio et al. 2017; Kim et al. 2016a; Li et al. 2017).

The presence of nucleic acids is also a constant for Gram-positive bacteria MVs. PCR analysis demonstrated the genes for 16S ribosomal RNA, α-toxin, and perfringolysin O in MVs from *Clostridium perfringens* (Jiang et al. 2014). Similarly, *Streptococcus mutans* produced MVs containing extracellular DNA during biofilm formation. The secretion of extracellular DNA during biofilm formation plays an important role in bacterial adhesion and the stabilization and maturation of the biofilm structure (Liao et al. 2014).

Another critical component of MVs is their lipids and some studies have started to explore the lipid composition of Gram-positive MVs. The lipid membrane composition of Gram-positive MVs is not entirely dissimilar from that of the bacterial cell membrane, but lipidomic analysis of *B. anthracis* (Rivera et al. 2010) and *S. pneumoniae* (Olaya-Abril et al. 2014) MVs revealed some significant differences as well. In comparison to the bacterial cell membrane, the isolated MVs were highly enriched in short-chain saturated fatty acids (such as myristic, palmitic, lauric, and pentadecanoic acids). In contrast, in *L. monocytogenes* (Coelho et al. 2019) MVs were enriched in unsaturated fatty acids compared to bacterial cells. To date only one study characterized metabolites in Gram-positive MVs. In *L. monocytogenes* some small metabolites such as ornithine, pyruvic acid, and sugars such as xylitol and D-mannose were found in MVs extract (Coelho et al. 2019).

### *3.5.3 Mycobacteria*

Compositional proteomic analysis of *M. tuberculosis* MVs identified 42 MV-associated proteins, where the putative TLR2 ligands such as the lipoproteins 19 kDa, LppX, and LprG, a well-known group of TLR2 ligands that interfere with the antigen presentation process in dendritic cells and macrophages, were the most abundant proteins (Prados-Rosales et al. 2011). This enrichment phenomenon was also observed in BCG MVs but not in *M. smegmatis* MVs. Results from this report were validated by a second study, which implemented a label-free mass spectrometry quantification of Mtb MVs, identifying 287 proteins. Among the most prevalent proteins of MVs were SodB, EsxN, LppX, PstS1, LpqH, KatG, GlnA1, Apa, FbpA, FadA3, Mtc28, AcpM, Fba, and Prs. Notably, only Lppx, PstS1, and LpqH accounted for more than 6% of the total vesicular proteins (Lee et al. 2015). In another study, it was revealed that Mtb enhances the release of MVs upon iron starvation and that such low iron MVs are loaded with siderophores (Prados-Rosales et al. 2014c).

## 3.6 Cell-Walled Organisms-Derived MV Functions

### *3.6.1 Fungi*

In the sections below we briefly summarize some functions and topics involving fungal MVs. For recent reviews in this topic see also de Toledo Martins et al. (2019); Joffe et al. (2016); and Rodrigues et al. (2015).

MVs function as vehicles for non-classical secretion of many different components including proteins, small molecules, lipids, and nucleic acids. Packaging multiple components into MVs has the obvious advantage that these are delivered as a concentrated bolus and thus avoid the dilution that would occur had they been released at the cell membrane and diffused away. Extracellular vesicles from *C. neoformans*, *S. cerevisiae*, and *P. brasiliensis* have each been shown to package small RNAs (Peres da Silva et al. 2015). The function of these RNAs is not known but their presence in MVs raises the possibility that they are used in communication and/or modulation of target cells. Proteomic analysis of MV protein content from *C. neoformans*, *H. capsulatum*, and *P. brasiliensis* has shown great variability in composition with the cargo including structural proteins and enzymes (Rodrigues et al. 2014; Vallejo et al. 2012b). Many of the proteins found in MV lack signal peptides associated with conventional secretion and vesicles represent a mechanism for unconventional secretion (Rodrigues et al. 2014).

For *C. neoformans* MVs transport components associated with virulence such as polysaccharides and laccase (Rodrigues et al. 2008a, 2007) that can modulate macrophage functions (Oliveira et al. 2010). Lipid staining was used to document punctate structures in the *C. neoformans* capsule that could represent MVs (Nicola et al. 2009). MVs from the highly pathogenic *Cryptococcus gattii* responsible for the Vancouver outbreak are taken up by macrophages where they impair cellular functions and facilitate the growth of less pathogenic strains (Bielska et al. 2018). *C. albicans*-derived vesicles activate NF-κB in murine macrophages and this effect is highly dependent on MV lipid composition (Wolf et al. 2015). Host defenses against MVs include Galectin-3, which is lytic to *C. neoformans* MVs (Almeida et al. 2017). *C. neoformans* MVs have also been shown to be disrupted by other host proteins such as albumin (Wolf et al. 2012). Antibodies to epitopes in the fungal cell wall of *H. capsulatum* interfere with vesicle release (Matos Baltazar et al. 2016), suggesting that adaptive immune responses during infection can modulate fungal physiology to reduce the secretion of vesicle-associated virulence factors.

### *3.6.2 Gram-Positive Bacteria*

Some of the Gram-positive bacterial MVs proteins are related to virulence and are involved in drug resistance, host cell invasion, immune system evasion, cytotoxicity, and pathogenicity. Proteomic analysis showed that *S. aureus* produces MVs containing superantigens that can result in nonspecific activation of host T-cells and thus MV may help the bacterium to evade host immune system (Lee et al. 2009). Other virulence factors involved in immune evasion included IgG-binding protein SbI, lactamase, coagulase, lipase, and N-acetylmuramoyl-1-alanine amidase (Lee et al. 2009). MVs isolated from Group B *Streptococcus* contained hyaluronate lyase and possessed collagenase activity that can result in the breakdown of extracellular structures (Surve et al. 2016). Some other virulence factors contributing to the spread and cellular invasion of bacteria included cysteine protease (Staphopain A) and serine protease (exfoliative toxins) identified in *S. aureus* MVs (Jeon et al. 2016; Lee et al. 2009).

MV-associated toxins play an important role in pathogenesis by forming pores in the host membrane and leading to cytotoxicity and apoptosis, consistent with the virulence functions of Gram-positive MVs. Importantly, toxins carried in MVs may explain secretion of toxins that lack any export or secretion signal sequences, an example of which are pneumolysin-containing MVs produced by *S. pneumoniae* (Hirst et al. 2008; Olaya-Abril et al. 2014). To further support an MV-associated secretion, activity of these toxins is usually dependent upon its association with vesicles. For instance, in the case of *S. aureus* the MV-associated α-hemolysin proved to be more cytotoxic and induced necrosis in comparable amounts of soluble α-hemolysin (Hong et al. 2014); or the mycolactone toxin present in the *M. ulcerans* MVs showed a more potent toxic effect on non-phagocytic cells relative to pure toxin (Marsollier et al. 2007).

MVs serve not only as attack weapons but also in defensive roles. MVs can bind to antibiotics (Lee et al. 2013a; Rossi et al. 2003) or help in collecting micronutrients from the extracellular environment, such as iron (Jeon et al. 2016; Schrempf et al. 2011).

While there have been several studies on the response triggered by bacterial MVs in host cells, including the toxic and immunomodulatory effects of MVs, one area that is still relatively unexplored is the role of MVs in intercellular communication among bacterial communities. MVs have been shown to be critical for the formation of biofilms, through export of DNA and proteins, but certainly other intercellular communications take advantage of MVs. For example, *B. subtilis* produces MVs for striking different compositions in different stages of growth (Kim et al. 2016b). Are these serving as communication devices, for example, in quorum sensing, as is proposed for the Gram-negative OMVs? If in mixed communities, MVs have been shown as defenses from antibiotics produced by other bacterial species, it seems logical that MVs serve as interspecies communication devices in symbiotic relationships to help both species thrive.

### *3.6.3 Mycobacteria*

Production of MVs by Mtb was shown to occur during an ongoing infection (Prados-Rosales et al. 2011), suggesting that released MVs and their corresponding cargo have the capacity of interfering with the surrounding environment. Considering that these MVs are enriched in lipoproteins which, via TLR2, are able to reduce the capacity of dendritic cells and macrophages to present antigen, it is reasonable to hypothesize that Mtb might use these structures to modify the nearest environment to its benefit. Supporting this notion was the observation of an increase in bacterial burden in lungs and spleen, as well as lung inflammation, in Mtb-infected mice, which were treated with intratracheal administration of Mtb MVs (Prados-Rosales et al. 2011). Mtb MVs have gained interest in the context of antigen transfer during the prime immune response to Mtb infection. In an attempt to provide an explanation to the observed robust immune response to Mtb beside the impaired capacity of infected DC for antigen presentation, the hypothesis of antigen export to bystander uninfected DC was proposed. A major premise for antigen export from infected phagocytes is that Mtb-related antigens associate with host membranes and are incorporated into the host endocytic machinery resulting in eukaryotic vesicles harboring bacterial antigens. Two recent reports indicate that the landscape of antigen export by Mtb-infected phagocytes is more complex than expected (Srivastava and Ernst 2014; Srivastava et al. 2016). Strikingly, they showed that antigen transfer was not dependent on eukaryotic exosomes or apoptotic bodies derived from macrophages. Another recent report exploring the origin of MVs in Mtb-infected macrophages showed that exosomes and bacterial-derived extracellular vesicles might represent two independent populations of vesicles, suggesting a scenario where antigen transfer could be mediated by bacterial derived MVs (Athman et al. 2015).

## 3.7 MVs in Medicine

### *3.7.1 Clinical Implications of MVs*

Examples of both detrimental and beneficial effects of MVs from cell-walled microorganisms have been provided. The production of MVs by many pathogenic species of Gram-positive bacteria, mycobacteria, and fungi suggests a connection between microbial pathogenesis and vesicles. Studies where isolated MVs were administered locally before challenge with the corresponding pathogenic species translated into exacerbation of the disease. In *C. neoformans*, this approach resulted in an enhanced efficiency in crossing the blood–brain barrier by the fungus (Huang et al. 2012). In *M. tuberculosis*-infected mice a Koch's phenomenon was observed in mice previously infused with Mtb MVs (Prados-Rosales et al. 2011). In fact, some of these findings could help to explain the occurrence of such detrimental effects at distant locations from the infection site. Similarly, it was reported that there was an association between preterm pregnancy termination and exposure of mice to *S. agalatiae* (GBS) MVs. In this particular case, documentation of MVs traveling through the uterine reproductive tract was reported (Surve et al. 2016). Conversely, MVs produced by nonpathogenic variants of cell-walled microorganisms have been shown to provide a benefit or at least no negative effect for the host. The best examples of this type of effect are MVs from probiotic microorganisms such as those from the *Bifiobacterium* and *Lactobacillus* genus. Importantly, such probiotic effects were observed with isolated MVs but not with whole cells, highlighting again the potential of these structures in reaching distant sites of the host more efficiently than a whole bacterial cell (Liu et al. 2018).

### 3.7.2 MVs as Vaccine

The observation that the administration of MVs prior to infection with pathogenic cell-walled microorganisms promote disease, contrasts with scenarios where naturally produced MVs have been shown to serve as potential vaccines. The potential advantages of using MVs as vaccines are *(i)* they are easy to obtain; *(ii)* they have intrinsic adjuvant properties; and *(iii)* they lack the ability to replicate like live bacteria. Protective effects of isolated MVs have been shown for *M. tuberculosis* (Prados-Rosales et al. 2014b), *B. anthracis* (Rivera et al. 2010), *S. pneumoniae* (Olaya-Abril et al. 2014), *C. perfringes* (Jiang et al. 2014), *S. aureus* (Choi et al. 2015) and *C. albicans* (Vargas et al. 2015). Among beneficial effects derived from vaccination with MV are the extended life of the infected host, reduced inflammation, or lower bacterial or fungal burden in organs. One of the major issues associated with the use of naturally produced MVs from bacteria is their potential toxicity. Although this problem is greater with Gram-negative bacteria because of the presence of LPS in OMVs, toxicity issues may also appear in Gram-positive MVs. A recent study generated genetically engineered *S. aureus* strains to produce MVs with limited toxicity (Wang et al. 2018) and propose them as a vaccine platform.

### 3.7.3 MVs in Biofilm Production

It is believed that environmental microbial communities are primarily organized in biofilms, which creates one of the most recalcitrant issues in clinical settings and represent a leading source of antibiotic-resistant infections. The biofilm matrix constituted by exopolysaccharides, proteins, and DNA, among other molecules, protects the bacterial and or fungal communities from external insults. MVs have been implicated in biofilm formation, presumably due to their role in transporting the cargo used in biofilm formation. Association between MVs and the biofilm formation was studied in *M. ulcerans* (Marsollier et al. 2007), *B. subtilis* (Brown et al. 2014), *S. mutants* (Liao et al. 2014), and *C. neoformans* (Robertson et al. 2012). MVs were visualized in association with the outermost layer of the extracellular matrix of *B. subtilis* and *M. ulcerans* biofilms. For *C. albicans,* MVs produced in biofilm conditions carry components used in matrix construction and differ from those produced by planktonic cells (Zarnowski et al. 2018). Such biofilm-promoting vesicles reduce the antifungal susceptibility of *C. albicans* cells and thus represent an example of community sharing of resources (Zarnowski et al. 2018). The addition of subinhibitory concentrations of EDTA to *C. neoformans* interfered with both MV release and biofilm formation (Robertson et al. 2012).

### *3.7.4 MVs in Diagnosis*

The literature on the use of MVs from cell-walled microorganisms as a platform for the diagnosis of infectious diseases is scarce. In a previous report, the human serologic response to isolated *M. tuberculosis* MVs was evaluated. A signature for TB was identified and consisted of three proteins that were specifically recognized by both smear-positive and smear-negative TB patients but not in healthy controls with or without latent tuberculosis infection (Ziegenbalg et al. 2013). These studies supported the use of Mtb MVs as a source of biomarkers. More recently, the finding that microbiota-derived MVs could be detected in urine or blood opens the door to the use of these components as a surrogate of the health status of the bacterial communities of our gut (Kang et al. 2013; Lee et al. 2017; Park et al. 2017).

## 3.8 Unsolved Problems and Concluding Remarks

The discovery of MVs in cell-walled microorganisms such as fungi, Gram-positive bacteria, and mycobacteria has opened a whole new set of questions and avenues of investigation that were not considered in microbiology until very recently. Some of the new problems are common to both fungi and bacteria, while others are specific to each class of microorganisms.

### *3.8.1 Cell Wall Transport*

Perhaps the most intriguing question raised by the discovery of MVs in cell-walled microorganisms is: how do these structures get through the cell wall? This question is intriguing because it challenges long held impressions of cell walls and their functions. In fact, the discovery of MVs in this class of microorganisms was probably delayed by the fact that their existence did not fit within the paradigm that cell walls were rigid and permeable only to small molecules. This view of cell walls meant that there was no need to consider extracellular vesicles and when such structures were occasionally noted in electron microscopy images these were dismissed as artifacts from the association of lipids released from dead and dying cells. In microbiology, fields are organized based on the species of the microbe such that communities working on different microbes constitute different fields

(Casadevall and Fang 2015). Hence, different fields have demonstrated different levels of acceptance of MVs. At this present time, the fungal and mycobacterial fields have wholeheartedly accepted the existence of MVs while the same notion has found different acceptance in various Gram-positive bacterial fields. Nevertheless, the increasing accumulation of papers from independent groups reporting MVs in different Gram-positive bacteria means that it will hopefully be only a matter of time before there is universal acceptance of this phenomenon.

### 3.8.2 Mechanism of Vesicle Production and Sorting Contents

The discovery of MV provides a transport system for great myriad of biomolecules that is not dependent on classical secretion mechanisms. MVs have been shown to carry proteins, polysaccharides, lipids, and even small molecules such as siderophores. Vesicle production poses different questions in fungi, Gram-positive bacteria, and mycobacteria. In fungi, the current view is that MVs are produced in vesicular bodies that then release these structures into the space between the cell membrane and cell wall for the crossing of the latter (Casadevall et al. 2009). For Gram-positive and mycobacteria, these vesicles are presumably produced at cell membranes for their transit across the cell wall. At this time there is little information regarding how OMVs and MVs are made and loaded with their cargo. Given that MVs have numerous roles in the extracellular space from biofilm formation, to communication, and promoting microbial virulence, one can imagine that there must exist a mechanism that sorts the cargo into vesicles depending on their intended role.

### 3.8.3 Role of Vesicles in Pathogenesis and Vaccines

The finding that many microbial MVs packs a suite of virulence factors means that vesicles have a role in microbial pathogenesis. For example, the finding that such toxigenic Gram-positive bacteria as *B. anthracis* (Rivera et al. 2010), *L. monocytogens* (Coelho et al. 2019), and *C. perfringes* (Jiang et al. 2014) pack their toxins into MVs is now an important factor to consider when incorporating vesicles into concepts of pathogenesis. MVs allow microbes to deliver a concentrated punch to a target such as predatory amoeba or a macrophage without concern about dilution during diffusion from the microbial surface. MVs also affect and induce immune responses (Kuipers et al. 2018). MVs from mycobacteria (Prados-Rosales et al. 2011, 2014b), *S. aureus* (Choi et al. 2015; Wang et al. 2018), and *S. pneumoniae* (Choi et al. 2017) elicit immune responses with vaccine potential. Hence, MV secretion poses new questions on how microbes damage host cells as well as providing a new set of components that can be used in vaccine design.

## 3.9 Concluding Remarks

A decade plus since the first reports of MVs in cell-walled microorganisms, this topic has emerged as an exciting area of study that poses new scientific problems that will undoubtedly produce new insights into prokaryotic and eukaryotic cell biology when solved. The production of MVs by cell-walled microorganisms places them in a continuum that includes plant and animal cells and Gram-negative bacteria, each of which uses vesicles to package cellular components for use in a wide array of functions that range from communication to modification of their environment. In fact, MV production may be a universal feature of cells irrespective of whether they have cell walls. One can anticipate that as additional investigators spin the culture supernatants of their favorite cell wall microorganisms, additional species will be associated with MV production, which in turn will lead to new research projects.

## References

Albuquerque PC et al (2008) Vesicular transport in *Histoplasma capsulatum*: an effective mechanism for trans-cell wall transfer of proteins and lipids in Ascomycetes. Cell Microbiol 10:1695–1710

Almeida F et al (2017) Galectin-3 impacts *Cryptococcus neoformans* infection through direct antifungal effects. Nat Commun 8:1968

Andreoni F et al (2019) Antibiotics stimulate vesicles formation in *Staphylococcus aureus* in a phage-dependent and independent fashion and via different routes. Antimicrob Agents Chemother 63(2):e01439-18

Athman JJ, Wang Y, McDonald DJ, Boom WH, Harding CV, Wearsch PA (2015) Bacterial membrane vesicles mediate the release of *Mycobacterium tuberculosis* lipoglycans and lipoproteins from infected macrophages. J Immunol 195:1044–1053

Behzadi E, Mahmoodzadeh Hosseini H, Imani Fooladi AA (2017) The inhibitory impacts of *Lactobacillus rhamnosus* GG-derived extracellular vesicles on the growth of hepatic cancer cells. Microb Pathog 110:1–6

Bielska E, Sisquella MA, Aldeieg M, Birch C, O'Donoghue EJ, May RC (2018) Pathogen-derived extracellular vesicles mediate virulence in the fatal human pathogen *Cryptococcus gattii*. Nat Commun 9:1556

Brown L, Kessler A, Cabezas-Sanchez P, Luque-Garcia JL, Casadevall A (2014) Extracellular vesicles produced by the Gram-positive bacterium *Bacillus subtilis* are disrupted by the lipopeptide surfactin. Mol Microbiol 93:183–198

Brown L, Wolf JM, Prados-Rosales R, Casadevall A (2015) Through the wall: extracellular vesicles in Gram-positive bacteria, mycobacteria and fungi. Nat Rev Microbiol 13:620–630

Casadevall A, Fang FC (2015) Field science—the nature and utility of scientific fields. MBio 6: e01259–e01215

Casadevall A, Nosanchuk JD, Williamson P, Rodrigues ML (2009) Vesicular transport across the fungal cell wall. Trends Microbiol 17:158–162

Choi SJ et al (2015) Active immunization with extracellular vesicles derived from *Staphylococcus aureus* effectively protects against staphylococcal lung infections, mainly via Th1 cell-mediated immunity. PLoS One 10:e0136021. https://doi.org/10.1371/journal.pone.0136021

Choi CW, Park EC, Yun SH, Lee SY, Kim SI, Kim GH (2017) Potential usefulness of *Streptococcus pneumoniae* extracellular membrane vesicles as antibacterial vaccines. J Immunol Res 2017:7931982

Chutkan H, Macdonald I, Manning A, Kuehn MJ (2013) Quantitative and qualitative preparations of bacterial outer membrane vesicles. Methods Mol Biol 966:259–272

Coelho C et al (2019) *Listeria monocytogenes* virulence factors, including Listeriolysin O, are secreted in biologically active extracellular vesicles. J Biol Chem 294(4):1202–1217. https://doi.org/10.1074/jbc.RA118.006472

de Souza Pereira R, Geibel J (1999) Direct observation of oxidative stress on the cell wall of *Saccharomyces cerevisiae* strains with atomic force microscopy. Mol Cell Biochem 201:17–24

de Toledo Martins S, Szwarc P, Goldenberg S, Alves LR (2019) Extracellular vesicles in fungi: composition and functions. Curr Top Microbiol Immunol 422:45–59. https://doi.org/10.1007/82_2018_141

Dominguez Rubio AP, Martinez JH, Martinez Casillas DC, Coluccio Leskow F, Piuri M, Perez OE (2017) *Lactobacillus casei* BL23 produces microvesicles carrying proteins that have been associated with its probiotic effect. Front Microbiol 8:1783

Dorward DW, Garon CF (1990) DNA is packaged within membrane-derived vesicles of Gram-negative but not Gram-positive bacteria. Appl Environ Microbiol 56:1960–1962

Dubois JY et al (2009) Immunity to the bacteriocin sublancin 168 is determined by the SunI (YolF) protein of *Bacillus subtilis*. Antimicrob Agents Chemother 53:651–661

Fulton SA, Reba SM, Pai RK, Pennini M, Torres M, Harding CV, Boom WH (2004) Inhibition of major histocompatibility complex II expression and antigen processing in murine alveolar macrophages by *Mycobacterium bovis* BCG and the 19-kilodalton mycobacterial lipoprotein. Infect Immun 72:2101–2110

Gehrmann U et al (2011) Nanovesicles from *Malassezia sympodialis* and host exosomes induce cytokine responses – novel mechanisms for host-microbe interactions in atopic eczema. PLoS One 6(7):e21480. https://doi.org/10.1371/journal.pone.0021480

Gil-Bona A, Monteoliva L, Gil C (2015) Global proteomic profiling of the secretome of *Candida albicans* ecm33 cell wall mutant reveals the involvement of Ecm33 in Sap2 secretion. J Proteome Res 14:4270–4281

Glaeser RM, Hall RJ (2011) Reaching the information limit in cryo-EM of biological macromolecules: experimental aspects. Biophys J 100:2331–2337

Grande R et al (2017) Detection and physicochemical characterization of membrane vesicles (MVs) of *Lactobacillus reuteri* DSM 17938. Front Microbiol 8:1040. https://doi.org/10.3389/fmicb.2017.01040

Gurung M et al (2011) *Staphylococcus aureus* produces membrane-derived vesicles that induce host cell death. PLoS One 6:e27958. https://doi.org/10.1371/journal.pone.0027958

Haas B, Grenier D (2015) Isolation, characterization and biological properties of membrane vesicles produced by the swine pathogen *Streptococcus suis*. PLoS One 10:e0130528. https://doi.org/10.1371/journal.pone.0130528

Hirst RA, Gosai B, Rutman A, Guerin CJ, Nicotera P, Andrew PW, O'Callaghan C (2008) *Streptococcus pneumoniae* deficient in pneumolysin or autolysin has reduced virulence in meningitis. J Infect Dis 197:744–751

Hong SW et al (2011) Extracellular vesicles derived from *Staphylococcus aureus* induce atopic dermatitis-like skin inflammation. Allergy 66:351–359

Hong SW et al (2014) An important role of alpha-hemolysin in extracellular vesicles on the development of atopic dermatitis induced by *Staphylococcus aureus*. PLoS One 9:e100499. https://doi.org/10.1371/journal.pone.0100499

Huang SH, Wu CH, Chang YC, Kwon-Chung KJ, Brown RJ, Jong A (2012) *Cryptococcus neoformans*-derived microvesicles enhance the pathogenesis of fungal brain infection. PLoS One 7:e48570. https://doi.org/10.1371/journal.pone.0048570

Ikeda MAK et al (2018) Extracellular vesicles from *Sporothrix brasiliensis* are an important virulence factor that induce an increase in fungal burden in experimental sporotrichosis. Front Microbiol 9:2286
Jacobson ES, Ikeda R (2005) Effect of melanization upon porosity of the cryptococcal cell wall. Med Mycol 43:327–333
Jeon H et al (2016) Variation among *Staphylococcus aureus* membrane vesicle proteomes affects cytotoxicity of host cells. Microb Pathog 93:185–193
Jeon J et al (2017) Proteomic analysis of extracellular vesicles derived from *Propionibacterium acnes*. Proteomics Clin Appl 11:1–2. https://doi.org/10.1002/prca.201600040
Jiang Y, Kong Q, Roland KL, Curtiss R 3rd (2014) Membrane vesicles of *Clostridium perfringens* type a strains induce innate and adaptive immunity. Int J Med Microbiol 304:431–443
Joffe LS, Nimrichter L, Rodrigues ML, Del Poeta M (2016) Potential roles of fungal extracellular vesicles during infection. mSphere 1:e00099-16. https://doi.org/10.1128/mSphere.00099-16
Johansson HJ et al (2018) Extracellular nanovesicles released from the commensal yeast *Malassezia sympodialis* are enriched in allergens and interact with cells in human skin. Sci Rep 8:9182. https://doi.org/10.1038/s41598-018-27451-9
Kang CS et al (2013) Extracellular vesicles derived from gut microbiota, especially *Akkermansia muciniphila*, protect the progression of dextran sulfate sodium-induced colitis. PLoS One 8: e76520. https://doi.org/10.1371/journal.pone.0076520
Kim MR et al (2012) *Staphylococcus aureus*-derived extracellular vesicles induce neutrophilic pulmonary inflammation via both Th1 and Th17 cell responses. Allergy 67:1271–1281
Kim JH, Lee J, Park J, Gho YS (2015) Gram-negative and Gram-positive bacterial extracellular vesicles. Semin Cell Dev Biol 40:97–104
Kim JH et al (2016a) Extracellular vesicle-derived protein from *Bifidobacterium longum* alleviates food allergy through mast cell suppression. J Allergy Clin Immunol 137:507–516 e508
Kim Y, Edwards N, Fenselau C (2016b) Extracellular vesicle proteomes reflect developmental phases of *Bacillus subtilis*. Clin Proteomics 13(6):6
Kmetzsch L et al (2011) Role for Golgi reassembly and stacking protein (GRASP) in polysaccharide secretion and fungal virulence. Mol Microbiol 81:206–218
Kowal J et al (2016) Proteomic comparison defines novel markers to characterize heterogeneous populations of extracellular vesicle subtypes. Proc Natl Acad Sci U S A 113:E968–E977
Kuipers ME, Hokke CH, Smits HH, Nolte-'t Hoen ENM (2018) Pathogen-derived extracellular vesicle-associated molecules that affect the host immune system: an overview. Front Microbiol 9:2182
Lai CP et al (2014) Dynamic biodistribution of extracellular vesicles in vivo using a multimodal imaging reporter. ACS Nano 8:483–494
Lai CP, Kim EY, Badr CE, Weissleder R, Mempel TR, Tannous BA, Breakefield XO (2015) Visualization and tracking of tumour extracellular vesicle delivery and RNA translation using multiplexed reporters. Nat Commun 6:7029
Lee EY et al (2009) Gram-positive bacteria produce membrane vesicles: proteomics-based characterization of *Staphylococcus aureus*-derived membrane vesicles. Proteomics 9:5425–5436
Lee J et al (2013a) *Staphylococcus aureus* extracellular vesicles carry biologically active beta-lactamase. Antimicrob Agents Chemother 57:2589–2595
Lee JH, Choi CW, Lee T, Kim SI, Lee JC, Shin JH (2013b) Transcription factor sigmaB plays an important role in the production of extracellular membrane-derived vesicles in *Listeria monocytogenes*. PLoS One 8:e73196
Lee J et al (2015) Proteomic analysis of extracellular vesicles derived from *Mycobacterium tuberculosis*. Proteomics 15:3331–3337
Lee Y et al (2017) Rapid assessment of microbiota changes in individuals with autism spectrum disorder using bacteria-derived membrane vesicles in urine. Exp Neurobiol 26:307–317
Li M, Lee K, Hsu M, Nau G, Mylonakis E, Ramratnam B (2017) Lactobacillus-derived extracellular vesicles enhance host immune responses against vancomycin-resistant enterococci. BMC Microbiol 17:66

Liao S et al (2014) *Streptococcus mutans* extracellular DNA is upregulated during growth in biofilms, actively released via membrane vesicles, and influenced by components of the protein secretion machinery. J Bacteriol 196:2355–2366
Liu Y, Defourny KAY, Smid EJ, Abee T (2018) Gram-positive bacterial extracellular vesicles and their impact on health and disease. Front Microbiol 9:1502
Maas SL et al (2015) Possibilities and limitations of current technologies for quantification of biological extracellular vesicles and synthetic mimics. J Control Release 200:87–96
Marsollier L et al (2007) Impact of *Mycobacterium ulcerans* biofilm on transmissibility to ecological niches and Buruli ulcer pathogenesis. PLoS Pathog 3:e62. https://doi.org/10.1371/journal.ppat.0030062
Mateescu B et al (2017) Obstacles and opportunities in the functional analysis of extracellular vesicle RNA—an ISEV position paper. J Extracell Vesicles 6:1286095
Matos Baltazar L, Nakayasu ES, Sobreira TJ, Choi H, Casadevall A, Nimrichter L, Nosanchuk JD (2016) Antibody binding alters the characteristics and contents of extracellular vesicles released by *Histoplasma capsulatum*. mSphere 1:e00085-15. https://doi.org/10.1128/mSphere.00085-15
Morales-Kastresana A, Jones JC (2017) Flow cytometric analysis of extracellular vesicles. Methods Mol Biol 1545:215–225
Morales-Kastresana A et al (2017) Labeling extracellular vesicles for nanoscale flow cytometry. Sci Rep 7:1878. https://doi.org/10.1038/s41598-017-01731-2
Muniz M, Riezman H (2000) Intracellular transport of GPI-anchored proteins. EMBO J 19:10–15
Nicola AM, Frases S, Casadevall A (2009) Lipophilic dye staining of *Cryptococcus neoformans* extracellular vesicles and capsule. Eukaryot Cell 8:1373–1380
Nimrichter L, de Souza MM, Del Poeta M, Nosanchuk JD, Joffe L, Tavares Pde M, Rodrigues ML (2016) Extracellular vesicle-associated transitory cell wall components and their impact on the interaction of fungi with host cells. Front Microbiol 7:1034
Olaya-Abril A et al (2014) Characterization of protective extracellular membrane-derived vesicles produced by *Streptococcus pneumoniae*. J Proteome 106:46–60
Oliveira DL et al (2010) Characterization of yeast extracellular vesicles: evidence for the participation of different pathways of cellular traffic in vesicle biogenesis. PLoS One 5:e11113
Park JY et al (2017) Metagenome analysis of bodily microbiota in a mouse model of Alzheimer disease using bacteria-derived membrane vesicles in blood. Exp Neurobiol 26:369–379
Peres da Silva R et al (2015) Extracellular vesicle-mediated export of fungal RNA. Sci Rep 5:7763
Peres da Silva R et al (2018) Golgi reassembly and stacking protein (GRASP) participates in vesicle-mediated RNA export in *Cryptococcus Neoformans*. Genes (Basel) 9:E400
Prados-Rosales R et al (2011) Mycobacteria release active membrane vesicles that modulate immune responses in a TLR2-dependent manner in mice. J Clin Invest 121:1471–1483
Prados-Rosales R, Brown L, Casadevall A, Montalvo-Quiros S, Luque-Garcia JL (2014a) Isolation and identification of membrane vesicle-associated proteins in Gram-positive bacteria and mycobacteria. MethodsX 1:124–129
Prados-Rosales R et al (2014b) Mycobacterial membrane vesicles administered systemically in mice induce a protective immune response to surface compartments of *Mycobacterium tuberculosis*. MBio 5:e01921–e01914
Prados-Rosales R, Weinrick BC, Pique DG, Jacobs WR Jr, Casadevall A, Rodriguez GM (2014c) Role for *Mycobacterium tuberculosis* membrane vesicles in iron acquisition. J Bacteriol 196:1250–1256
Rath P et al (2013) Genetic regulation of vesiculogenesis and immunomodulation in *Mycobacterium tuberculosis*. Proc Natl Acad Sci U S A 110:E4790–E4797
Rayner S, Bruhn S, Vallhov H, Andersson A, Billmyre RB, Scheynius A (2017) Identification of small RNAs in extracellular vesicles from the commensal yeast *Malassezia sympodialis*. Sci Rep 7:39742
Resch U et al (2016) A two-component regulatory system impacts extracellular membrane-derived vesicle production in group a streptococcus. MBio 7:e00207–e00216. https://doi.org/10.1128/mBio.00207-16

Rivera J, Cordero RJ, Nakouzi AS, Frases S, Nicola A, Casadevall A (2010) *Bacillus anthracis* produces membrane-derived vesicles containing biologically active toxins. Proc Natl Acad Sci U S A 107:19002–19007

Robertson EJ, Wolf JM, Casadevall A (2012) EDTA inhibits biofilm formation, extracellular vesicular secretion, and shedding of the capsular polysaccharide glucuronoxylomannan by *Cryptococcus neoformans*. Appl Environ Microbiol 78:7977–7984

Rodrigues ML, Casadevall A (2018) A two-way road: novel roles for fungal extracellular vesicles. Mol Microbiol 110:11–15

Rodrigues ML et al (2007) Vesicular polysaccharide export in *Cryptococcus neoformans* is a eukaryotic solution to the problem of fungal trans-cell wall transport. Eukaryot Cell 6:48–59

Rodrigues ML, Nakayasu ES, Oliveira DL, Nimrichter L, Nosanchuk JD, Almeida IC, Casadevall A (2008a) Extracellular vesicles produced by *Cryptococcus neoformans* contain protein components associated with virulence. Eukaryot Cell 7:58–67

Rodrigues ML, Nimrichter L, Oliveira DL, Nosanchuk JD, Casadevall A (2008b) Vesicular trans-cell wall transport in fungi: a mechanism for the delivery of virulence-associated macromolecules? Lipid Insights 2:27–40

Rodrigues ML, Nakayasu ES, Almeida IC, Nimrichter L (2014) The impact of proteomics on the understanding of functions and biogenesis of fungal extracellular vesicles. J Proteome 97:177–186

Rodrigues ML, Godinho RM, Zamith-Miranda D, Nimrichter L (2015) Traveling into outer space: unanswered questions about fungal extracellular vesicles. PLoS Pathog 11:e1005240. https://doi.org/10.1371/journal.ppat.1005240

Rossi J, Bischoff M, Wada A, Berger-Bachi B (2003) MsrR, a putative cell envelope-associated element involved in *Staphylococcus aureus* sarA attenuation. Antimicrob Agents Chemother 47:2558–2564

Schlatterer K et al (2018) The mechanism behind bacterial lipoprotein release: phenol-soluble modulins mediate toll-like receptor 2 activation via extracellular vesicle release from *Staphylococcus aureus*. MBio 9:e01851-18. https://doi.org/10.1128/mBio.01851-18

Schrempf H, Koebsch I, Walter S, Engelhardt H, Meschke H (2011) Extracellular Streptomyces vesicles: amphorae for survival and defence. Microb Biotechnol 4:286–299

Silva BM, Prados-Rosales R, Espadas-Moreno J, Wolf JM, Luque-Garcia JL, Goncalves T, Casadevall A (2014) Characterization of *Alternaria infectoria* extracellular vesicles. Med Mycol 52:202–210

Srivastava S, Ernst JD (2014) Cell-to-cell transfer of *M. tuberculosis* antigens optimizes CD4 T cell priming. Cell Host Microbe 15:741–752

Srivastava S, Grace PS, Ernst JD (2016) Antigen export reduces antigen presentation and limits T cell control of *M. tuberculosis*. Cell Host Microbe 19:44–54

Surve MV et al (2016) Membrane vesicles of group B streptococcus disrupt Feto-maternal barrier leading to preterm birth. PLoS Pathog 12:e1005816. https://doi.org/10.1371/journal.ppat.1005816

Szatanek R, Baj-Krzyworzeka M, Zimoch J, Lekka M, Siedlar M, Baran J (2017) The methods of choice for extracellular vesicles (EVs) characterization. Int J Mol Sci 18:E1153. https://doi.org/10.3390/ijms18061153

Tartaglia NR et al (2018) *Staphylococcus aureus* extracellular vesicles elicit an immunostimulatory response in vivo on the murine mammary gland. Front Cell Infect Microbiol 8:277. https://doi.org/10.3389/fcimb.2018.00277

Thay B, Wai SN, Oscarsson J (2013) *Staphylococcus aureus* alpha-toxin-dependent induction of host cell death by membrane-derived vesicles. PLoS One 8:e54661. https://doi.org/10.1371/journal.pone.0054661

Toyofuku M et al (2017) Prophage-triggered membrane vesicle formation through peptidoglycan damage in *Bacillus subtilis*. Nat Commun 8:481. https://doi.org/10.1038/s41467-017-00492-w

Toyofuku M, Nomura N, Eberl L (2019) Types and origins of bacterial membrane vesicles. Nat Rev Microbiol 17:13–24

Turnbull L et al (2016) Explosive cell lysis as a mechanism for the biogenesis of bacterial membrane vesicles and biofilms. Nat Commun 7:11220. https://doi.org/10.1038/ncomms11220

Vallejo MC et al (2011) The pathogenic fungus *Paracoccidioides brasiliensis* exports extracellular vesicles containing highly immunogenic alpha-galactosyl epitopes. Eukaryot Cell 10:343–351

Vallejo MC et al (2012a) Lipidomic analysis of extracellular vesicles from the pathogenic phase of *Paracoccidioides brasiliensis*. PLoS One 7:e39463. https://doi.org/10.1371/journal.pone.0039463

Vallejo MC et al (2012b) Vesicle and vesicle-free extracellular proteome of *Paracoccidioides brasiliensis*: comparative analysis with other pathogenic fungi. J Proteome Res 11:1676–1685

Vargas G et al (2015) Compositional and immunobiological analyses of extracellular vesicles released by *Candida albicans*. Cell Microbiol 17:389–407

Walker L et al (2018) The viscoelastic properties of the fungal cell wall allow traffic of AmBisome as intact liposome vesicles. MBio 9. https://doi.org/10.1128/mBio.02383-17

Wang X, Thompson CD, Weidenmaier C, Lee JC (2018) Release of *Staphylococcus aureus* extracellular vesicles and their application as a vaccine platform. Nat Commun 9:1379. https://doi.org/10.1038/s41467-018-03847-z

White DW, Elliott SR, Odean E, Bemis LT, Tischler AD (2018) *Mycobacterium tuberculosis* Pst/SenX3-RegX3 regulates membrane vesicle production independently of ESX-5 activity. MBio 9:e00778–e00718. https://doi.org/10.1128/mBio.00778-18

Wolf JM, Rivera J, Casadevall A (2012) Serum albumin disrupts *Cryptococcus neoformans* and *Bacillus anthracis* extracellular vesicles. Cell Microbiol 14:762–773

Wolf JM, Espadas-Moreno J, Luque-Garcia JL, Casadevall A (2014) Interaction of *Cryptococcus neoformans* extracellular vesicles with the cell wall. Eukaryot Cell 13:1484–1493

Wolf JM, Espadas J, Luque-Garcia J, Reynolds T, Casadevall A (2015) Lipid biosynthetic genes affect *Candida albicans* extracellular vesicle morphology, cargo, and immunostimulatory properties. Eukaryot Cell 14:745–754

Zarnowski R et al (2018) *Candida albicans* biofilm-induced vesicles confer drug resistance through matrix biogenesis. PLoS Biol 16:e2006872. https://doi.org/10.1371/journal.pbio.2006872

Ziegenbalg A, Prados-Rosales R, Jenny-Avital ER, Kim RS, Casadevall A, Achkar JM (2013) Immunogenicity of mycobacterial vesicles in humans: identification of a new tuberculosis antibody biomarker. Tuberculosis (Edinb) 93:448–455. https://doi.org/10.1016/j.tube.2013.03.001

# Chapter 4
# Extracellular Vesicles in the Environment

**Steven J. Biller**

**Abstract** Extracellular vesicles are small membrane-bound structures released by cells from all domains of life. Diverse populations of vesicles are found in many natural ecosystems, where they mediate complex networks of interactions between microbes and their local environment. Vesicles can serve numerous functions, including transporting and delivering compounds such as lipids, proteins, nucleic acids, and small molecules, between organisms and across both spatial and temporal dimensions. In this review I consider extracellular vesicles from an ecological perspective, exploring their influence on both the biotic and abiotic environment. I summarize our current understanding of vesicle contents and distributions in various microbial habitats, their potential contributions to nutrient pools and food webs, and the many ways in which vesicles can influence the physiology, ecology, and evolution of microbial communities. While many questions concerning the ecological impact of extracellular vesicles remain to be answered, it is becoming increasingly evident that these particles play important roles in the global ecosystem.

## 4.1 Introduction

Bacteria are found in essentially every habitat on Earth. Due to their remarkable abundance, small size, and fast metabolisms, microbes have the ability to profoundly impact both their local environment and, ultimately, drive the major biogeochemical cycles that sustain life on the planet. Bacteria evolve and function within complex ecological communities, and the physiology and behavior of these globally important microbial consortia can be affected by interactions with other cells and the surrounding abiotic environment (Azam and Malfatti 2007; Hibbing et al. 2010). In recent years, our understanding of the mechanisms underlying microbial interactions has been expanded by the discovery of extracellular vesicles (EVs). EVs are small,

---

S. J. Biller (✉)
Department of Biological Sciences, Wellesley College, Wellesley, MA, USA
e-mail: sbiller@wellesley.edu

 
M. Kaparakis-Liaskos, T. A. Kufer (eds.), *Bacterial Membrane Vesicles*,
https://doi.org/10.1007/978-3-030-36331-4_4

membrane-bound structures released by cells from all domains of life, with numerous studies documenting vesicle release by archaea, bacteria, and eukaryotes (Deatherage and Cookson 2012). These structures are now known to be prevalent in a variety of environments, raising questions about their roles within natural ecosystems.

This chapter discusses extracellular vesicles from an ecological and environmental standpoint, examining vesicle contents, distributions, contributions to global nutrient pools, and putative functions within microbial ecosystems. An ecological perspective is important when considering the roles of EVs produced by any organism and within any environment, whether in the ocean or within the human body; however, this chapter will focus primarily on bacterial membrane vesicles (MVs) within nonhost-associated communities and in nonpathogenic contexts. The emerging body of research on extracellular vesicles indicates that these submicron particles likely influence many aspects of microbial community function and evolution, ultimately affecting the global ecosystem in ways that we are just now beginning to identify.

## 4.2 Formation and Contents of Membrane Vesicles

Extracellular vesicles can be formed through a variety of processes in prokaryotic and eukaryotic cells, but all vesicles share a common set of basic properties: they are small, spherical structures, typically between 20 and 250 nm in diameter, which are bounded by a lipid bilayer membrane (Schwechheimer and Kuehn 2015). In bacteria, MVs are primarily thought to derive from local regions of the cell's outermost membrane that begin to protrude away, eventually expanding (i.e., "blebbing" out) until they separate from the rest of the cell (Schertzer and Whiteley 2012). The putative mechanisms of vesicle formation in bacteria have been reviewed extensively (Brown et al. 2015; Schwechheimer and Kuehn 2015; Pathirana and Kaparakis-Liaskos 2016; Toyofuku et al. 2019), including elsewhere in this book (see also Chaps. 2 and 3), and will not be detailed here.

The biological impacts of vesicles are largely attributable to the fact that they serve as a versatile secretion, transport, and delivery mechanism for cells (Kulp and Kuehn 2010; Guerrero-Mandujano et al. 2017). As MVs are formed and released by a bacterium, they can take a variety of compounds with them into the extracellular milieu (Brown et al. 2015; Schwechheimer and Kuehn 2015). Besides the lipids that comprise the vesicle membrane, MVs can contain an array of proteins originating from all compartments of a bacterial cell, though the degree to which specific proteins are preferentially "packaged" as cargo into vesicles remains an open question (Bonnington and Kuehn 2014). MVs carry a variety of small molecules as well, and vesicles may be a particularly useful vehicle for exporting hydrophobic compounds such as quorum-sensing quinolones into an aqueous extracellular environment (Mashburn-Warren and Whiteley 2006).

A remarkable diversity of nucleic acids can be found in EVs, including chromosomal DNA fragments (ranging from hundreds to many thousands of bp long), plasmids, viral genomes, and a variety of messenger, transfer, ribosomal, and small RNAs (Soler et al. 2008; Biller et al. 2014; Gaudin et al. 2014; Sjöström et al. 2015; Blenkiron et al. 2016). In bacterial and archaeal vesicles, these nucleic acids are found either within MVs and/or associated with their outer surface (Renelli et al. 2004; Bitto et al. 2017). The mechanisms responsible for moving nucleic acids from the cytosol into a MV remain unclear. In the Gram-positive bacterium *Streptococcus mutans*, the protein secretion machinery may play a role in moving DNA into MVs (Liao et al. 2014). In Gram-negative bacteria, the discovery that some outer membrane vesicles can contain both outer and inner membranes provides an alternate explanation for DNA export based on cellular topology (Perez-Cruz et al. 2013): if both inner and outer membranes were to simultaneously "bleb" out of the cell and form a vesicle, then a sub-compartment within the vesicle would originate directly from the cytosol, avoiding the need for DNA to cross the inner membrane. It is not yet clear, however, whether such "outer-inner membrane vesicles" are the only ones to contain DNA, nor how the chromosomal DNA fragments found within vesicles are generated in the first place.

The ability of bacteria, archaea, and eukaryotes to export portions of their genetic information via extracellular vesicles has a number of functional implications (discussed in more detail below), and also provides a means to identify which cells in a given environment produce these structures. Based on the assumption that the majority of DNA contained within a vesicle likely originated from the cell that produced it, metagenomic sequencing of vesicle-enclosed DNA collected from both coastal and open-ocean water samples has shown that organisms from all three domains, including representatives of at least 33 different phyla, produce vesicles in marine environments (Biller et al. 2014). These results further support the notion that most, if not all, microbes release extracellular vesicles, and indicate that vesicle production is a common occurrence in the natural environment—not just in laboratory cultures (Fig. 4.1a).

## 4.3 Variation Among Vesicles

Studies of MVs from *Pseudomonas aeruginosa* and *Escherichia coli* have long noted that these model organisms produce heterogeneous vesicle populations varying in physical properties including particle size (Kadurugamuwa and Beveridge 1995) and buoyant density (Kesty and Kuehn 2004; Renelli et al. 2004) in culture. Such heterogeneity has been observed in vesicles from the marine cyanobacteria *Prochlorococcus* and *Synechococcus* as well (Biller et al. 2014). The concept of vesicle heterogeneity further applies to their contents. For example, one investigation into the DNA distribution among vesicles isolated from various marine

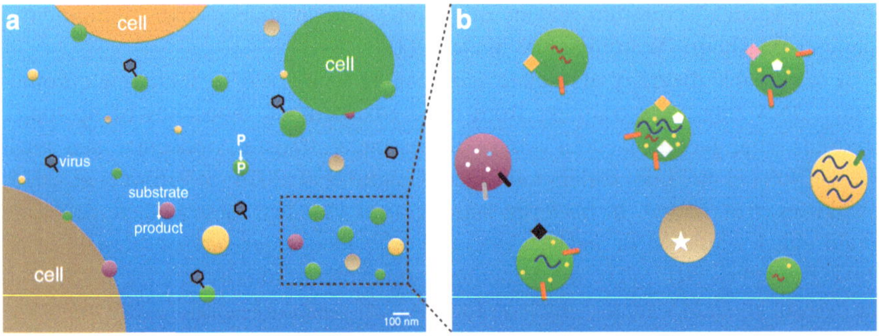


**Fig. 4.1** Extracellular vesicles in the marine environment. (**a**) Vesicles (small circles) originating from different cells (larger circles, distinguished by colors) can freely diffuse and interact with cells, viruses, or other components of the environment. Vesicles can impact the local ecosystem through activities such as mediating extracellular enzymatic reactions, gathering nutrients like phosphorus (P), or delivering materials among cells. Cells and vesicles are shown at a greater relative abundance (and closer proximity) than would be expected at this scale for illustrative purposes. Vesicles are colored to match the cell that produced them; cells and vesicles associated with particle surfaces are not depicted. (**b**) Individual extracellular vesicles, whether produced by the same or different microbes, can vary in size and the composition of contents such as DNA, RNA, proteins and metabolites (depicted by different symbols on the surface or within the vesicles)

microbes showed that long (>5 kb) DNA fragments were found in fewer than 1% of vesicles, with a larger (but unknown) fraction of vesicles presumably containing shorter DNA fragments (Biller et al. 2017). Similarly, most DNA in *P. aeruginosa* MVs is associated with only the smallest vesicles (Bitto et al. 2017). Following from these data, it seems reasonable to speculate that the distribution of proteins, RNAs, and other components will almost certainly vary among individual vesicles as well (Kulp and Kuehn 2010). Thus, while vesicles can collectively contain a huge diversity of compounds, it is important to consider vesicles as a population of heterogeneous individual particles, each with potentially distinct functional capabilities (Fig. 4.1b). Further complicating the picture is that the vesicles produced by a single strain can also vary with changes in growth phase and environmental conditions (Orench-Rivera and Kuehn 2016; Zavan et al. 2019) or due to different mechanisms of vesicle formation (Turnbull et al. 2016).

## 4.4 Distribution of Vesicles in the Environment

Where are vesicles found, and how many are there? Diverse microbes isolated from both aquatic and terrestrial environments release MVs in culture, implying that vesicles are likely to be widespread throughout nature. Recent studies have started to shed some light on the abundance and distribution of vesicles in the field, highlighting the need to consider how the physical context of different habitats may influence MV functions.

### 4.4.1 Vesicles in Aquatic Systems

Extracellular vesicle abundances in the environment are best understood in the oceans. Bacteria are a widespread and essential part of the marine ecosystem, where they can be found at densities on the order of $10^5$–$10^6$ cells $mL^{-1}$ in surface waters; archaea and eukaryotes are important components of ocean microbiomes as well (Moran 2015). Membrane vesicle release has been observed in laboratory cultures of numerous marine bacteria, ranging from globally abundant cyanobacteria to heterotrophs isolated from both mid-latitude and polar regions (Frias et al. 2010; Biller et al. 2014, 2017; Li et al. 2016). The presence of vesicles in seawater has also been directly confirmed by electron microscopy in samples from both coastal and open-ocean environments (Biller et al. 2014). Though measurements of vesicle abundances have been traditionally based on measurements of lipid abundance or vesicle protein content, newer technologies for nanoparticle characterization are beginning to provide more quantitative estimates of vesicle numbers. Measurements based on nanoparticle tracking analysis revealed that in surface waters, EVs are found at concentrations comparable to that of bacterial cells, with ~$10^6$ vesicles $mL^{-1}$ identified in seawater samples taken from the coast of Massachusetts and ~$10^5$ vesicles $mL^{-1}$ in samples collected in the Sargasso Sea (Biller et al. 2014). In the Sargasso, vesicle abundance gradually decreased with depth, mirroring the overall change in bacterial abundance. Since vesicles were almost certainly lost during many of the sample collection and processing steps used to complete those measurements, it is important to note that those concentrations should be considered lower bounds—the actual vesicle concentration in these environments is likely greater. While direct measurements of vesicle abundance have only been made in samples from the mid-latitudes to date, vesicle release has also been noted in cultures of Antarctic microbes, suggesting that vesicles are a cosmopolitan feature of marine environments globally (Frias et al. 2010).

Vesicles are components of freshwater ecosystems as well. Electron microscopy evidence indicates that microbes naturally produce MVs within a variety of tropical freshwater habitats (Silva et al. 2014), and vesicles are found in cultures of diverse freshwater bacteria such as *Cylindrospermopsis raciborskii* (Zarantonello et al. 2018), *Shewanella* spp. (Gorby et al. 2008), and *Synechocystis* PCC6803 (Pardo et al. 2015). The abundance and distribution of vesicles in fresh or brackish water samples remains to be characterized. The presence of endotoxins in drinking water has been suggested as additional indirect evidence for vesicles in freshwater (Toyofuku et al. 2015), though whether this material is truly associated with vesicles is unclear. Vesicles, or vesicle-like liposomes, have also been implicated as a potential source of material clogging filtration membranes used in water treatment processes (Barry et al. 2014).

Vesicles are generally assumed to diffuse throughout a well-mixed aqueous system, allowing them to randomly encounter other components of the environment (Fig. 4.1). There are, however, potential conditions under which EVs may not be able to freely move away from the cells that produced them, resulting in microscale "patchiness" of vesicle distributions. *Shewanella oneidensis* vesicles, for example, can be tethered to cells by thin nanowires (Gorby et al. 2008), which may provide some restraint on the ability of a vesicle to move within low-shear settings. More broadly, considerations of microscale physics show that bacteria live in a world of low Reynold's numbers and high viscosity, where flows around a cell can be laminar instead of generally turbulent (Lauga 2016). This may impact the ability of diffusion to separate vesicles from cell surfaces under some conditions, where vesicle movement will be subject to constraints of the local diffusion boundary layer surrounding a cell (Stocker and Seymour 2012). Over time, and in a relatively turbulent aqueous environment, this boundary layer may not provide much of an impediment to vesicle release away from a microbial cell; by contrast, in a low flow, poorly mixed environment, vesicles may have greater difficulty moving away from the cell surface. Such conditions may lead to vesicles remaining locally restricted near the cell, potentially influencing their ecological function(s).

### 4.4.2 Vesicles in Surface-Associated Communities

Many microbes do not live free-floating in a liquid environment, but are instead attached to surfaces. Such communities are frequently found in the form of biofilms: collections of bacteria bound to each other and fastened to a surface by secreted extracellular polymeric substances. Electron microscopy surveys have demonstrated that vesicles are an abundant feature of natural biofilms isolated from such diverse sources as domestic water drains, sewage and water treatment plants, pulp and paper manufacturers, freshwater fish aquariums, water storage tanks, and riverbed sediments (Schooling and Beveridge 2006). Large organic particles (μm to cm scale), such as "marine snow" and transparent exopolymer particles, are also plentiful in marine (and freshwater) environments (Simon et al. 2002). These aggregates are colonized by diverse microbial biofilm communities, and presumably should contain vesicles as well. Marine and freshwater sediments also harbor complex communities of bacteria (Nealson 1997), but the concentration and distribution of vesicles on sediments are not yet known. Most vesicles released within a biofilm or other aggregate will likely remain confined within the biofilm matrix and thus would not be included in measurements of planktonic vesicles, which typically remove particles of this larger size class. These considerations suggest that vesicles in aquatic environments may be even more abundant, and perhaps more spatially structured, than is reflected in current data. Abundant bacteria are also associated with particles in the atmosphere (Barberán et al. 2015), and work on indoor dust particles has revealed data consistent with the idea that particle-associated vesicles are found in the air (Kim et al. 2013).

Terrestrial bacteria, such as the model organism *Bacillus subtilis*, release MVs in liquid cultures grown in the lab (Brown et al. 2014). While these organisms presumably also produce vesicles in their natural environment, vesicle concentrations in soils are unknown. The physical properties of soil particles impose a distinct set of constraints on vesicle movements and interactions as compared to aqueous environments. Soil composition can vary across numerous dimensions, including the composition of the soil particles (organics, sand, clay, etc.), the charge of these particles, and their hydration levels. Water films surrounding soil particles can be less than 10 nm thick (Or et al. 2007); since this is markedly thinner than a typical MV, vesicles would not be expected to freely diffuse in the same way that they could in a well-mixed, turbulent aquatic habitat (Shetty et al. 2011). Vesicle movement in soils may be further restricted by charge interactions between vesicles and soil particles.

Since dry or nearly desiccated particle surfaces are not necessarily conducive to vesicle diffusion, biofilms represent a likely ecological context for many vesicle activities in soils. Additionally, terrestrial microbes can utilize other mechanisms to overcome the physical constraints associated with soil particles and enable vesicle-based interactions. *B. subtilis*, for instance, can produce "nanotubes" that extend between cells and mediate intercellular exchanges (Dubey and Ben-Yehuda 2011), possibly via vesicles. Another related mechanism was identified in the soil microbe *Delftia*, which forms tubular surface structures termed "nanopods" that can extend more than 6 μm away from the cell surface (Shetty et al. 2011). These nanopods are composed of surface layer proteins and membrane vesicles, providing a means for vesicles to be exchanged directly over long distances. Vesicles have also been seen to form long "chains" reaching away from the surface of *Myxococcus xanthus* (Remis et al. 2014). The potential use of intercellular tubes to exchange vesicles may not be limited to soil microbes, as the human pathogen *Francisella tularensis* also produces tubular outer membrane vesicles from its surface when grown in liquid culture and on agar plates (McCaig et al. 2013).

### *4.4.3 What Modulates Vesicle Distributions in the Environment?*

The abundance of vesicles at any given time and place represents the net balance between the rate of vesicle production by the community and the rate of vesicle loss. The relevant rates on both sides of this equation are currently poorly constrained. Vesicle production in a given environment will fluctuate as a function of many factors: first, MV release varies among bacterial strains as measured on a per-cell, per-generation basis (Biller et al. 2014), so community composition and growth rates of individual taxa will affect vesicle abundance. MV production can be further influenced by environmental conditions such as nutrient availability (Prados-Rosales et al. 2014a; Orench-Rivera and Kuehn 2016; Sampath et al. 2018; Gerritzen et al. 2019), temperature (Frias et al. 2010; MacDonald and Kuehn 2013), UV exposure (Gamalier et al. 2017, Zarantonello et al. 2018), and oxidative stress (MacDonald and Kuehn 2013). The rotation of sheathed flagella has also been linked to vesicle release (Aschtgen et al. 2016a), suggesting that conditions increasing flagellar activity may also lead to higher local MV concentrations. There is evidence that vesicle production is under at least some degree of genetic control as well, since mutations have been identified in multiple Gram-negative bacteria that either increase or decrease the rate of vesicle formation (McBroom et al. 2006; Rath et al. 2013; Kulp et al. 2015; Nakayama-Imaohji et al. 2016; Resch et al. 2016). It is not yet clear how many of these genetic mechanisms specifically regulate vesicle production as opposed to indirectly influencing vesicle release through impacts on cellular envelope structure and/or other physiological factors, including oxidative stress levels or lipid production (Gerritzen et al. 2019). Regardless, these data indicate that the specific genetic composition of a microbial community will also affect overall vesicle production rates.

Even less is known about the rates and mechanisms of vesicle loss. Studies of cyanobacterial extracellular vesicles showed that average vesicle size and abundance did not significantly change over the span of 2 weeks when kept in sterile seawater (Biller et al. 2014). Though these data cannot rule out microscale changes occurring among individual vesicles, they indicate that vesicles are inherently stable structures—at least in a high salinity environment where lipids will be most thermodynamically stable in a spherical form. Dispersal rates, whether through random diffusion or active mixing, will further influence vesicle concentrations within a given region of an aqueous environment. Other processes surely contribute to vesicle loss, including breakdown by microbes or extracellular enzymes, consumption of vesicles, uptake/fusion of vesicles into the surfaces of other cells, abiotic degradation, or adsorption and subsequent sequestering of vesicles onto surfaces. The relative contributions of these different vesicle removal mechanisms are not yet known, but together such factors will affect not only the number of vesicles in a given place and time but also the average vesicle half-life and distance a vesicle might be able to travel.

## 4.5 Vesicles as Discrete, Structured Packets of "Dissolved" Nutrients

At a basic level, extracellular vesicles can be viewed as simply secreted packets of organic molecules. Given that vesicles can contain lipids, proteins, small molecules, and nucleic acids, these structures are therefore a potential source of biologically important nutrients including organic carbon, nitrogen, oxygen, phosphorous, sulfur, and trace metals. Vesicles thereby represent components of global nutrient pools and microbial food webs.

### *4.5.1 Vesicles Are an Investment of Cellular Resources*

The bulk of our current understanding of MVs comes from laboratory studies of relatively large, fast-growing bacteria grown in nutrient-rich media, where the release of proteins and other material into vesicles may represent a negligible loss to the cell. By contrast, most bacteria in the environment grow slowly, with doubling times on the order of days to weeks or even longer (Kirchman 2016). The oligotrophic oceans, for example, are extremely nutrient-poor and dilute environments, where essential nutrients are found at picomolar concentrations and bacterial cells are at least 100–200 body lengths away from each other on average (Biller et al. 2015; Moran 2015). These conditions impose a distinct set of selective pressures on marine microbes as compared to those experienced within other environments such as the human gut, where organisms generally live at much higher local densities and experience "feast and famine" type regimes. Oligotrophic microbes have a number of adaptations that appear to help them survive in a world of limited nutrients. One evolutionary approach is to reduce cell size: *Prochlorococcus* and *Pelagibacter*—the most abundant autotrophic and heterotrophic bacteria, respectively, in the oceans—are both tiny cells, with diameters less than 1 μm (Chisholm et al. 1988; Rappé et al. 2002). This means that the production of a single 100 nm diameter vesicle by these organisms represents a proportionally greater amount of cellular resources than it would in a larger bacterium such as *E. coli*. *Prochlorococcus* has evolved numerous ways to reduce its nutrient requirements, such as using sulfolipids instead of phospholipids in its membrane to conserve phosphorous, which is a limiting nutrient in some ocean regions (Van Mooy et al. 2009). Given all of this, it is therefore perhaps surprising that *Prochlorococcus* cells would release potentially critical limiting nutrients within a vesicle. For bacteria producing on the order of 1–10 MVs per cell per generation (Biller et al. 2014), this might represent a nontrivial amount of material to export—particularly when considering the nutrients that are lost are in the form of energetically and chemically "expensive" materials like lipids, proteins, and nucleic acids.

Why, then, might cells—especially those growing in nutrient-poor environments—release extracellular vesicles in the first place? Do EVs simply perform sufficient beneficial functions to justify the investment of resources, meaning vesicle production is maintained through natural selection? Or are other factors at play? One hypothesis for the apparent ubiquity of EV production is that vesicle release reflects a vestige of the earliest forms of life on Earth, wherein lipid vesicles could have provided an environment for metabolic reactions to occur and facilitated exchange of RNA and other compounds among ancient cells (Gill and Forterre 2016). Vesicles could also be, at least to some degree, an unavoidable consequence of having a lipid bilayer membrane, since bits of membrane protruding from the cell will be thermodynamically favored to self-assemble into a spherical vesicle under various conditions (Huang et al. 2017). Structuring the cell envelope in such a way as to prevent membrane "blebs" from forming, such as by increasing membrane crosslinking, may have too many other deleterious consequences for the cell, making vesicle production essentially an accepted loss term. Consistent with this idea is the observation that while some genetic mutations either increase or decrease vesicle production in *E. coli*, none have yet been found that can completely abolish vesicle release (McBroom et al. 2006; Kulp et al. 2015). Evidence from *Salmonella enterica* also indicates that vesicle release may be tied, at least in part, to the processes required to remodel bacterial outer membranes in response to environmental changes (Elhenawy et al. 2016). Regardless of the original evolutionary and mechanistic origins of vesicle release, the secretion of these particles and their associated contents represents a means through which bacteria release potentially valuable nutrients and energy.

### 4.5.2 How Much "Dissolved" Material in the Environment Is Enclosed Within Vesicles?

In aquatic sciences, measurements of "dissolved" versus "particulate" nutrients have historically been based on operational criteria wherein materials are considered "dissolved" if they pass through a 0.2 μm diameter filter, and "particulate" if they are >0.2 μm (Azam and Malfatti 2007). Within this "dissolved" fraction are found not only truly dissolved monomeric molecules, but also an entire size continuum of particles—ranging from nm-scale inorganic colloids to larger, biologically-derived organic structures like viruses and vesicles (Azam and Malfatti 2007). Of perhaps particular relevance in the open ocean is the fact that vesicles represent a discrete, colloidal structure that is relatively concentrated with nutrients within an otherwise dilute environment. Though many common methods for determining the concentration of important nutrients like nitrite, nitrate, and organic carbon in seawater likely already include vesicle materials, it is not yet known what fraction of these nutrients are truly "dissolved" versus associated with vesicles.

The realization that seawater contains abundant extracellular vesicles may partly explain a number of previous oceanographic observations. For instance, early proteomic studies of seawater revealed that some of the most abundant "dissolved" peptides were membrane proteins (Tanoue et al. 1995), and it is quite likely that this includes membrane proteins associated with extracellular vesicles. Similarly, lipids were found to comprise a notable component of total dissolved organic carbon (DOC) in seawater (Aluwihare et al. 1997), and lipid membrane-bound extracellular vesicles likely represent a fraction of this. Extracellular DNA has been repeatedly observed in the oceans (DeFlaun et al. 1987; Brum 2005), which is somewhat surprising given that microbes can rapidly utilize free DNA as a nutrient source (Jørgensen and Jacobsen 1996; Lennon 2007). Given what we now know, at least some of this dissolved DNA may not be truly "free" but instead afforded a degree of protection inside vesicles. As a final example, ATP has been identified in bacterial MVs (Pérez-Cruz et al. 2015); ATP is also found within the <0.2 μm size fraction of seawater, where it is utilized by microbes as a source of either phosphorous or purines for biosynthesis (Azam and Hodson 1977). Chemical analyses of this "dissolved" ATP have shown that it likely comes from grazing and/or cellular excretion (Nawrocki and Karl 1989; Björkman and Karl 2001), consistent with the hypothesis that at least some fraction of seawater ATP is associated with MVs.

### 4.5.3 Vesicles as a Component of Global Dissolved Organic Carbon Pools

From a food web perspective, vesicles represent a "snack pack"—a discrete, locally concentrated bundle of bioavailable nutrients that could be utilized by single-celled or multicellular organisms. It is therefore worth considering the place of vesicles within global food webs, focusing here again on the ocean ecosystem. Marine primary producers (phytoplankton) use sunlight to fix inorganic carbon into organic molecules, which are later released and then utilized by the heterotrophic community for either energy or biosynthesis. These organic molecules are typically thought to be released by phytoplankton through direct excretion of the compounds, leakage across membranes, or as a consequence of cell lysis (Azam and Malfatti 2007). Recent findings indicate that vesicle secretion should be considered in this context as well. Experiments with the common marine heterotroph *Alteromonas* demonstrated that they can utilize purified extracellular vesicles from *Prochlorococcus* as their sole organic carbon source (Biller et al. 2014). This result highlights the potential for vesicles to contribute to marine food web interactions as part of "dissolved" organic carbon pools. The identity of the vesicle-associated biomolecule(s) consumed by *Alteromonas* or other microbes is not yet known, but could potentially include the membrane lipids as well as vesicle-associated proteins and/or small molecules. Extracellular DNA can serve as a nutrient source in planktonic environments and marine sediments (Jørgensen and Jacobsen 1996; Dell'Anno and Danovaro 2005), leading to the hypothesis that some cells might utilize vesicle-associated DNA for food.

Dissolved organic carbon pools in the global oceans are massive, with the deep ocean containing roughly as much carbon as there is $CO_2$ in the atmosphere (Hansell and Carlson 2015). DOC can generally be divided into two fractions: "labile" carbon, which is quickly utilized by microbes within the ocean food web, and "refractory" carbon which is less accessible. While any individual extracellular vesicle likely contains only $50–100 \times 10^{-18}$ g of carbon (Biller et al. 2014), their apparent ubiquity in the marine environment and ability to be consumed by other bacteria indicates that they may have a large combined impact on labile DOC pools. A back-of-the-envelope calculation shows that if an average marine bacterium releases 1 vesicle per day, with each vesicle containing 0.1–1% of a typical marine bacterial cell's mass (Fukuda et al. 1998; Biller et al. 2014; Bar-On et al. 2018), then the $\sim 1.3 \times 10^{15}$ g of bacterial carbon biomass in the pelagic oceans (Bar-On et al. 2018) would produce on the order of $\sim 4.7 \times 10^{14}–4.7 \times 10^{15}$ g of "dissolved" vesicle C biomass each year. Assuming that vesicles are composed of primarily labile material, this suggests that bacterial vesicle biomass represents a potentially notable fraction of the estimated $15–25 \times 10^{15}$ g of total marine labile DOC produced annually (Hansell 2013). There is, of course, considerable uncertainty in these estimates, but the exercise emphasizes the potential for vesicle-associated carbon to be an important part of organic carbon flux in the oceans. Future work will hopefully permit better estimates of the contribution of vesicles to global DOC pools and fluxes, and improve our understanding of the variation in both vesicle production and consumption among different bacterial, archaeal, and eukaryotic taxa. In addition to the role of vesicles in carbon flux, the protein and DNA within vesicles implies that vesicles may also comprise part of the dissolved nitrogen, sulfur, and phosphorous pools (Lynch and Alegado 2017).

### 4.5.4 Vesicles as Reservoirs and Scavengers of Inorganic Nutrients

Nutrient-binding proteins and transporters have been identified within vesicles, raising the possibility that some vesicles might contain materials they have "scavenged" extracellularly, in addition to any nutrients carried out of the producing cell (Schwechheimer and Kuehn 2015). For example, MVs released by the marine cyanobacterium *Prochlorococcus* contain putative phosphate-binding proteins (Biller et al. 2014), suggesting that these vesicles might be able to bind extracellular phosphate as they diffuse through the ocean and shuttle those concentrated materials to another cell (Fig. 4.1a). Binding and transport of trace metals represent another potential function for vesicles; to date these mechanisms have been primarily studied in pathogens and host-associated microbes, but the concepts are likely broadly relevant. For instance, the proteome of *Neisseria meningitidis* vesicles contains both iron- and zinc-binding proteins (Lappann et al. 2013). When *Mycobacterium tuberculosis* is grown under low-iron conditions, it releases extracellular vesicles containing mycobactin, a compound that can bind iron and help these organisms obtain this nutrient (Prados-Rosales et al. 2014b). In *P. aeruginosa*, the *Pseudomonas* quinolone signal (PQS) is both a quorum-sensing molecule trafficked within MVs (Mashburn-Warren et al. 2008; Schertzer and Whiteley 2012) as well as an iron-binding compound (Bredenbruch et al. 2006; Schertzer et al. 2009), providing yet another avenue for vesicles to contribute to metal uptake. From an evolutionary perspective, the ultimate benefit of releasing vesicles into the environment where they can serve as a nutrient "snack pack" for others may depend on a complicated set of tradeoffs that considers the amount of cellular resources invested into each vesicle, the rates of vesicle–cell encounters, the identity of the limiting resource(s) in that environment, any mutualistic interactions between organisms, and the degree of specificity with which that vesicle will interact with closely or distantly related cells.

## 4.6 Ecological Roles of Vesicles

The variety of biological compounds that can be transported by extracellular vesicles underlies the remarkable diversity of their potential functional abilities. Studies have revealed contributions of vesicles to an ever-increasing number of processes, suggesting that vesicles may play many varied ecological roles within natural communities. While the exact impacts of any individual vesicle interaction may be small and heavily context dependent, collectively such vesicle-mediated processes could have a marked impact on the structure, function, and evolution of microbial communities globally.

### 4.6.1 Manipulators of the Local Environment

Studies of MVs released by cultured bacteria have repeatedly shown that vesicles can be associated with a variety of proteins, including functional enzymes (Kadurugamuwa and Beveridge 1995). Extracellular vesicles thus provide a mechanism for cells to affect their local environment through enzymatic activity, facilitating microbial niche construction. Exporting proteins in or on a vesicle as opposed to directly releasing an exoenzyme into the extracellular milieu could provide benefits including affording the enzyme protection from environmental damage and allowing for simultaneous co-secretion of multiple proteins involved in a complex or pathway (Borch and Kirchman 1999; Lee et al. 2013). Vesicles could also provide a local structure wherein substrates are kept in close proximity to enzymes that act on them, avoiding issues with diffusion in aqueous extracellular environments (Bonnington and Kuehn 2014).

Vesicle proteomes have revealed not only the diversity of vesicle-associated proteins, but also the fascinating observation that these proteins can originate from a variety of cellular compartments. For instance, MVs from Gram-negative bacteria contain not just outer membrane and periplasmic proteins—as would be expected based on the common models of vesicle formation—but also cytoplasmic proteins (Pérez-Cruz et al. 2015; Yun et al. 2017; Zakharzhevskaya et al. 2017). These repeated observations, across distantly related bacteria, suggest that the export of cytoplasmic proteins is a common biological feature of vesicles and not simply a technical artifact. Extracellular vesicles, therefore, provide a mechanism through which enzyme-mediated activities can occur away from the cell, perhaps allowing these proteins access to substrates that they might not otherwise encounter in their typical subcellular location (Ebner and Götz 2019).

Vesicle-associated enzymes are responsible for many of the functional roles currently described for these structures (Schwechheimer and Kuehn 2015). Pioneering studies in *P. aeruginosa* noted that enzymes contributing to pathogenesis were found in vesicles (Kadurugamuwa and Beveridge 1995). Vesicles were also implicated as agents of microbial "warfare," based on experimental demonstrations that vesicle-associated hydrolases and endopeptidases could lyse other microbes (Kadurugamuwa and Beveridge 1996; Li et al. 1998; Vasilyeva et al. 2008). Vesicles are also capable of catalyzing reactions that enable cells to broadly manipulate their local chemical environment. For example, *S. oneidensis* vesicles contain active reductases that facilitate electron transfer in these cells, allowing them to reduce terminal electron accepters located away from the cell surface (Gorby et al. 2008). *Staphylococcus aureus* vesicles contain functional β-lactamases that can degrade the antibiotic ampicillin (Lee et al. 2013). Other findings indicate that MV-associated enzymes can contribute to nutrient acquisition. Vesicles released by human gut *Bacteroides* contain active hydrolases that can break down extracellular substrates, releasing soluble nutrients that can then be utilized by the cells (Rakoff-Nahoum et al. 2014; Li et al. 2016). Relatedly, the rumen-associated microbe *Fibrobacter succinogenes* produces vesicles containing carbohydrate-degrading enzymes that can depolymerize various plant polysaccharides, likely facilitating further degradation and utilization of these nutrients (Arntzen et al. 2017). MVs from coral-associated *Vibrio* strains also contain active proteases, glucosidases, lipases, and chitinases, which could contribute to pathogenesis or nutrient acquisition as well (Li et al. 2016).

These enzymatic data raise additional questions concerning how much of total bacterial exoenzyme activity is associated with individually secreted, truly "dissolved" proteins as compared to enzymes associated with vesicles. For instance, alkaline phosphatase is an exoenzyme long known to play important roles in the marine phosphorous cycle (Hoppe 2003). The majority of marine alkaline phosphatase activity is typically found in the "dissolved" seawater fraction, but at least some of this activity may come from vesicle-associated enzymes, as suggested by the finding that MVs from a marine *Vibrio* contain active alkaline phosphatase activity (Li et al. 2016). Bacteria from other environments, such as *P. aeruginosa* and the soil bacterium *Myxococcus xanthus*, also secrete active alkaline phosphatase in vesicles (Kadurugamuwa and Beveridge 1995; Evans et al. 2012). While these data lead to a number of compelling hypotheses concerning potential vesicle functions, to date they are only proof of concept. We still lack quantitative information about the degree to which vesicle-associated enzymes contribute to processes within natural microbial ecosystems, either extracellularly or within cells that interact with vesicles.

### 4.6.2 Vectors of Intercellular Exchange and Signaling

Extracellular vesicles can carry a diverse suite of compounds released by one cell across spatial and temporal distances to another cell. Such intercellular exchanges may lead to a variety of potential outcomes, the exact nature of which will depend on the organisms involved as well as the contents of those vesicles. Some examples of vesicle-mediated intercellular exchange include previously discussed roles in nutrient exchange and protein delivery. In addition, vesicles can serve as vehicles for transferring signaling molecules. For instance, MVs from *P. aeruginosa* have been shown to traffic the hydrophobic bacterial quorum-sensing signal PQS; interestingly, PQS can also directly contribute to vesicle formation and signal packaging in this bacterium (Mashburn and Whiteley 2005; Schertzer and Whiteley 2012). By enclosing a relatively high local concentration of such molecules, vesicles likely prevent secreted signaling molecules from diluting to the point where the signal concentration would be insufficient to elicit a response. This can, however, change the nature of these signaling interactions. Individual vesicles from *Paracoccus denitrificans* carry sufficient quorum signal to elicit a response in recipient cells, suggesting that vesicle-mediated signaling may lead to a binary response that differs from the more canonical density-dependent quorum sensing (Toyofuku et al. 2017). MV-mediated signaling contributes to cross-domain interactions as well. In one example, the marine Bacteroidetes *Algoriphagus machipongonensis* was found to induce multicellular colony development of a choanoflagellate, *Salpingoeca rosetta*, via sulfonolipids trafficked via MVs (Alegado et al. 2012; Lynch and Alegado 2017). MVs also participate in the mutualistic symbiosis between *Vibrio fischeri* and the Hawaiian bobtail squid, *Euprymna scolopes*. Here, *Vibrio* MVs convey developmental signals to the squid, helping to drive host developmental changes required for successful colonization (Aschtgen et al. 2016b).

Vesicle-based intercellular delivery could occur through various mechanisms. Some models of vesicle delivery implicate the fusion of vesicles with the outer membrane of a cell, releasing all of the vesicle's contents into the recipient (Kadurugamuwa and Beveridge 1996, 1999). Alternatively, vesicles either attached to or held in stable proximity to the surface of a cell through charge interactions (Kadurugamuwa and Beveridge 1996) could deliver material via "flipping" of molecules from the vesicle membrane into the cell (Remis et al. 2014). Cells could perhaps also use enzymes to degrade a vesicle extracellularly and subsequently acquire specifically desired components through standard import pathways. The uptake of vesicle material into eukaryotic cells can occur via any one of multiple endocytic processes, which have been reviewed elsewhere (Mulcahy et al. 2014).

The factors that determine whether a given vesicle will associate with and deliver material to another cell remain unclear. Vesicles can clearly mediate transfer between different microbes (Yaron et al. 2000), but experimental data is emerging indicating that species/strain boundaries can exist between vesicles and cells in some cases (MacDonald and Beveridge 2002; Toyofuku et al. 2017; Tashiro et al. 2017). The factors mediating such strain specificity could include surface protein interactions, zeta potential (Tashiro et al. 2017), hydrophobicity (MacDonald and Beveridge 2002), envelope and boundary layer structure of the cells, or other properties yet to be identified.

### 4.6.3 Reservoirs of Genetic Information and Vectors of Horizontal Gene Transfer

Transformation, transduction, and conjugation have historically been viewed as the primary mechanisms of bacterial horizontal gene transfer (HGT), but extracellular vesicles can mediate HGT as well. Bacterial and archaeal vesicles can enclose DNA ranging in size from entire plasmids to both short ($<100$ bp) and long (anywhere from hundreds of bp to $>20$ kb) fragments of chromosomal DNA (Hagemann et al. 2013; Biller et al. 2014, 2017; Gaudin et al. 2014; Erdmann et al. 2017). Multiple studies have shown that bacterial MVs can successfully deliver this DNA cargo into other bacterial or eukaryotic cells, supporting the likely contribution of vesicle-mediated delivery to HGT in natural systems (Dorward et al. 1989; Kolling and Matthews 1999; Yaron et al. 2000; Chatterjee et al. 2017; Bitto et al. 2017; Grüll et al. 2018).

The extensive diversity of DNA associated with vesicles in the marine environment (Biller et al. 2014) highlights not only the vast potential for vesicles to mediate cross-species HGT, but also points to vesicles as an environmental genetic "reservoir" from which any cell could sample. Vesicles provide a degree of protection to DNA in the environment that a free-floating DNA molecule, which could be consumed or degraded, would not receive. In one study, vesicle DNA could still be transformed into cells following nearly 2 years of storage at 4 °C—even with nuclease enzymes present in the same solution (Blesa and Berenguer 2015). Thus, the "dissolved information" contained within vesicles represents an expansive and possibly long-lived pool, with the vesicle-associated DNA potentially outlasting the strain that released it. The transfer of RNA, including small RNAs, among cells also suggests potential roles for vesicle-mediated exchange impacting genetic regulation within microbial communities (Dauros-Singorenko et al. 2018; Tsatsaronis et al. 2018; Cai et al. 2018).

Some DNA may be able to facilitate its own transfer via vesicles in a quasi-viral manner. The plasmid pR1SE, identified in the Antarctic haloarchaeon *Halorubrum lacusprofundi* R1S1, encodes proteins that facilitate the formation of plasmid-containing extracellular vesicles. These vesicles can then "infect" cells lacking pR1SE, causing those hosts to produce plasmid-containing vesicles themselves (Erdmann et al. 2017). This finding mirrors previous observations of "viral-like particles" (likely membrane vesicles) in *E. coli* that can move DNA between cells and induce the recipients to produce more of these DNA-bearing structures; whether a similar mechanism mediates this phenomenon remains unclear (Chiura et al. 2011; Velimirov and Ranftler 2018).

The relative contribution of vesicle-mediated transfer, conjugation, transformation, and transduction to overall rates of HGT in different environments, or among different strains, is not known, though each mechanism has unique tradeoffs that likely influences the rate of successful HGT under different conditions (Nazarian et al. 2018). In laboratory cultures, rates of vesicle-mediated plasmid transfer between different Gram-negative strains have been shown to vary as a function of the specific donor strain, recipient strain, plasmid characteristics, and the genes being transferred; these rates were not correlated with the genetic relatedness of the donor and recipient (Tran and Boedicker 2017, 2019). Vesicle-mediated HGT is also subject to the previously discussed factors possibly influencing vesicle–cell interaction rates, combined with a consideration of the heterogeneity of DNA contained within any one individual vesicle (Biller et al. 2017). While the extent of vesicle HGT still needs to be elucidated, it is exciting to consider that vesicles could provide cells with a means to acquire DNA from a broader diversity of sources than transduction or conjugation. Many viruses, particularly the tailed viruses, exhibit quite specific and narrow host ranges (Kauffman et al. 2018), whereas the breadth of vesicle transfers between disparate cell types demonstrated to date hint that there may be fewer potential barriers for EVs. Ultimately, the amount of HGT mediated by vesicles as compared to viruses will depend not only on the relative abundance of these structures in a given environment, but also the fraction of each particle that contains host DNA, and the differences between their encounter dynamics, host specificity, and delivery efficiency (Nazarian et al. 2018).

### 4.6.4 Impacts on Cell–Surface Interactions

Production of extracellular vesicles affects the physical interaction of cells with surfaces, thus influencing cellular distributions and motility. In some instances, membrane vesicles promote attachment of bacteria by contributing to the formation of biofilms (Schooling and Beveridge 2006; Yonezawa et al. 2009; Grande et al. 2015), possibly through a structural role for MVs and their associated DNA. Vesicles can influence the physical attachment of microbes to other cells, as has been noted in studies of the oral cavity where vesicles can promote bacterial attachment to both host epithelial cells (Meyer and Fives-Taylor 1994; Inagaki et al. 2006) as well as to other microbes (Kamaguchi et al. 2003). In different contexts, vesicles inhibit cell–surface interactions. For example, vesicle production by *Xylella fastidiosa* facilitates the ability of this plant pathogen to move throughout the plant by preventing bacteria from sticking to surfaces (Ionescu et al. 2014).

### 4.6.5 Defensive Roles and Vesicle–Virus Interactions

Vesicles can play a variety of defensive and protective roles for cells. Some bacterial MVs bind toxic compounds such as antimicrobial peptides or hemin, thereby reducing the local concentration of that molecule and promoting cell survival (Manning and Kuehn 2011; Roden et al. 2012). MVs have also been proposed to provide a means for cells to remove damaged molecules (Schwechheimer and Kuehn 2015). Vesicles are further able to contribute to bacterial defenses against viral infections. Viruses and vesicles have a complex and fascinating set of interrelationships, and EVs are known to impact viral infection dynamics in multiple systems. Phage recognizes potential target cells through specific interactions with molecules on cell surfaces; since bacterial MVs contain material from the outermost membrane, any vesicle with the appropriate phage receptor molecule could be bound by that virus (Manning and Kuehn 2011). In this way, vesicles can serve as a "decoy" of sorts for the cell, which would lead to nonproductive infections and a reduction of the infectious viral population. Phage can bind vesicles released by marine cyanobacteria (Biller et al. 2014), and the presence of vesicles has been shown to reduce phage infection of both *E. coli* and *Vibrio cholerae* (Manning and Kuehn 2011; Reyes-Robles et al. 2018).

Vesicles do not, however, only act to inhibit viral infection. In the marine alga *Emiliania huxleyi*, viruses instead appear to use EVs to promote their own infection cycle (Schatz et al. 2017). Infected *E. huxleyi* cells release many EVs containing both small RNAs as well as a putative small signaling molecule; when these vesicles are taken up by uninfected *E. huxleyi* cells, the vesicle contents induce some currently unknown changes in the recipient cells that speed up subsequent viral infection cycles. In this system, the presence of EVs also led to a marked increase in viral half-life through an unknown mechanism (Schatz et al. 2017).

The fact that vesicles can transport surface molecules between cells provides a means for MVs to expand the host range of a phage. In *B. subtilis*, pre-treatment of a strain resistant to a particular phage with MVs released by a phage-sensitive strain led to the infection of the previously resistant strain, mediated by vesicle delivery of the phage receptor onto resistant bacterial cells (Tzipilevich et al. 2017). Changes in the phage host range could also occur via vesicles which contain viral genomes (either from cellular sources or perhaps an extracellular phage infection) and then deliver that DNA into cells (Yaron et al. 2000; Gaudin et al. 2014). In this way, vesicles containing viral DNA might facilitate viral "infection" in a manner that is not subject to the same barriers experienced by the virus itself. Since viruses and vesicles co-occur in marine (Biller et al. 2017) and other environments, future work will be required to untangle the many ways in which EVs impact phage dynamics—and vice versa.

## 4.7 The Future of Vesicle Research: Challenges and Opportunities

Despite the many advances we have made in understanding extracellular vesicles from bacteria, archaea, and eukaryotes, the extracellular vesicle field is, in many ways, still in its infancy. While there are countless questions to address, much of EV research is currently hindered by the simple fact that these structures are extremely difficult to work with. Nanoparticle analysis and imaging technologies are rapidly improving, but isolating and quantifying vesicles, particularly from the environment, remains a particular challenge. EVs from many types of natural samples can be found at concentrations close to or below some instrument detection limits, necessitating extensive sample concentration and processing; in addition, different isolation protocols can greatly influence study results (Singorenko et al. 2017). Separating extracellular vesicles from other types of small particles like viruses and inorganic colloids is still an inexact science, dependent on differences in charge, density, and other properties that do not always sufficiently discriminate among particle types. Whereas studies of eukaryotic extracellular vesicles (exosomes) frequently utilize antibodies to isolate or identify specific exosome populations of interest, there does not currently appear to be anything close to a universal epitope shared among all bacterial MVs produced by diverse communities. These technical considerations are further complicated by one of the fascinating properties of vesicles, namely their heterogeneity. This diversity is likely an important contributor to the functional capabilities of vesicles, and raises questions concerning when and where it is appropriate to study EVs at the level of individual structures as compared to populations. Regardless, the field has made rapid progress over the last few years, and continued technical advances will undoubtedly help us to overcome some of these challenges.

It is now clear that vesicle production is widespread in the natural environment, and that they likely mediate a diverse network of microbial interactions. This is an exciting time in which we are beginning to unveil an entirely new dimension of complexity within natural ecosystems. Despite the technical and conceptual challenges that remain, I believe that the field is well poised to take on these challenges and develop a basic understanding of vesicle ecology—to study the processes that determine the abundance and distribution of vesicles, determine how vesicles interact with the biotic and abiotic components of the environment, and quantify the influence of vesicle-mediated processes within natural systems. To this end, we need to advance our knowledge of the basic "natural history" of vesicles in disparate habitats: How many are there? When are they produced? By which cells? How long do they last, and where do they go? On top of this, many questions remain to be answered concerning vesicle functions. For example, what is the relative rate of vesicle-mediated HGT in the environment as compared to other mechanisms such as phage transduction? When and where do vesicle-associated enzymes function, and how can we quantify their impact? How much of a role do vesicles play in organic carbon cycling? In other nutrient cycles?

Given the abundance of vesicles in the environment and the diversity of their cargo, it seems that exchange among bacteria, archaea, and eukaryotes could be much more frequent than is commonly appreciated; this, in turn, raises important questions as to how cells handle potentially frequent encounters with vesicles from either related or dissimilar organisms in the wild. What are the encounter dynamics between vesicles and cells in different environments, and how does this compare to cell–cell encounter rates? What factors influence the ability of a vesicle to interact with a cell? To what degree does an average bacterium contain some number of biomolecules produced by a different organism and delivered by vesicles? What are the consequences of this? The ability of cells to discriminate among vesicles, or not, will also influence the structure of microbial interaction networks and the degree to which extracellular vesicles should be considered a true 'public good' (Hasegawa et al. 2015). While many functions of extracellular vesicles have been described to date, it seems likely that we have only begun to uncover the ways in which EVs affect the global ecosystem.

**Acknowledgments** My apologies to all the authors whose vesicle research I was not able to discuss here or highlight to the degree that it deserved. I wish to thank Julie Roden, Paul Berube, Jamie Becker, and Libusha Kelly for providing helpful comments on the manuscript.

## References

Alegado RA, Brown LW, Cao S et al (2012) A bacterial sulfonolipid triggers multicellular development in the closest living relatives of animals. eLife 1:e00013

Aluwihare L, Repeta D, Chen R (1997) A major biopolymeric component to dissolved organic carbon in surface sea water. Nature 387:166–169

Arntzen MØ, Várnai A, Mackie RI et al (2017) Outer membrane vesicles from *Fibrobacter succinogenes* S85 contain an array of carbohydrate-active enzymes with versatile polysaccharide-degrading capacity. Environ Microbiol 19:2701–2714
Aschtgen M-S, Lynch JB, Koch E et al (2016a) Rotation of *Vibrio fischeri* flagella produces outer membrane vesicles that induce host development. J Bacteriol 198:2156–2165
Aschtgen M-S, Wetzel K, Goldman W et al (2016b) *Vibrio fischeri*-derived outer membrane vesicles trigger host development. Cell Microbiol 18:488–499
Azam F, Hodson RE (1977) Dissolved ATP in the sea and its utilisation by marine bacteria. Nature 267:696–698
Azam F, Malfatti F (2007) Microbial structuring of marine ecosystems. Nat Rev Microbiol 5:782–791
Bar-On YM, Phillips R, Milo R (2018) The biomass distribution on Earth. Proc Natl Acad Sci 115:6506–6511
Barberán A, Ladau J, Leff JW et al (2015) Continental-scale distributions of dust-associated bacteria and fungi. Proc Natl Acad Sci 112:5756–5761
Barry MC, Hristovski K, Westerhoff P (2014) Membrane fouling by vesicles and prevention through ozonation. Environ Sci Technol 48:7349–7356
Biller SJ, Berube PM, Lindell D, Chisholm SW (2015) *Prochlorococcus*: the structure and function of collective diversity. Nat Rev Microbiol 13:13–27
Biller SJ, McDaniel LD, Breitbart M et al (2017) Membrane vesicles in sea water: heterogeneous DNA content and implications for viral abundance estimates. ISME J 11:394–404
Biller SJ, Schubotz F, Roggensack SE et al (2014) Bacterial vesicles in marine ecosystems. Science 343:183–186
Bitto NJ, Chapman R, Pidot S et al (2017) Bacterial membrane vesicles transport their DNA cargo into host cells. Sci Rep 7:7072
Björkman KM, Karl DM (2001) A novel method for the measurement of dissolved adenosine and guanosine triphosphate in aquatic habitats: applications to marine microbial ecology. J Microbiol Methods 47:159–167
Blenkiron C, Simonov D, Muthukaruppan A et al (2016) Uropathogenic *Escherichia coli* releases extracellular vesicles that are associated with RNA. PLoS One 11:e0160440
Blesa A, Berenguer J (2015) Contribution of vesicle-protected extracellular DNA to horizontal gene transfer in *Thermus* spp. Int Microbiol 18:177–187
Bonnington KE, Kuehn MJ (2014) Protein selection and export via outer membrane vesicles. Biochim Biophys Acta 1843:1612–1619
Borch N, Kirchman D (1999) Protection of protein from bacterial degradation by submicron particles. Aquat Microb Ecol 16:265–272
Bredenbruch F, Geffers R, Nimtz M et al (2006) The *Pseudomonas aeruginosa* quinolone signal (PQS) has an iron-chelating activity. Environ Microbiol 8:1318–1329
Brown L, Kessler A, Cabezas-Sanchez P et al (2014) Extracellular vesicles produced by the Gram-positive bacterium *Bacillus subtilis* are disrupted by the lipopeptide surfactin. Mol Microbiol 93:183–198
Brown L, Wolf JM, Prados-Rosales R, Casadevall A (2015) Through the wall: extracellular vesicles in Gram-positive bacteria, mycobacteria and fungi. Nat Rev Microbiol 13:620–630
Brum JR (2005) Concentration, production and turnover of viruses and dissolved DNA pools at Stn ALOHA, North Pacific Subtropical Gyre. Aquat Microb Ecol 41:103–113
Cai Q, Qiao L, Wang M et al (2018) Plants send small RNAs in extracellular vesicles to fungal pathogen to silence virulence genes. Science 360:1126–1129
Chatterjee S, Mondal A, Mitra S, Basu S (2017) *Acinetobacter baumannii* transfers the $bla_{\mathrm{NDM-1}}$ gene via outer membrane vesicles. J Antimicrob Chemother 72:2201–2207
Chisholm S, Olson RJ, Zettler ER et al (1988) A novel free-living prochlorophyte abundant in the oceanic euphotic zone. Nature 334:340–343
Chiura HX, Kogure K, Hagemann S et al (2011) Evidence for particle-induced horizontal gene transfer and serial transduction between bacteria. FEMS Microbiol Ecol 76:576–591

Dauros-Singorenko P, Blenkiron C, Phillips A, Swift S (2018) The functional RNA cargo of bacterial membrane vesicles. FEMS Microbiol Lett 365:fny023
Deatherage BL, Cookson BT (2012) Membrane vesicle release in bacteria, eukaryotes, and archaea: a conserved yet underappreciated aspect of microbial life. Infect Immun 80:1948–1957
DeFlaun MF, Paul JH, Jeffrey WH (1987) Distribution and molecular weight of dissolved DNA in subtropical estuarine and oceanic environments. Mar Ecol Prog Ser 38:65–73
Dell'Anno A, Danovaro R (2005) Extracellular DNA plays a key role in deep-sea ecosystem functioning. Science 309:2179
Dorward DW, Garon CF, Judd RC (1989) Export and intercellular transfer of DNA via membrane blebs of *Neisseria gonorrhoeae*. J Bacteriol 171:2499–2505
Dubey GP, Ben-Yehuda S (2011) Intercellular nanotubes mediate bacterial communication. Cell 144:590–600
Ebner P, Götz F (2019) Bacterial excretion of cytoplasmic proteins (ECP): occurrence, mechanism, and function. Trends Microbiol 27:176–187
Elhenawy W, Bording-Jorgensen M, Valguarnera E et al (2016) LPS remodeling triggers formation of outer membrane vesicles in *Salmonella*. MBio 7:e00940–16
Erdmann S, Tschitschko B, Zhong L et al (2017) A plasmid from an Antarctic haloarchaeon uses specialized membrane vesicles to disseminate and infect plasmid-free cells. Nat Microbiol 2:1446–1455
Evans AGL, Davey HM, Cookson A et al (2012) Predatory activity of *Myxococcus xanthus* outer-membrane vesicles and properties of their hydrolase cargo. Microbiology 158:2742–2752
Frias A, Manresa A, de Oliveira E et al (2010) Membrane vesicles: a common feature in the extracellular matter of cold-adapted Antarctic bacteria. Microb Ecol 59:476–486
Fukuda R, Ogawa H, Nagata T, Koike I (1998) Direct determination of carbon and nitrogen contents of natural bacterial assemblages in marine environments. Appl Environ Microbiol 64:3352–3358
Gamalier JP, Silva TP, Zarantonello V et al (2017) Increased production of outer membrane vesicles by cultured freshwater bacteria in response to ultraviolet radiation. Microbiol Res 194:38–46
Gaudin M, Krupovic M, Marguet E et al (2014) Extracellular membrane vesicles harbouring viral genomes. Environ Microbiol 16:1167–1175
Gerritzen MJH, Martens DE, Uittenbogaard JP et al (2019) Sulfate depletion triggers overproduction of phospholipids and the release of outer membrane vesicles by *Neisseria meningitidis*. Sci Rep 9:4716
Gill S, Forterre P (2016) Origin of life: LUCA and extracellular membrane vesicles (EMVs). Int J Astrobiol 15:7–15
Gorby Y, McLean J, Korenevsky A et al (2008) Gram-negative outer membrane vesicles: beyond the cell surface. Geobiology 6:214–219
Grande R, Di Marcantonio MC, Robuffo I et al (2015) *Helicobacter pylori* ATCC 43629/NCTC 11639 outer membrane vesicles (OMVs) from biofilm and planktonic phase associated with extracellular DNA (eDNA). Front Microbiol 6:1369
Grüll MP, Mulligan ME, Lang AS (2018) Small extracellular particles with big potential for horizontal gene transfer: membrane vesicles and gene transfer agents. FEMS Microbiol Lett 365:67
Guerrero-Mandujano A, Hernández-Cortez C, Ibarra JA, Castro-Escarpulli G (2017) The outer membrane vesicles: secretion system type zero. Traffic 18:425–432
Hagemann S, Stöger L, Kappelmann M et al (2013) DNA-bearing membrane vesicles produced by *Ahrensia kielensis* and *Pseudoalteromonas marina*. J Basic Microbiol 54:1062–1072
Hansell D, Carlson C (2015) Dissolved organic matter in the ocean carbon cycle. Eos 96
Hansell DA (2013) Recalcitrant dissolved organic carbon fractions. Ann Rev Mar Sci 5:421–445
Hasegawa Y, Futamata H, Tashiro Y (2015) Complexities of cell-to-cell communication through membrane vesicles: implications for selective interaction of membrane vesicles with microbial cells. Front Microbiol 6:633

Hibbing ME, Fuqua C, Parsek MR, Peterson SB (2010) Bacterial competition: surviving and thriving in the microbial jungle. Nat Rev Microbiol 8:15–25
Hoppe H-G (2003) Phosphatase activity in the sea. Hydrobiologia 493:187–200
Huang C, Quinn D, Sadovsky Y et al (2017) Formation and size distribution of self-assembled vesicles. Proc Natl Acad Sci 114:2910–2915
Inagaki SS, Onishi SS, Kuramitsu HKH, Sharma AA (2006) *Porphyromonas gingivalis* vesicles enhance attachment, and the leucine-rich repeat BspA protein is required for invasion of epithelial cells by *Tannerella forsythia*. Infect Immun 74:5023–5028
Ionescu M, Zaini PA, Baccari C et al (2014) *Xylella fastidiosa* outer membrane vesicles modulate plant colonization by blocking attachment to surfaces. Proc Natl Acad Sci 111:E3910–E3918
Jørgensen NO, Jacobsen CS (1996) Bacterial uptake and utilization of dissolved DNA. Aquat Microb Ecol 11:263–270
Kadurugamuwa JL, Beveridge TJ (1995) Virulence factors are released from *Pseudomonas aeruginosa* in association with membrane vesicles during normal growth and exposure to gentamicin: a novel mechanism of enzyme secretion. J Bacteriol 177:3998–4008
Kadurugamuwa JL, Beveridge TJ (1996) Bacteriolytic effect of membrane vesicles from *Pseudomonas aeruginosa* on other bacteria including pathogens: conceptually new antibiotics. J Bacteriol 178:2767–2774
Kadurugamuwa JL, Beveridge TJ (1999) Membrane vesicles derived from *Pseudomonas aeruginosa* and *Shigella flexneri* can be integrated into the surfaces of other Gram-negative bacteria. Microbiology 145:2051–2060
Kamaguchi A, Nakayama K, Ichiyama S et al (2003) Effect of *Porphyromonas gingivalis* vesicles on coaggregation of *Staphylococcus aureus* to oral microorganisms. Curr Microbiol 47:485–491
Kauffman KM, Hussain FA, Yang J et al (2018) A major lineage of non-tailed dsDNA viruses as unrecognized killers of marine bacteria. Nature 554:118–122
Kesty NC, Kuehn MJ (2004) Incorporation of heterologous outer membrane and periplasmic proteins into *Escherichia coli* outer membrane vesicles. J Biol Chem 279:2069–2076
Kim YS, Choi EJ, Lee WH et al (2013) Extracellular vesicles, especially derived from Gram-negative bacteria, in indoor dust induce neutrophilic pulmonary inflammation associated with both Th1 and Th17 cell responses. Clin Exp Allergy 43:443–454
Kirchman DL (2016) Growth rates of microbes in the oceans. Ann Rev Mar Sci 8:285–309
Kolling GL, Matthews KR (1999) Export of virulence genes and Shiga toxin by membrane vesicles of *Escherichia coli* O157:H7. Appl Environ Microbiol 65:1843–1848
Kulp A, Kuehn MJ (2010) Biological functions and biogenesis of secreted bacterial outer membrane vesicles. Annu Rev Microbiol 64:163–184
Kulp AJ, Sun B, Ai T et al (2015) Genome-wide assessment of outer membrane vesicle production in *Escherichia coli*. PLoS One 10:e0139200–e0139216
Lappann M, Otto A, Becher D, Vogel U (2013) Comparative proteome analysis of spontaneous outer membrane vesicles and purified outer membranes of *Neisseria meningitidis*. J Bacteriol 195:4425–4435
Lauga E (2016) Bacterial hydrodynamics. Annu Rev Fluid Mech 48:105–130
Lee J, Lee EY, Kim SH et al (2013) *Staphylococcus aureus* extracellular vesicles carry biologically active β-lactamase. Antimicrob Agents Chemother 57:2589–2595
Lennon JT (2007) Diversity and metabolism of marine bacteria cultivated on dissolved DNA. Appl Environ Microbiol 73:2799–2805
Li J, Azam F, Zhang S (2016) Outer membrane vesicles containing signaling molecules and active hydrolytic enzymes released by a coral pathogen *Vibrio shilonii* AK1. Environ Microbiol 18:3850–3866
Li Z, Clarke AJ, Beveridge TJ (1998) Gram-negative bacteria produce membrane vesicles which are capable of killing other bacteria. J Bacteriol 180:5478–5483

Liao S, Klein MI, Heim KP et al (2014) *Streptococcus mutans* extracellular DNA is upregulated during growth in biofilms, actively released via membrane vesicles, and influenced by components of the protein secretion machinery. J Bacteriol 196:2355–2366
Lynch JB, Alegado RA (2017) Spheres of hope, packets of doom: the good and bad of outer membrane vesicles in interspecies and ecological dynamics. J Bacteriol 199:e00012–e00017
MacDonald IA, Kuehn MJ (2013) Stress-induced outer membrane vesicle production by *Pseudomonas aeruginosa*. J Bacteriol 195:2971–2981
MacDonald KL, Beveridge TJ (2002) Bactericidal effect of gentamicin-induced membrane vesicles derived from *Pseudomonas aeruginosa* PAO1 on Gram-positive bacteria. Can J Microbiol 48:810–820
Manning AJ, Kuehn MJ (2011) Contribution of bacterial outer membrane vesicles to innate bacterial defense. BMC Microbiol 11:258
Mashburn LM, Whiteley M (2005) Membrane vesicles traffic signals and facilitate group activities in a prokaryote. Nature 437:422–425
Mashburn-Warren L, Howe J, Garidel P et al (2008) Interaction of quorum signals with outer membrane lipids: insights into prokaryotic membrane vesicle formation. Mol Microbiol 69:491–502
Mashburn-Warren LM, Whiteley M (2006) Special delivery: vesicle trafficking in prokaryotes. Mol Microbiol 61:839–846
McBroom AJ, Johnson AP, Vemulapalli S, Kuehn MJ (2006) Outer membrane vesicle production by *Escherichia coli* is independent of membrane instability. J Bacteriol 188:5385–5392
McCaig WD, Koller A, Thanassi DG (2013) Production of outer membrane vesicles and outer membrane tubes by *Francisella novicida*. J Bacteriol 195:1120–1132
Meyer DH, Fives-Taylor PM (1994) Characteristics of adherence of *Actinobacillus actinomycetemcomitans* to epithelial cells. Infect Immun 62:928–935
Moran MA (2015) The global ocean microbiome. Science 350:aac8455
Mulcahy LA, Pink RC, Carter DRF (2014) Routes and mechanisms of extracellular vesicle uptake. J Extracell Vesicles 3:1093
Nakayama-Imaohji H, Hirota K, Yamasaki H et al (2016) DNA inversion regulates outer membrane vesicle production in *Bacteroides fragilis*. PLoS One 11:e0148887–e0148821
Nawrocki MP, Karl D (1989) Dissolved ATP turnover in the Bransfield Strait, Antarctica during a spring bloom. Mar Ecol Prog Ser 57:35–44
Nazarian P, Tran F, Boedicker JQ (2018) Modeling multispecies gene flow dynamics reveals the unique roles of different horizontal gene transfer mechanisms. Front Microbiol 9:2978
Nealson KH (1997) Sediment bacteria: who's there, what are they doing, and what's new? Annu Rev Earth Planet Sci 25:403–434
Or D, Smets BF, Wraith JM et al (2007) Physical constraints affecting bacterial habitats and activity in unsaturated porous media—a review. Adv Water Resour 30:1505–1527
Orench-Rivera N, Kuehn MJ (2016) Environmentally controlled bacterial vesicle-mediated export. Cell Microbiol 18:1525–1536
Pardo YA, Florez C, Baker KM et al (2015) Detection of outer membrane vesicles in *Synechocystis* PCC 6803. FEMS Microbiol Lett 362:fnv163
Pathirana R, Kaparakis-Liaskos M (2016) Bacterial membrane vesicles: biogenesis, immune regulation and pathogenesis. Cell Microbiol 18:1518–1524
Perez-Cruz C, Carrion O, Delgado L et al (2013) New type of outer membrane vesicle produced by the Gram-negative bacterium *Shewanella vesiculosa* M7T: implications for DNA content. Appl Environ Microbiol 79:1874–1881
Pérez-Cruz C, Delgado L, López-Iglesias C, Mercade E (2015) Outer-inner membrane vesicles naturally secreted by Gram-negative pathogenic Bacteria. PLoS One 10:e0116896
Prados-Rosales R, Carreno LJ, Batista-Gonzalez A et al (2014a) Mycobacterial membrane vesicles administered systemically in mice induce a protective immune response to surface compartments of *Mycobacterium tuberculosis*. MBio 5:e01921–e01914

Prados-Rosales R, Weinrick BC, Pique DG et al (2014b) A role for *Mycobacterium tuberculosis* membrane vesicles in iron acquisition. J Bacteriol 196:1250–1256
Rakoff-Nahoum S, Coyne MJ, Comstock LE (2014) An ecological network of polysaccharide utilization among human intestinal symbionts. Curr Biol 24:40–49
Rappé MS, Connon SA, Vergin KL, Giovannoni SJ (2002) Cultivation of the ubiquitous SAR11 marine bacterioplankton clade. Nature 418:630–633
Rath P, Huang C, Wang T et al (2013) Genetic regulation of vesiculogenesis and immunomodulation in *Mycobacterium tuberculosis*. Proc Natl Acad Sci 110:E4790–E4797
Remis JP, Wei D, Gorur A et al (2014) Bacterial social networks: structure and composition of *Myxococcus xanthus* outer membrane vesicle chains. Environ Microbiol 16:598–610
Renelli M, Matias V, Lo RY, Beveridge TJ (2004) DNA-containing membrane vesicles of *Pseudomonas aeruginosa* PAO1 and their genetic transformation potential. Microbiology 150:2161–2169
Resch U, Tsatsaronis JA, Le Rhun A et al (2016) A two-component regulatory system impacts extracellular membrane-derived vesicle production in group A Streptococcus. MBio 7: e00207–e00216
Reyes-Robles T, Dillard RS, Cairns LS et al (2018) *Vibrio cholerae* outer membrane vesicles inhibit bacteriophage infection. J Bacteriol 200:e00792–e00717
Roden JA, Wells DH, Chomel BB et al (2012) Hemin binding protein C is found in outer membrane vesicles and protects *Bartonella henselae* against toxic concentrations of hemin. Infect Immun 80:929–942
Sampath V, McCaig WD, Thanassi DG (2018) Amino acid deprivation and central carbon metabolism regulate the production of outer membrane vesicles and tubes by *Francisella*. Mol Microbiol 107:523–541
Schatz D, Rosenwasser S, Malitsky S et al (2017) Communication via extracellular vesicles enhances viral infection of a cosmopolitan alga. Nat Microbiol 2:1485–1492
Schertzer JW, Boulette ML, Whiteley M (2009) More than a signal: non-signaling properties of quorum sensing molecules. Trends Microbiol 17:189–195
Schertzer JW, Whiteley M (2012) A bilayer-couple model of bacterial outer membrane vesicle biogenesis. MBio 3:e00297–e00211
Schooling SR, Beveridge TJ (2006) Membrane vesicles: an overlooked component of the matrices of biofilms. J Bacteriol 188:5945–5957
Schwechheimer C, Kuehn MJ (2015) Outer-membrane vesicles from Gram-negative bacteria: biogenesis and functions. Nat Rev Microbiol 13:605–619
Shetty A, Chen S, Tocheva EI et al (2011) Nanopods: a new bacterial structure and mechanism for deployment of outer membrane vesicles. PLoS One 6:e20725
Silva TP, Noyma NP, Duque TLA et al (2014) Visualizing aquatic bacteria by light and transmission electron microscopy. Antonie Van Leeuwenhoek 105:1–14
Simon M, Grossart H-P, Schweitzer B, Ploug H (2002) Microbial ecology of organic aggregates in aquatic ecosystems. Aquat Microb Ecol 28:175–211
Singorenko PD, Chang V, Whitcombe A et al (2017) Isolation of membrane vesicles from prokaryotes: a technical and biological comparison reveals heterogeneity. J Extracell Vesicles 6:1324731
Sjöström AE, Sandblad L, Uhlin BE, Wai SN (2015) Membrane vesicle-mediated release of bacterial RNA. Sci Rep 5:15329
Soler NN, Marguet EE, Verbavatz J-MJ, Forterre PP (2008) Virus-like vesicles and extracellular DNA produced by hyperthermophilic archaea of the order *Thermococcales*. Res Microbiol 159:390–399
Stocker R, Seymour JR (2012) Ecology and physics of bacterial chemotaxis in the ocean. Microbiol Mol Biol Rev 76:792–812
Tanoue E, Nishiyama S, Kamo M, Tsugita A (1995) Bacterial membranes: possible source of a major dissolved protein in seawater. Geochim Cosmochim Acta 59:2643–2648

Tashiro Y, Hasegawa Y, Shintani M et al (2017) Interaction of bacterial membrane vesicles with specific species and their potential for delivery to target cells. Front Microbiol 8:873–813
Toyofuku M, Morinaga K, Hashimoto Y et al (2017) Membrane vesicle-mediated bacterial communication. ISME J 11:1504–1509
Toyofuku M, Nomura N, Eberl L (2019) Types and origins of bacterial membrane vesicles. Nat Rev Microbiol 17:13–24
Toyofuku M, Tashiro Y, Hasegawa Y et al (2015) Bacterial membrane vesicles, an overlooked environmental colloid: biology, environmental perspectives and applications. Adv Colloid Interf Sci 266:65–77
Tran F, Boedicker JQ (2017) Genetic cargo and bacterial species set the rate of vesicle-mediated horizontal gene transfer. Sci Rep 7:8813
Tran F, Boedicker JQ (2019) Plasmid characteristics modulate the propensity of gene exchange in bacterial vesicles. J Bacteriol 201:e00430–e00418
Tsatsaronis JA, Franch-Arroyo S, Resch U, Charpentier E (2018) Extracellular vesicle RNA: a universal mediator of microbial communication? Trends Microbiol 26:401–410
Turnbull L, Toyofuku M, Hynen AL et al (2016) Explosive cell lysis as a mechanism for the biogenesis of bacterial membrane vesicles and biofilms. Nat Commun 7:11220–11213
Tzipilevich E, Habusha M, Ben-Yehuda S (2017) Acquisition of phage sensitivity by bacteria through exchange of phage receptors. Cell 168:186–199
Van Mooy BAS, Fredricks HF, Pedler BE et al (2009) Phytoplankton in the ocean use non-phosphorus lipids in response to phosphorus scarcity. Nature 458:69–72
Vasilyeva NV, Tsfasman IM, Suzina NE et al (2008) Secretion of bacteriolytic endopeptidase L5 of *Lysobacter* sp. XL1 into the medium by means of outer membrane vesicles. FEBS J 275:3827–3835
Velimirov B, Ranftler C (2018) Unexpected aspects in the dynamics of horizontal gene transfer of prokaryotes: the impact of outer membrane vesicles. Wien Med Wochenschr 168:307–313
Yaron S, Kolling GL, Simon L, Matthews KR (2000) Vesicle-mediated transfer of virulence genes from *Escherichia coli* O157:H7 to other enteric bacteria. Appl Environ Microbiol 66:4414–4420
Yonezawa H, Osaki T, Kurata S et al (2009) Outer membrane vesicles of *Helicobacter pylori* TK1402 are involved in biofilm formation. BMC Microbiol 9:197
Yun SH, Lee S-Y, Choi CW et al (2017) Proteomic characterization of the outer membrane vesicle of the halophilic marine bacterium *Novosphingobium pentaromativorans* US6-1. J Microbiol 55:56–62
Zarantonello V, Silva TP, Noyma NP et al (2018) The cyanobacterium *Cylindrospermopsis raciborskii* (CYRF-01) responds to environmental stresses with increased vesiculation detected at single-cell resolution. Front Microbiol 9:272
Zavan L, Bitto NJ, Johnston EL et al (2019) *Helicobacter pylori* growth stage determines the size, protein composition, and preferential cargo packaging of outer membrane vesicles. Proteomics 19:1800209
Zakharzhevskaya NB, Vanyushkina AA, Altukhov IA et al (2017) Outer membrane vesicles secreted by pathogenic and nonpathogenic *Bacteroides fragilis* represent different metabolic activities. Sci Rep 7:65–16

# Chapter 5
# Functions of MVs in Inter-Bacterial Communication

**Masanori Toyofuku, Yosuke Tashiro, Nobuhiko Nomura, and Leo Eberl**

**Abstract** Bacterial communication depends on small molecules that are released into the environment and are perceived by other cells. Many of the bacterial communication molecules are hydrophobic and thus have a poor solubility in water. While it is well established that such molecules serve as bona fide signal molecules, very little is known on how these molecules travel in aqueous environments. In this chapter we will summarize the evidence that hydrophobic signals can be released by bacterial membrane vesicles (MVs), which also serve as vehicles for signal dispersal. Given that the signals are concentrated in MVs, which can target specific cell types, a new binary signaling mechanism has been proposed that is different from the classic diffusion-based signaling model. This has important implications on how bacteria communicate in natural aqueous habitats.

## 5.1 Introduction

Bacteria, including Gram-positive and Gram-negative bacteria such as *Bacillus subtilis* and *Pseudomonas aeruginosa*, can communicate by the aid of signal molecules with each other to synchronize gene expression within the population, a phenomenon which forms the basis of coordinated group behaviors and is commonly referred to as quorum sensing (QS) (Miller and Bassler 2001; Whiteley et al. 2017). Many of the signals employed by bacteria are very hydrophobic and their solubility in water is very limited, raising the question of how these molecules can

M. Toyofuku (✉) · N. Nomura
Faculty of Life and Environmental Sciences, University of Tsukuba, Ibaraki, Japan
e-mail: toyofuku.masanori.gf@u.tsukuba.ac.jp; nomura.nobuhiko.ge@u.tsukuba.ac.jp

Y. Tashiro
Applied Chemistry and Biochemical Engineering Course, Department of Engineering, Graduate School of Integrated Science and Technology, Shizuoka University, Hamamatsu, Japan
e-mail: tashiro.yosuke@shizuoka.ac.jp

L. Eberl
Department of Plant and Microbial Biology, University of Zürich, Zürich, Switzerland
e-mail: leberl@botinst.uzh.ch

 
M. Kaparakis-Liaskos, T. A. Kufer (eds.), *Bacterial Membrane Vesicles*,
https://doi.org/10.1007/978-3-030-36331-4_5

travel between cells (Decho et al. 2011). In a seminal paper, Marvin Whiteley's group (Mashburn and Whiteley 2005; Mashburn-Warren et al. 2009) showed that the *Pseudomonas aeruginosa* quinolone signal (PQS) is not only associated with membrane vesicles (MVs) but also stimulates MV formation through PQS intercalation into the outer membrane. Subsequent studies have provided compelling evidence that many different hydrophobic signal molecules, including long-chain *N*-acyl homoserine lactone (AHL) signals, are associated with and transported via MVs (Brameyer et al. 2018; Li et al. 2016; Toyofuku et al. 2017b). Given that the signal molecules are highly concentrated in MVs, the fusion of a single MV with a bacterial cell is often sufficient to trigger its quorum sensing (QS) response (Toyofuku et al. 2017b). Such a binary signaling mechanism is fundamentally different from the classic diffusion-based QS model both in terms of its kinetics and their socioecological consequences.

In this chapter we will give an overview of different bacterial signals such as PQS, *N*-hexadecanoyl homoserine lactone (C16-HSL), (z)-3-aminoundec-2-en-4-one (Ea-$C_8$-CAI-1), which have been reported to be associated with MVs, together with diffusible signal factor (DSF; cis-11-methyl-2-dodecenoic acid) and 2-cis-tetradecenoic acid (*Xylella fastidiosa* DSF; *Xf*DSF) that regulate MV production (Fig. 5.1).

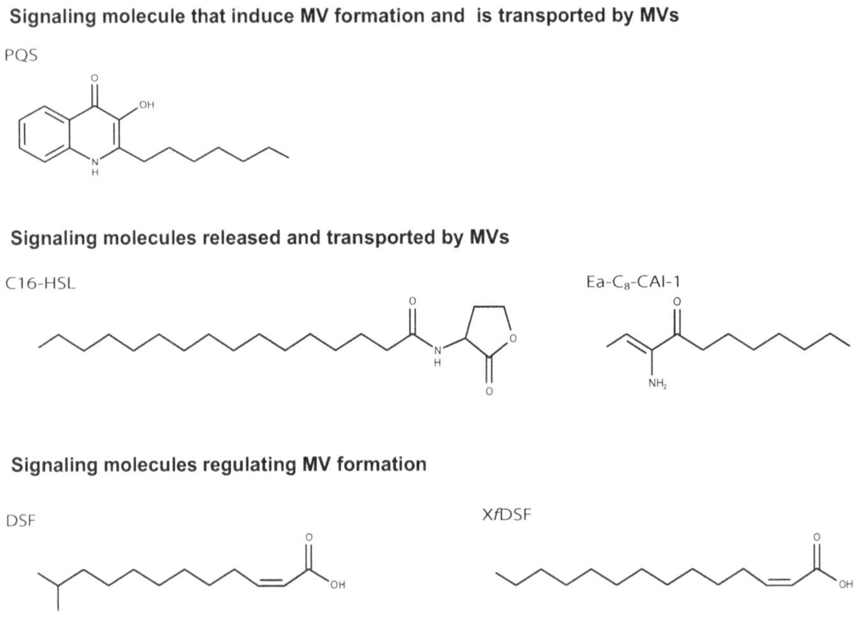


**Fig. 5.1** Bacterial signal molecules associated with MVs. PQS can induce MV formation and is transported by MVs. Likewise, C16HSL and Ea-C8-CAI-1 are released and transported by MVs. DSF and *Xf*DSF regulate MV formation by an unknown mechanism

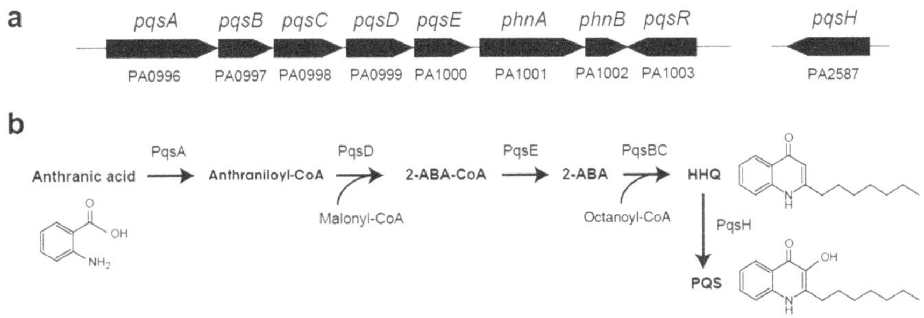


**Fig. 5.2** The synthesis of *Pseudomonas* quinolone signal (PQS) in *P. aeruginosa*. (**a**) Genetic organization of the PQS gene clusters. (**b**) Proposed pathway for PQS biosynthesis from anthranilate. Biosynthesis of PQS requires PqsABCDE proteins and PqsH. 2-ABA-CoA, 2-aminobenzoylacetyl-CoA; 2-ABA, 2-aminobenzoylacetate; HHQ, 2-heptyl-4-hydroxyquinoline; PQS, 2-heptyl-3-hydroxy-4-quinolone

## 5.2 *Pseudomonas* Quinolone Signal

A wide variety of quinolone compounds are synthesized by animals, plants, and microorganisms, and those compounds are of medical interest because of their antiallergenic, anticancer, and antimicrobial activities (Heeb et al. 2011). *P. aeruginosa* produces at least fifty-five 2-alkyl-4(*1H*)-quinolones (AQs) and some of those compounds exhibit antimicrobial activities (Déziel et al. 2004). 2-heptyl-3-hydroxy-4-quinolone (known as PQS) is used as a signal in the quinolone-based QS system in *P. aeruginosa* (Pesci et al. 1999), in addition to the two main *N*-acyl homoserine lactone (AHL) signals *N*-3-oxododecanoyl homoserine lactone (3-oxo-C12-HSL) and *N*-butanoyl homoserine lactone (C4-HSL). PQS and the two AHLs signals control expression of overlapping sets of hundreds of genes, many of which encode for virulence factors (Jimenez et al. 2012). PQS is synthesized from anthranilic acid by the products of the *pqsABCDE* operon and *pqsH* (Fig. 5.2) (Lin et al. 2018). The *pqsABCDE* operon is adjacent to the anthranilate synthase genes *phnAB* and the gene *pqsR*, which encodes the PQS receptor (Gallagher et al. 2002). The first step of PQS biosynthesis from anthranilate is conducted by PqsA, an anthranilate coenzyme A ligase (Coleman et al. 2008). Then, PqsD, which belongs to the FabH (β-ketoacyl-(acyl carrier protein) synthase III) protein family, synthesizes 2-aminobenzoylacetyl-CoA (2-ABA-CoA) from anthraniloyl-coenzyme A and malonyl-coenzyme A (Dulcey et al. 2013; Zhang et al. 2008). PqsE functions as a pathway-specific thioesterase, hydrolyzing 2-ABA-CoA to 2-aminobenzoylacetate (2-ABA) (Drees and Fetzner 2015). This reaction can be processed by the broad-specificity thioesterase TesB, explaining why the depletion of PqsE does not block the synthesis of 2-heptyl-4-hydroxyquinoline (HHQ) and PQS (Drees and Fetzner 2015). PqsBC are involved in adding an octanoyl moiety to 2-ABA to produce HHQ (Dulcey et al. 2013). Finally, the FAD-dependent mono-oxygenase PqsH converts HHQ into PQS (Déziel et al. 2004; Gallagher et al. 2002). Therefore, the deletion of *pqsH* abolishes PQS synthesis but the production of other AQs is continued.

### 5.2.1 Multifunctional PQS

PQS is the main signal among the AQs in *P. aeruginosa*. PQS binds to the LysR-type transcriptional regulator PqsR (also called MvfR) and its affinity to PqsR is higher than that of its precursor HHQ (Wade et al. 2005; Xiao et al. 2006). The PQS–PqsR complex regulates the expression of many genes, many of which encode functions related to motility, virulence, and biofilm formation (Déziel et al. 2005; Schuster et al. 2003). Interestingly, PQS-mediated transcriptional regulation can occur in a PqsR-dependent and PqsR-independent manner (Rampioni et al. 2010). In addition to PqsR, PQS directly interacts with other proteins, including MexG and RhlR as well as proteins involved in respiration (Baker et al. 2017; Dandela et al. 2018; Hodgkinson et al. 2016; Toyofuku et al. 2008).

It has been demonstrated that PQS can chelate ferric iron ($Fe^{3+}$) whereas HHQ, which is lacking the 3-hydroxy group of PQS, cannot (Bredenbruch et al. 2006; Diggle et al. 2007). PQS has a high affinity for iron and induces the expression of genes involved in the synthesis of the siderophores pyoverdine and pyochelin (Bredenbruch et al. 2006; Diggle et al. 2007). Iron-chelating activity of PQS also represses denitrification in *P. aeruginosa* as well as growth of some bacteria (Toyofuku et al. 2010; Toyofuku et al. 2008). In addition, PQS has been shown to exert immune modulatory and cytotoxic activities (Lin et al. 2018).

### 5.2.2 PQS Delivery Through MVs

AQs have generally a limited aqueous solubility. For example, the solubility of PQS is only 1 mg/L water (Lépine et al. 2003) and the octanol–water partition coefficient (logP) is 3.60 (Mashburn and Whiteley 2005). Despite its hydrophobic nature, PQS functions as a cell-to-cell communication signal in the aqueous solution. An elegant solution to this problem was offered by Mashburn and Whitley, who showed that about 80% of the total PQS is associated with MVs, in contrast to the homoserine lactone signals where only 1% of the signal was found to be within MVs (Mashburn and Whiteley 2005). The PQS carried by MVs was shown to be biologically active and can increase *pqs* gene expression, virulence factor production, and antimicrobial activities (Mashburn and Whiteley 2005; Tashiro et al. 2010b).

### 5.2.3 Vesiculation Is Stimulated by PQS

PQS is not only a constituent of MVs, but also stimulates MV formation in *P. aeruginosa* (Mashburn and Whiteley 2005). Consequently, MV production is decreased in mutants with inactivated *pqsA*, *pqsH*, or *pqsR* genes but can be restored by the exogenous addition of PQS (Mashburn and Whiteley 2005). The increase of

MV production in the *pqsR* mutant indicates that PQS-mediated MV formation is not due to PQS-dependent gene regulation. The structure of PQS is important for MV formation as HHQ is less effective (Mashburn-Warren et al. 2009). Interestingly, PQS can induce MV production also in many other bacterial strains (Horspool and Schertzer 2018; Mashburn-Warren et al. 2008; Tashiro et al. 2010a). More detailed biophysical experiments revealed that PQS stimulates outer membrane blebbing through intercalation into the outer membrane. When PQS is present at the extracellular milieu, it specifically interacts with the 4′-phosphate and acyl chains of lipid A of the lipopolysaccharide (LPS) (Mashburn-Warren et al. 2008). Salt bridges containing divalent cations ($Mg^{2+}$ and $Ca^{2+}$) normally stabilize the negatively charged phosphates between adjacent LPS molecules (Kadurugamuwa and Beveridge 1996). PQS sequesters these cations and thereby increases the anionic repulsion between LPS molecules and creates membrane curvature (Mashburn-Warren and Whiteley 2006; Tashiro et al. 2012). Adding exogenous $Mg^{2+}$ abolishes the effect of PQS on *Escherichia coli* MV formation, suggesting that excess cations can neutralize the repulsion of negatively charged LPS molecules (Tashiro et al. 2010a). Subsequent work resulted in a refined model, the so-called bilayer-couple model (Schertzer and Whiteley 2012), in which the specific interaction of PQS with LPS contributes to a low rate of flip-flops between the leaflets of the outer membrane. This leads to an expansion of the outer leaflet relative to the inner leaflet, which eventually gives rise to membrane curvature and MV formation. PQS has to be initially secreted from the cells via an unknown export mechanism to interact with the outer leaflet of the outer membrane (Lin et al. 2018) (Fig. 5.3). When strains PAO1 and PA14 were compared, PQS export and MV formation were significantly higher in the latter strain, even though equivalent amounts of total PQS were synthesized in both strains (Florez et al. 2017). While PQS accumulates in the outer membrane of PA14, in PAO1 most of the PQS accumulates in the inner membrane, presumably due to early saturation of the PQS export pathway (Florez et al. 2017). The differences in PQS localization among strains and its dependence on culture conditions may explain reports that could not confirm a role of PQS in MV formation (Macdonald and Kuehn 2013; Turnbull et al. 2016). Furthermore, PQS is not synthesized under anaerobic conditions as PqsH requires oxygen for PQS synthesis (Schertzer et al. 2010; Toyofuku et al. 2008). Molecules such as indole and its derivatives also repress MV formation in *P. aeruginosa* by inhibiting PQS biosynthesis (Tashiro et al. 2010c).

While PQS synthesis does not start before early stationary phase, MVs are already released during the exponential phase, suggesting that other routes for MV biogenesis exist (Tashiro et al. 2010b). The alternative sigma factor AlgU and the periplasmic protease MucD are related to the envelope stress pathway, and MV formation was shown to be increased when either of the genes were deleted (Tashiro et al. 2009). A mutation of *mucD* stimulates MV formation in PQS-deficient mutants, indicating that periplasmic stress-mediated MV formation is independent from PQS pathway. Similarly, MV formation is also induced in a *pqsA* mutant by other stresses, such as exposure to D-cycloserine, polymyxin B, or $H_2O_2$ (Macdonald and Kuehn 2013). Moreover, the peptidoglycan-associated outer membrane proteins

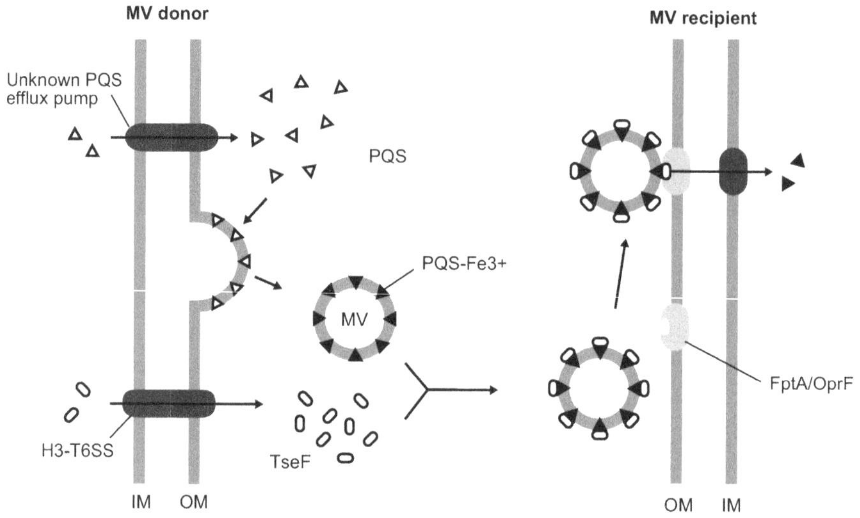


**Fig. 5.3** Membrane vesicle-mediated PQS transfer and iron uptake. *Pseudomonas* quinolone signal (PQS) is synthesized in *P. aeruginosa* cytoplasm and moved out of the cell via a yet-to-be-defined export mechanism. PQS integrates the outer membrane of donor *P. aeruginosa*, resulting in membrane curvature and MV secretion. PQS chelates ferric iron and an extracellular protein TseF, which is secreted by the Type VI Secretion System H3 (H3-T6SS) interacts with PQS-$Fe^{3+}$. The complex of TseF and PQS-$Fe^{3+}$ localized in MVs recognizes the pyochelin receptor FptA and the porin OprF on the recipient bacterial cell membrane, to facilitate the uptake of iron. OM, outer membrane; IM, inner membrane. The figure is modified from (Tashiro et al. 2019)

OprF and OprI affect MV formation (Wessel et al. 2013). Increased MV production by the *oprF* mutant was shown to be caused by upregulation of PQS production, while the effect of OprI on vesiculation is independent of PQS. Endolysin-mediated explosive cell lysis appears to be the main route for vesicle biogenesis in biofilms as well as under anoxic conditions in *P. aeruginosa* (Toyofuku et al. 2014; Turnbull et al. 2016).

## 5.3 *N*-Acyl Homoserine Lactone (AHL) Signals

The most common signaling molecules produced by Gram-negative bacteria are the AHLs, which are produced by hundreds of species mainly of the Proteobacteria. AHLs, which are typically synthesized by the LuxI-family enzymes, consist of a homoserine lactone ring and a fatty chain that contains 4–20 carbons that can have additional modifications (Arashida et al. 2018). These variations confer specificity to the signals, which are generally recognized by LuxR-type receptors. In general, the longer the fatty acid chain the more hydrophobic the signal is. The classic QS model assumes that the signals diffuse freely between cells to synchronize their gene expression when a particular population density has been attained. While the short-chain AHLs such as C4-HSL were shown to freely diffuse in and out of bacterial cells, AHLs containing longer fatty acid chains often accumulate in the cell envelope and require transporters to be released (Buroni et al. 2009; Chan et al. 2007; Pearson et al. 1999). Given the low solubility of long-chain AHLs in water it is unclear how they are released and can travel to other cells.

### 5.3.1 Binary Signaling Involving MVs

In a recent study it has been demonstrated that *Paracoccus denitrificans* release C16-HSL through MVs (Toyofuku et al. 2017b). In this bacterium, cell aggregation and biofilm formation were inhibited in the presence of the signal. Of the extracellular C16-HSL, half of the molecules were found to be tightly associated with MVs, which were able to trigger the QS response in a *P. denitrificans* mutant deficient in signal production. Stimulation of MV formation increased C16-HSL release without affecting C16-HSL production. Importantly, the amount of C16-HSL associated with one MV was shown to be higher than the threshold concentration required to trigger the QS response of a single cell (Toyofuku et al. 2017b). This mechanism for cell–cell communication is fundamentally different from the classical QS model, which assumes the homogenous distribution of the signal in the medium (Miller and Bassler 2001). MVs deliver the signal molecules in high concentrations to target cells such that their QS response will inevitably be induced. For this reason, this MV-dependent signal dissemination mechanism was referred to as binary signaling (Toyofuku et al. 2017b) (Fig. 5.4).

This study provided evidence that hydrophobic AHL signals may be mainly released by MVs, which will also solubilize the signals in water (Toyofuku et al. 2017b). Likewise, the coral-associated bacterium *Vibrio shilonii* was shown to produce MVs that contain AHLs, although the structure of this signal remains to be identified (Li et al. 2016). The MV-dependent binary QS mechanism could be particularly relevant for trafficking hydrophobic signal molecules in open aqueous environments where free signals would be infinitely diluted.

### 5.3.2 Specific Signal Delivery Through MVs

One of the advantages of packaging signals into MVs is the possibility of a specific delivery to certain target cells, particularly relevant in polymicrobial communities. In such a situation, MVs would traffic the signals with high precision to the target cells and at the same time ensure that the signal is concentrated high enough that the quorum threshold of the receiver cell is attained.

Specific delivery was reported for *P. denitrificans*, for which it has been shown that MVs derived from this species fuse with a higher frequency with cells of their own species than those of other species (Toyofuku et al. 2017b). Furthermore, by using

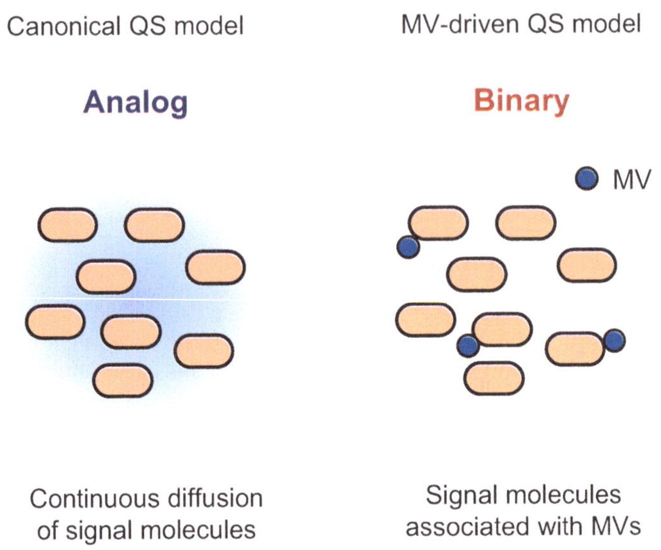


**Fig. 5.4** Bacterial binary signaling involving MVs. *Left panel:* The canonical QS model assumes free diffusion of the signal molecule that will synchronize the gene expression within the bacterial population when its concentration has reached a certain threshold (analog signaling). *Right panel:* The MV-driven QS model involves MV-mediated delivery of the signal molecules in sufficiently high concentrations to target cells that their QS response will inevitably be induced (binary signaling)

AHL-reporter strains it was demonstrated that *P. denitrificans* responds equally well to free C16-HSL than to MV-associated C16-HSL, while *P. aeruginosa* does not respond well to MV-associated C16-HSL compared to free C16-HSL. It is currently not clear how MVs can be specifically delivered to target cells. Similar results of self-recognition by MVs have been observed in *Buttiauxella agrestis* for which gene transfer through MVs was only observed between cells of the same species (Tashiro et al. 2017). In this bacterium, the physiochemical properties of surfaces were critical for self-recognition by MVs. *B. agrestis* has a lower zeta-potential compared to other bacteria leading to a low energy barrier for fusion between MVs and *B. agrestis*. In the case of PQS of *P. aeruginosa*, TseF, a protein secreted by the type VI secretion system H3 (H3-T6SS), is involved in MV delivery (Fig. 5.3) (Lin et al. 2017). TseF is incorporated into MVs due to its direct interaction with iron-bound PQS (Lin et al. 2017). As TseF is captured by the Fe(III)-pyochelin receptor FptA and the porin OprF that is localized on the cell surface, PQS-$Fe^{3+}$ loaded MVs are specifically delivered to *P. aeruginosa* cells. However, the mechanism of how PQS-$Fe^{3+}$ is unloaded from MVs and is taken up by the cells is unknown. The presence of a protein like TseF on MVs that is recognized by surface receptors allows the specific delivery of the MV cargo to bacteria expressing the cognate receptor protein (Tashiro et al. 2019). These studies imply that the MV composition will influence the specificity of cargo delivery.

### 5.3.3 Signal Piracy by MVs

Besides releasing signals from the cell, MVs can also collect signal molecules from the environment and deliver them to a target bacterium (Morinaga et al. 2018). For example, *P. denitrificans* can sequester long-chain AHLs and exploit them to trigger its QS response. Although this bacterium primarily produces C16-HSL, it can respond to other long-chain AHLs such as C12-, C14- and C18-HSLs. When each signal was incubated with MVs derived from a *pdnI* mutant of *P. denitrificans*, which does not produce C16-HSL, the signals were adsorbed by the MVs. These AHL-loaded MVs were able to induce the QS response of the *P. denitrificans pdnI* mutant (Morinaga et al. 2018). It has been hypothesized that this signal piracy allows *P. denitrificans* to trigger its QS response in the presence of cooperating bacteria.

## 5.4 *cis*-2-Unsaturated Fatty Acids Signaling

Another group of quorum sensing signals are *cis*-2-unsaturated fatty acids. The first molecule belonging to this class of signals is the DSF, which was first reported to be produced by the plant pathogen *Xanthomonas campestris* pv. *campestris* (*Xcc*) (Barber et al. 1997). The biosynthesis and perception of the DSF signal molecule involves protein encoded by the *rpf* (regulation of pathogenicity factors) gene cluster (Dow 2017; Tang et al. 1991). DSF biosynthesis is dependent on the putative enoyl CoA hydratase Rpf. RpfB, a long-chain fatty acyl CoA ligase, appears to be involved in DSF turnover (Zhou et al. 2015). Cells perceive the DSF signal through RpfR or a two-component system, comprising the sensor histidine kinase RpfC and the HD-GYP domain response regulator RpfG (Deng et al. 2012; Suppiger et al. 2016). The HD-GYP domain of RpfG is a phosphodiesterase that hydrolyses the second messenger cyclic di-GMP. Upon binding of DSF to its receptor, the intracellular level of c-di-GMP is reduced by RpfC, which in turn regulates a broad range of biological functions, many of which are associated with virulence, biofilm formation, and stress tolerance. Subsequent work showed that DSF-family signals are produced by bacteria, which can use their signals even for interspecies communication (Boon et al. 2008; Deng et al. 2010). In addition, *cis*-2-dodecenoic acid, referred to as BDSF (***B****urkholderia* **d**iffusible **s**ignal **f**actor), was shown to inhibit the yeast-hyphal transition of *Candida albicans*, suggesting that this class of signals also have a role in interkingdom interactions (Boon et al. 2008).

### 5.4.1 The Role of DSF Family Signals in MV Formation

DSF family molecules are hydrophobic and little is known on how these signals are released from the cells and travel in aqueous environments. *X. fastidiosa* is a bacterium that causes disease in a wide range of important crop plants. They can colonize specific areas of the foreguts of insect vectors that transmit the pathogen to their host plants, where they migrate and proliferate within xylem vessels. *X. fastidiosa* uses the signal *Xf*DSF to suppress motility while stimulating the production of cell-surface adhesins, and thus is required for cell aggregation, surface attachment, and biofilm formation (Chatterjee et al. 2008). An *rpfF* mutant of *X. fastidiosa* that no longer produces *Xf*DSF is hyper virulent to grapevine but the *rpfF* mutant is impaired in insect colonization and transmission. MV production in *X. fastidiosa* is suppressed by *Xf*DSF (Ionescu et al. 2013). Interestingly, MV production was shown to influence bacteria–surface interactions (Ionescu et al. 2014). At a low population density, *X. fastidiosa* produces large numbers of MVs that inhibit attachment of the cells to the walls of xylem vessels and as a consequence the bacterium spreads in the plant host. On the other hand, when *X. fastidiosa* population is high, *Xf*DSF accumulates and MV formation is suppressed and *X. fastidiosa* form a biofilm on the plant surface. Another study showed that the DSF also influences MV production in *Stenotrophomonas maltophilia,* an emerging multidrug resistant pathogen that is associated with bacteremia, pneumonia, and soft-tissue infections (Devos et al. 2015). In contrast to *X. fastidiosa,* however, DSF signal stimulates MV production in *S. maltophilia*. Vesiculation in this organism was also stimulated by BDSF but not by *cis*-2-decenoic acid (PDSF), a signal produced by *P. aeruginosa*.

MVs derived from one species can benefit other species and can play important roles in polymicrobial communities. For example, it was shown that MV formation in *S. maltophilia* can be induced by the β-lactam antibiotic imipenem, presumably through perturbation of the cell wall (Devos et al. 2015). The imipenem-induced MVs contain a β-lactamase, which degrades the antibiotic. These vesicles can protect other co-residing bacteria, including *P. aeruginosa* and *Burkholderia cenocepacia* that frequently coexist with *S. maltophilia* in the lungs of chronically infected cystic fibrosis (CF) patients (Devos et al. 2016). Naturally, signal-induced MVs do not contain β-lactamase, whereas imipenem-induced MVs do.

It has been postulated that membrane proteins that are regulated by DSF may influence MV biogenesis (Ionescu et al. 2014). However, the exact molecular mechanism of DSF-controlled MV formation remains to be elucidated.

## 5.5 CAI-1 Signaling

Another signal released through MVs is CAI-1 [(s)-3-hydroxytridecan-4-one] that is produced by many *Vibrio* species and is the predominant QS molecule of the human pathogen *Vibrio cholerae* (Brameyer et al. 2018). Other *Vibrio* species produce signal molecules with different acyl chain length and modifications (Ng et al. 2011). *Vibrio harveyi* produces Ea-$C_8$-CAI-1 and two additional signals, HAI-1 [*N*-3-(hydroxybutyryl)-homoserine lactone], and AI-2, a furanosyl borate diester that is widely spread among bacteria and is known as an inter-species signaling molecule

(Defoirdt et al. 2008). Each signal is perceived by specific membrane-bound hybrid sensor kinase. CAI-1 is sensed by CqsS, HAI-1 is sensed by LuxN, and AI-2 is sensed by LuxPQ.

### *5.5.1 CAI-1 Delivery Through MVs*

While HAI-1 and AI-2 are both hydrophilic with LogP values of −0.94 and −1.25, respectively, CAI-1 is hydrophobic with a LogP of 3.05 (Brameyer et al. 2018). Due to its hydrophobicity, CAI-1 was thought to be partitioned into the lipid bilayer of the outer membrane. Indeed, CAI-1 was detected in MVs collected from stationary cultures of *V. harveyi* and was shown to activate the QS cascade in a CAI-1 mutant (Brameyer et al. 2018). In this study, MVs were fractionated by size and CAI-1 were detected using a CAI-1 reporter strain. Interestingly, only larger MVs (about 10% of all MVs) induced the QS response in the reporter strain. It is still unclear whether only the large MVs are able to fuse with neighboring cells or whether these MVs are formed through different routes and therefore have different contents and functions. *Vibrio* species are known to possess sheathed flagella at the pole of the cell. In *Aliivibrio fischeri*, rotation of the sheathed flagella generates MVs with smaller size than the ones produced by a mutant with a paralyzed flagellum (Aschtgen et al. 2016a). Such MVs may not contain signals and the different types of MVs generated from different routes may have different functions in *Vibrio* species.

## 5.6 Types of MVs and Their Role in Cell-to-Cell Communication

The main routes of MV formation in Gram-negative bacteria are blebbing of the outer membrane and cell lysis, and our knowledge regarding the production of Gram-positive MVs is still limited (Toyofuku et al. 2019). MVs were originally considered to be formed through controlled blebbing of the outer membrane without affecting the cell viability (Schwechheimer and Kuehn 2015). DNA damage was recently shown to induce MV formation through explosive cell lysis in *P. aeruginosa* and bubbling cell death in a Gram-positive bacterium, *B. subtilis* (Toyofuku et al. 2017a; Turnbull et al. 2016). Both processes are dependent on the enzymatic action of endolysins. These peptidoglycan (PG)-degrading enzymes are typically encoded in prophage regions and are required to lyse the host for phage release. In explosive cell lysis, the enzyme degrades the PG layer and consequently the cells round up and explode (Turnbull et al. 2016). The resulting shattered membrane fragments circularize and form MVs. It has been proposed that the composition and cargo of these MVs would differ from the ones that arise from blebbing and were therefore named EOMV (explosive outer membrane vesicles).

Outer-inner membrane vesicles (OIMV), which have recently been discovered in various bacteria, could also be generated through this route (Toyofuku et al. 2019). It is likely that the different genesis mechanisms lead to distinct MV types that serve particular functions. For example, evidence is accumulating that hydrophobic signal molecules have a high affinity to intercalate into the outer membrane, which may stimulate MV formation through blebbing (Horspool and Schertzer 2018). MVs generated via explosive cell lysis will randomly capture cellular materials such as DNA/RNA and cytoplasmic proteins as well as signal molecules. Small RNAs (sRNAs) are considered to be a universal "signal" across kingdoms (Cai et al. 2018), and MV types arising from explosive cell lysis and bubbling cell death appear to be the main carrier of this signal.

## 5.7 Interkingdom Signals Carried by MVs

Extracellular vesicles are released from cells of all domains of life and evidence is emerging that they play an important role in intra- and interkingdom interactions (Cai et al. 2018). An interesting example is the interaction between choanoflagellates and bacteria. Choanoflagellates are a group of eukaryotic microbes that are the closest living relatives of animals. *Salpongoeca rosetta* develops into multicellular rosettes from a single founding cell that undergoes serial rounds of oriented cell division. The initiation of rosette formation was found to be induced by bioactive lipids, including sulfonolidis and lysophosphatidylethanolamines, which are produced by the bacterium *Algoriphagus machipongonensis* (Alegado et al. 2012; Woznica et al. 2016). MVs isolated from *A. machipongonensis* cultures induce rosette formation, indicating that the inducing lipids are packed into MVs (Woznica et al. 2016). The involvement of bacterial MVs in host development is also well studied in the Hawaiian bobtail squid-*Vibrio fischeri* interaction. The LPS associated with MVs released by *V. fischeri* trigger morphogenesis of the light organ of the bobtail squid (Aschtgen et al. 2016a; Aschtgen et al. 2016b). In addition, recent work unraveled a role of MV-associated small bacterial RNAs in intercellular communication with eukaryotic cells (Tsatsaronis et al. 2018).

## 5.8 Concluding Remarks

In this chapter, we have reviewed recent studies that have demonstrated that hydrophobic signal molecules can be transported by MVs. The transport of signaling molecules by MVs has several advantages over simple diffusion, which in the case of highly hydrophobic molecules is not even possible. In MVs the signals are concentrated and this ensures that the QS response in recipient cells is induced (Toyofuku et al. 2017b). As fusion of vesicles with bacterial cells is a stochastic process, this may generate heterogeneous populations, with cells being either induced or not. By contrast, the classic diffusion-based QS model assumes that upon reaching a threshold concentration the signals synchronize gene expression within all cells of the population to allow coordinated group behaviors (Miller and Bassler 2001). Another important difference is that free signals are accessible to the entire bacterial community, while evidence has emerged that MVs can deliver their signal contents to certain bacterial species and thus allow targeted communication (Tashiro et al. 2017; Toyofuku et al. 2017b). It will be a highly interesting line of research to investigate how MVs specifically fuse with certain bacteria and how these MV-associated signals are then delivered into the cytoplasm of the recipient cell. In addition, how MVs can deliver their cargo through the thick cell wall of Gram-positive bacteria has become a challenging question, and whether membrane vesicles produced by Gram-positive bacteria have a role in bacterial cell-to-cell communication remains to be investigated. It has also been shown that MVs can also sequester signals from the environment and this signal piracy may allow eavesdropping for cooperation partners in polymicrobial communities (Morinaga et al. 2018). MVs may even serve as "time capsules," as they protect the signals and thus allow bacteria to disperse their messages beyond their life span. The fact that some MV formation mechanisms are in fact based on bacterial cell death adds another interesting aspect to this idea (Toyofuku et al. 2019).

## References


Alegado RA et al (2012) A bacterial sulfonolipid triggers multicellular development in the closest living relatives of animals. elife 1:e00013. https://doi.org/10.7554/eLife.00013

Arashida N et al (2018) Identification of novel long chain N-acylhomoserine lactones of chain length C20 from the marine phototrophic bacterium *Rhodovulum sulfidophilum*. Biosci Biotechnol Biochem 82:1683–1693

Aschtgen MS, Lynch JB, Koch E, Schwartzman J, McFall-Ngai M, Ruby E (2016a) Rotation of *Vibrio fischeri* flagella produces outer membrane vesicles that induce host development. J Bacteriol 198:2156–2165

Aschtgen MS, Wetzel K, Goldman W, McFall-Ngai M, Ruby E (2016b) *Vibrio fischeri*-derived outer membrane vesicles trigger host development. Cell Microbiol 18:488–499

Baker YR et al (2017) Identification of new quorum sensing autoinducer binding partners in *Pseudomonas aeruginosa* using photoaffinity probes. Chem Sci 8:7403–7411

Barber CE et al (1997) A novel regulatory system required for pathogenicity of *Xanthomonas campestris* is mediated by a small diffusible signal molecule. Mol Microbiol 24:555–566

Boon C et al (2008) A novel DSF-like signal from *Burkholderia cenocepacia* interferes with *Candida albicans* morphological transition. ISME J 2:27–36

Brameyer S, Plener L, Muller A, Klingl A, Wanner G, Jung K (2018) Outer membrane vesicles facilitate trafficking of the hydrophobic signaling molecule CAI-1 between *Vibrio harveyi* cells. J Bacteriol 200:e00740–e00717. https://doi.org/10.1128/JB.00740-17

Bredenbruch F, Geffers R, Nimtz M, Buer J, Häussler S (2006) The *Pseudomonas aeruginosa* quinolone signal (PQS) has an iron-chelating activity. Environ Microbiol 8:1318–1329

Buroni S et al (2009) Assessment of three resistance-nodulation-cell division drug efflux transporters of *Burkholderia cenocepacia* in intrinsic antibiotic resistance. BMC Microbiol 9:200. https://doi.org/10.1186/1471-2180-9-200

Cai Q et al (2018) Plants send small RNAs in extracellular vesicles to fungal pathogen to silence virulence genes. Science 360:1126–1129

Chan YY et al (2007) Control of quorum sensing by a *Burkholderia pseudomallei* multidrug efflux pump. J Bacteriol 189:4320–4324

Chatterjee S, Wistrom C, Lindow SE (2008) A cell-cell signaling sensor is required for virulence and insect transmission of *Xylella fastidiosa*. Proc Natl Acad Sci U S A 105:2670–2675

Coleman JP, Hudson LL, McKnight SL, Farrow JM, Calfee MW, Lindsey CA, Pesci EC (2008) *Pseudomonas aeruginosa* PqsA is an anthranilate-coenzyme a ligase. J Bacteriol 190:1247–1255

Dandela R, Mantin D, Cravatt BF, Rayo J, Meijler MM (2018) Proteome-wide mapping of PQS-interacting proteins in *Pseudomonas aeruginosa*. Chem Sci 9:2290–2294

Decho AW, Frey RL, Ferry JL (2011) Chemical challenges to bacterial AHL signaling in the environment. Chem Rev 111:86–99

Defoirdt T, Boon N, Sorgeloos P, Verstraete W, Bossier P (2008) Quorum sensing and quorum quenching in *Vibrio harveyi*: lessons learned from in vivo work. ISME 2:19–26

Deng Y, Wu J, Eberl L, Zhang LH (2010) Structural and functional characterization of diffusible signal factor family quorum-sensing signals produced by members of the *Burkholderia cepacia* complex. Appl Environ Microb 76:4675–4683

Deng Y et al (2012) Cis-2-dodecenoic acid receptor RpfR links quorum-sensing signal perception with regulation of virulence through cyclic dimeric guanosine monophosphate turnover. Proc Natl Acad Sci U S A 109:15479–15484

Devos S et al (2015) The effect of imipenem and diffusible signaling factors on the secretion of outer membrane vesicles and associated Ax21 proteins in *Stenotrophomonas maltophilia*. Front Microbiol 6:298. https://doi.org/10.3389/fmicb.2015.00298

Devos S, Stremersch S, Raemdonck K, Braeckmans K, Devreese B (2016) Intra- and interspecies effects of outer membrane vesicles from *Stenotrophomonas maltophilia* on beta-lactam resistance. Antimicrob Agents Chemother 60:2516–2518

Déziel E, Lépine F, Milot S, He J, Mindrinos M, Tompkins R, Rahme L (2004) Analysis of *Pseudomonas aeruginosa* 4-hydroxy-2-alkylquinolines (HAQs) reveals a role for 4-hydroxy-2-heptylquinoline in cell-to-cell communication. Proc Natl Acad Sci U S A 101:1339–1344

Déziel E et al (2005) The contribution of MvfR to *Pseudomonas aeruginosa* pathogenesis and quorum sensing circuitry regulation: multiple quorum sensing-regulated genes are modulated without affecting l*asRI*, *rhlRI* or the production of *N*-acyl-L-homoserine lactones. Mol Microbiol 55:998–1014

Diggle S et al (2007) The *Pseudomonas* aeruginosa 4-quinolone signal molecules HHQ and PQS play multifunctional roles in quorum sensing and iron entrapment. Chem Biol 14:87–96

Dow JM (2017) Diffusible signal factor-dependent quorum sensing in pathogenic bacteria and its exploitation for disease control. J Appl Microbiol 122:2–11

Drees Steffen L, Fetzner S (2015) PqsE of *Pseudomonas aeruginosa* acts as pathway-specific thioesterase in the biosynthesis of alkylquinolone signaling molecules. Chem Biol 22:611–618

Dulcey Carlos E et al (2013) The end of an old hypothesis: the *Pseudomonas* signaling molecules 4-hydroxy-2-alkylquinolines derive from fatty acids, not 3-ketofatty acids. Chem Biol 20:1481–1491

Florez C, Raab JE, Cooke AC, Schertzer JW (2017) Membrane distribution of the *Pseudomonas* quinolone signal modulates outer membrane vesicle production in *Pseudomonas aeruginosa*. mBio 8 doi:https://doi.org/10.1128/mBio.01034-17

Gallagher L, McKnight S, Kuznetsova M, Pesci E, Manoil C (2002) Functions required for extracellular quinolone signaling by *Pseudomonas aeruginosa*. J Bacteriol 184:6472–6480

Heeb S, Fletcher MP, Chhabra SR, Diggle SP, Williams P, Cámara M (2011) Quinolones: from antibiotics to autoinducers. FEMS Microbiol Rev 35:247–274

Hodgkinson JT, Gross J, Baker YR, Spring DR, Welch M (2016) A new *Pseudomonas* quinolone signal (PQS) binding partner: MexG. Chem Sci 7:2553–2562

Horspool AM, Schertzer JW (2018) Reciprocal cross-species induction of outer membrane vesicle biogenesis via secreted factors. Sci Rep 8:9873. https://doi.org/10.1038/s41598-018-28042-4

Ionescu M, Baccari C, Da Silva AM, Garcia A, Yokota K, Lindow SE (2013) Diffusible signal factor (DSF) synthase RpfF of *Xylella fastidiosa* is a multifunction protein also required for response to DSF. J Bacteriol 195:5273–5284

Ionescu M, Zaini PA, Baccari C, Tran S, da Silva AM, Lindow SE (2014) *Xylella fastidiosa* outer membrane vesicles modulate plant colonization by blocking attachment to surfaces. Proc Natl Acad Sci U S A 111:E3910–E3918

Jimenez PN, Koch G, Thompson JA, Xavier KB, Cool RH, Quax WJ (2012) The multiple signaling systems regulating virulence in *Pseudomonas aeruginosa*. Microbiol Mol Biol Rev 76:46–65

Kadurugamuwa JL, Beveridge TJ (1996) Bacteriolytic effect of membrane vesicles from *Pseudomonas aeruginosa* on other bacteria including pathogens: conceptually new antibiotics. J Bacteriol 178:2767–2774

Lépine F, Déziel E, Milot S, Rahme LG (2003) A stable isotope dilution assay for the quantification of the *Pseudomonas* quinolone signal in *Pseudomonas aeruginosa* cultures. Biochim Biophys Acta 1622:36–41

Li J, Azam F, Zhang S (2016) Outer membrane vesicles containing signaling molecules and active hydrolytic enzymes released by a coral pathogen *Vibrio shilonii** AK1. Environ Microbiol 18:3850–3866

Lin J et al (2017) A *Pseudomonas* T6SS effector recruits PQS-containing outer membrane vesicles for iron acquisition. Nat Commun 8:14888. https://doi.org/10.1038/ncomms14888

Lin J, Cheng J, Wang Y, Shen X (2018) The *Pseudomonas* quinolone signal (PQS): not just for quorum sensing anymore. Front Cell Infect Microbiol 8:230. https://doi.org/10.3389/fcimb.2018.00230

Macdonald IA, Kuehn MJ (2013) Stress-induced outer membrane vesicle production by *Pseudomonas aeruginosa*. J Bacteriol 195:2971–2981

Mashburn LM, Whiteley M (2005) Membrane vesicles traffic signals and facilitate group activities in a prokaryote. Nature 437:422–425

Mashburn-Warren LM, Whiteley M (2006) Special delivery: vesicle trafficking in prokaryotes. Mol Microbiol 61:839–846

Mashburn-Warren L et al (2008) Interaction of quorum signals with outer membrane lipids: insights into prokaryotic membrane vesicle formation. Mol Microbiol 69:491–502

Mashburn-Warren L, Howe J, Brandenburg K, Whiteley M (2009) Structural requirements of the *Pseudomonas* quinolone signal for membrane vesicle stimulation. J Bacteriol 191:3411–3414

Miller MB, Bassler BL (2001) Quorum sensing in bacteria. Annu Rev Microbiol 55:165–199

Morinaga K, Yamamoto T, Nomura N, Toyofuku M (2018) *Paracoccus denitrificans* can utilize various long-chain N-acyl homoserine lactones and sequester them in membrane vesicles. Environ Microbiol Rep 10:651–654

Ng WL, Perez LJ, Wei Y, Kraml C, Semmelhack MF, Bassler BL (2011) Signal production and detection specificity in *Vibrio* CqsA/CqsS quorum-sensing systems. Mol Microbiol 79:1407–1417

Pearson JP, Van Delden C, Iglewski BH (1999) Active efflux and diffusion are involved in transport of *Pseudomonas aeruginosa* cell-to-cell signals. J Bacteriol 181:1203–1210

Pesci E, Milbank J, Pearson J, McKnight S, Kende A, Greenberg E, Iglewski B (1999) Quinolone signaling in the cell-to-cell communication system of *Pseudomonas aeruginosa*. Proc Natl Acad Sci U S A 96:11229–11234

Rampioni G et al (2010) Transcriptomic analysis reveals a global alkyl-quinolone-independent regulatory role for PqsE in facilitating the environmental adaptation of *Pseudomonas aeruginosa* to plant and animal hosts. Environ Microbiol Rep 12:1659–1673

Schertzer JW, Whiteley M (2012) A bilayer-couple model of bacterial outer membrane vesicle biogenesis. mBio 3:e00297–e00211. https://doi.org/10.1128/mBio.00297-11

Schertzer JW, Brown SA, Whiteley M (2010) Oxygen levels rapidly modulate *Pseudomonas aeruginosa* social behaviours via substrate limitation of PqsH. Mol Microbiol 77:1527–1538

Schuster M, Lostroh CP, Ogi T, Greenberg EP (2003) Identification, timing, and signal specificity of *Pseudomonas aeruginosa* quorum-controlled genes: a transcriptome analysis. J Bacteriol 185:2066–2079

Schwechheimer C, Kuehn MJ (2015) Outer-membrane vesicles from Gram-negative bacteria: biogenesis and functions. Nat Rev Microbiol 13:605–619

Suppiger A, Eshwar AK, Stephan R, Kaever V, Eberl L, Lehner A (2016) The DSF type quorum sensing signalling system RpfF/R regulates diverse phenotypes in the opportunistic pathogen *Cronobacter*. Sci Rep 6:18753. https://doi.org/10.1038/srep18753

Tang JL, Liu YN, Barber CE, Dow JM, Wootton JC, Daniels MJ (1991) Genetic and molecular analysis of a cluster of rpf genes involved in positive regulation of synthesis of extracellular enzymes and polysaccharide in *Xanthomonas campestris* pathovar campestris. Mol Gen Genet 226:409–417

Tashiro Y, Sakai R, Toyofuku M, Sawada I, Nakajima-Kambe T, Uchiyama H, Nomura N (2009) Outer membrane machinery and alginate synthesis regulators control membrane vesicle production in *Pseudomonas aeruginosa*. J Bacteriol 191:7509–7519

Tashiro Y, Ichikawa S, Nakajima-Kambe T, Uchiyama H, Nomura N (2010a) *Pseudomonas* quinolone signal affects membrane vesicle production in not only Gram-negative but also Gram-positive bacteria. Microbes Environ 25:120–125

Tashiro Y et al (2010b) *Variation of physioche*mical properties and cell association activity of membrane vesicles with growth phase in *Pseudomonas aeruginosa*. Appl Environ Microbiol 76:3732–3739

Tashiro Y, Toyofuku M, Nakajima-Kambe T, Uchiyama H, Nomura N (2010c) Bicyclic compounds repress membrane vesicle production and *Pseudomonas* quinolone signal synthesis in *Pseudomonas aeruginosa*. FEMS Microbiol Lett 304:123–130

Tashiro Y, Uchiyama H, Nomura N (2012) Multifunctional membrane vesicles in *Pseudomonas aeruginosa*. Environ Microbiol 14:1349–1362

Tashiro Y, Hasegawa Y, Shintani M, Takaki K, Ohkuma M, Kimbara K, Futamata H (2017) Interaction of bacterial membrane vesicles with specific species and their potential for delivery to target cells. Front Microbiol 8:571. https://doi.org/10.3389/fmicb.2017.00571

Tashiro Y, Takaki K, Futamata H (2019) Targeted delivery using membrane vesicles in prokaryotes. Biophys Physicobiol 16:114–120

Toyofuku M, Nomura N, Kuno E, Tashiro Y, Nakajima T, Uchiyama H (2008) Influence of the *Pseudomonas* quinolone signal on denitrification in *Pseudomonas aeruginosa*. J Bacteriol 190:7947–7956

Toyofuku M, Nakajima-Kambe T, Uchiyama H, Nomura N (2010) The effect of a cell-to-cell communication molecule, *Pseudomonas* quinolone signal (PQS), produced by *P. aeruginosa* on other bacterial species. Microbes Environ 25:1–7

Toyofuku M, Zhou S, Sawada I, Takaya N, Uchiyama H, Nomura N (2014) Membrane vesicle formation is associated with pyocin production under denitrifying conditions in *Pseudomonas aeruginosa* PAO1. Environ Microbiol 16:2927–2938

Toyofuku M et al (2017a) Prophage-triggered membrane vesicle formation through peptidoglycan damage in *Bacillus subtilis*. Nat Commun 8:481. https://doi.org/10.1038/s41467-017-00492-w

Toyofuku M et al (2017b) Membrane vesicle-mediated bacterial communication. ISME J 11:1504–1509

Toyofuku M, Nomura N, Eberl L (2019) Types and origins of bacterial membrane vesicles. Nat Rev Microbiol 17:13–24

Tsatsaronis JA, Franch-Arroyo S, Resch U, Charpentier E (2018) Extracellular vesicle RNA: a universal mediator of microbial communication? Trends Microbiol 26:401–410

Turnbull L et al (2016) Explosive cell lysis as a mechanism for the biogenesis of bacterial membrane vesicles and biofilms. Nat Commun 7:11220. https://doi.org/10.1038/ncomms11220

Wade DS, Calfee MW, Rocha ER, Ling EA, Engstrom E, Coleman JP, Pesci EC (2005) Regulation of *Pseudomonas* quinolone signal synthesis in *Pseudomonas aeruginosa*. J Bacteriol 187:4372–4380

Wessel AK, Liew J, Kwon T, Marcotte EM, Whiteley M (2013) Role of *Pseudomonas aeruginosa* peptidoglycan-associated outer membrane proteins in vesicle formation. J Bacteriol 195:213–219

Whiteley M, Diggle SP, Greenberg EP (2017) Progress in and promise of bacterial quorum sensing research. Nature 551:313–320

Woznica A, Cantley AM, Beemelmanns C, Freinkman E, Clardy J, King N (2016) Bacterial lipids activate, synergize, and inhibit a developmental switch in choanoflagellates. Proc Natl Acad Sci U S A 113:7894–7899

Xiao G et al (2006) MvfR, a key *Pseudomonas aeruginosa* pathogenicity LTTR-class regulatory protein, has dual ligands. Mol Microbiol 62:1689–1699

Zhang YM, Frank MW, Zhu K, Mayasundari A, Rock CO (2008) PqsD is responsible for the synthesis of 2,4-dihydroxyquinoline, an extracellular metabolite produced by *Pseudomonas aeruginosa*. J Biol Chem 283:28788–28794

Zhou L, Wang XY, Sun S, Yang LC, Jiang BL, He YW (2015) Identification and characterization of naturally occurring DSF-family quorum sensing signal turnover system in the phytopathogen *Xanthomonas*. Environ Microbiol 17:4646–4658

# Chapter 6
# Membrane Vesicles from Plant Pathogenic Bacteria and Their Roles During Plant–Pathogen Interactions

**Ofir Bahar**

**Abstract** Membrane vesicle (MV) release occurs in all forms of life, including Gram-negative and Gram-positive bacteria. Bacterial MVs have been studied mostly in relation to the bacterial lifestyle and regarding their role during mammalian host interactions. Surprisingly, while plants are known to be colonized by pathogenic, mutualistic, and commensal bacteria, the functions of MVs produced by these plant colonizers have only begun to be studied in the past decade. In fact, only a handful of studies have been published on this topic. Nevertheless, it is apparent that this field is gaining increasing attention, as does the role of plant and fungal extracellular vesicles (EVs) during plant–pathogen interactions. In this chapter I will review the current literature on plant-associated bacterial MVs and their interactions with plants. I will focus on MV cargo with emphasis on virulence-related proteins and on MVs' function during host colonization including interactions with the plant immune system. I will further provide a view of the possible, yet unexplored, roles of MVs in plant–bacteria interactions, and highlight important questions and limitations in the study of MVs.

## 6.1 Background

Extracellular bacterial membrane vesicles (MVs) are spherical nanostructures originating from the cell envelope and released into the extracellular environment. The initial bulging of MVs can be observed using transmission electron microscopy, as small blebs projecting from the outer membrane (OM) along the cell periphery (Fig. 6.1). One of the first descriptions of bacterial MVs was over 50 years ago by Knox et al. (1966), yet, the following three decades saw only a handful of publications that further developed this topic. The slight interest in MVs in the first decades can be partly explained by the common conception of MVs being an artifact of cell growth or a result of cell breakage and death (Haurat et al. 2014). However, in recent

O. Bahar (✉)
Department of Plant Pathology and Weed Research, Agricultural Research Organization, Volcani Center, Rishon LeZion, Israel
e-mail: ofirb@agri.gov.il



M. Kaparakis-Liaskos, T. A. Kufer (eds.), *Bacterial Membrane Vesicles*,
https://doi.org/10.1007/978-3-030-36331-4_6

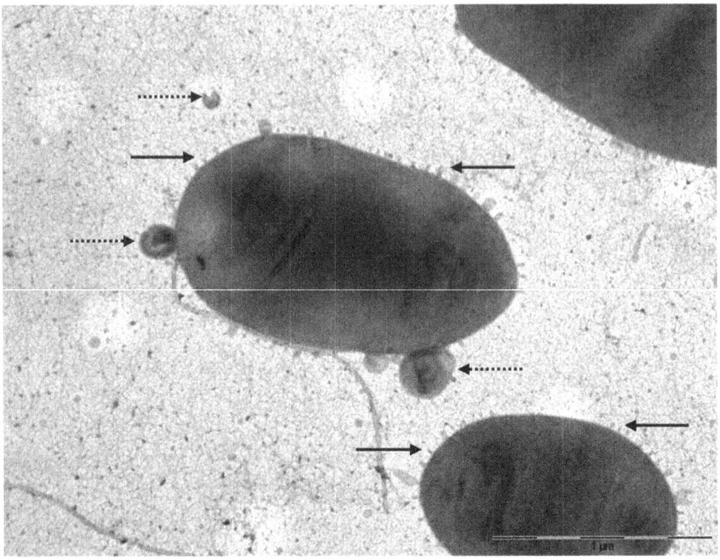

**Fig. 6.1** Membrane vesicle formation in the plant pathogen *Xanthomonas campestris* pv. *campestris* (*Xcc*). A *Xcc* (strain 33913) culture grown on nutrient agar plate for 48 h was washed off the plate, diluted and negatively stained using 4% uranyl acetate on Cu-400FC grids. Specimens were analyzed using a JEM-1400Plus Transmission Electron Microscope at 40 K magnification. In the micrograph, few *Xcc* cells are shown with small membrane blebs forming along the margins of the cell (solid-line arrow). A few larger membrane vesicles, that appear to have already formed and dissociate from the cells are also seen (dashed-line arrows). Size bar indicates 1 μm

years, there is a significant increase in the number of publications on extracellular vesicles (EVs) in general and on bacterial MVs specifically, indicative of the growing interest the scientific community has in this field.

Similar to the progress in the general field of MVs, the first studies on MVs released by plant pathogenic bacteria were late to arrive and were published more than 40 years after the work of Knox et al. (1966). Here too, the scientific community was slow to take up this topic and very few research papers have since been published on MVs of plant-associated bacteria. Nevertheless, the recent 5 years saw a significant increase in published papers studying and reviewing the role, or involvement, of MVs of important plant pathogenic bacteria such as *Xanthomonas campestris* pv. *campestris* (Xcc), *X. euvesicatoria*, *Pseudomonas syringae* pv. *tomato* (Pst), and *Xylella fastidiosa*, in plant–microbe interactions (Bahar et al. 2016; Baldrich et al. 2019; Ionescu et al. 2014; Katsir and Bahar 2017; Mendes et al. 2016; Nascimento et al. 2016; Regente et al. 2017; Rutter and Innes 2017, 2018; Rybak and Robatzek 2019; Solé et al. 2015).

Bacterial MVs have been implicated in multiple functions such as virulence, host immune modulation, surface adherence and biofilm formation, cell–cell communication, genetic material transfer and more (Schwechheimer and Kuehn 2015) (see also Chaps. 1, 2 and 5). Owing to this wide range of activities, the cargo of MV is rich and diverse, containing membrane lipids, lipopolysaccharides, membrane proteins, soluble proteins, nucleic acids, and peptidoglycan. Most of our knowledge on the cargo and functions of bacterial MVs comes from studies involving mammalian pathogens. In this chapter, I will try to summarize the main studies and findings related to MV cargo and function in plant pathogenic bacteria. More specifically, this chapter will focus on proteomic studies of plant pathogenic bacterial MVs, the presence of virulence factors in MVs and their possible functions, and the role of bacterial MVs during host colonization including MV interactions with the plant immune system.

## 6.2 Characterization of the Molecular Cargo of Bacterial Plant Pathogens MVs

Bacterial MVs are not de novo synthesized and are rather formed from preexisting cell structures such as the outer membrane (OM). While, it has become widely accepted that the formation of MVs is not a random process, the mechanisms that govern MV cargo sorting have remained largely elusive (Haurat et al. 2014). Considering that cargo sorting into MVs is a regulated and deliberate process, it is still reasonable to assume that at least some of the molecular cargo associated with MVs is a result of their presence in the preformed structures from which the MV originates, i.e., the OM and the periplasmic space. With this assumption in mind, one of the challenges en route to understanding the role MVs play during plant colonization is to be able to distinguish MV molecules that have specific roles in planta from molecules that are merely associated with MVs.

One way to address this challenge would be to purify bacterial MVs that are formed during plant colonization and compare their molecular cargo with MVs of the same organism, that were produced in a rich artificial medium. Such an approach has been used to identify genes and/or proteins that are specifically expressed in planta using transcriptomic and proteomic approaches (Andrade et al. 2008; Jacobs et al. 2012). However, since the study of plant pathogenic bacterial MVs is only in its infancy, methods to purify MVs from infected plants have not yet been optimized and published. To overcome this limitation, attempts were made to purify and characterize MV proteins following growth in culture media that mimics the plant environment.

*Xcc* is a Gram-negative bacterium that belongs to the family Xanthomonadaceae and is the causal agent of "black rot" disease of crucifers. Similar to some mammalian bacterial pathogens, *Xcc* depends on a functional type 3 secretion system (T3SS) for pathogenicity (Ryan et al. 2011). To identify MV proteins that are more likely to have a role during plant colonization, Sidhu et al. (Sidhu et al. 2008) purified MVs from *Xcc* cultures grown in two different minimal media: M9 and XVM2 and characterized their protein cargo using liquid chromatography mass spectrometry

(LC-MS/MS). XVM2 media was shown in the past to induce the expression of T3S-genes, which are thought to be induced strictly during plant colonization, and hence could serve as a proxy for the plant environment (Wengelnik and Bonas 1996). This first proteomic characterization of a plant pathogenic bacterium MVs revealed several interesting insights. First, the fact that certain proteins are enriched in the MV fraction compared with the OM fraction. This result suggests the existence of a protein sorting mechanism that directs proteins specifically to MVs and is also supported by previous studies with mammalian bacterial pathogens (Haurat et al. 2011). Second, that culture media affects the composition of MV proteins; and third, the association of virulence-related proteins with MVs. The association of virulence factors with MVs of plant pathogenic bacteria was also demonstrated for the tomato pathogen—*Pst* T1 strain (Chowdhury and Jagannadham 2013).

In the study by Sidhu et al. (2008), structural T3S-proteins, T3S-regulators, and T3S-effectors were found in association with secreted MVs. However, these structural proteins were not specifically expressed in the plant-mimicking media as they were also expressed in the control media, highlighting the limitation of this approach in finding plant-induced MV proteins. Additional virulence-related proteins that were found in the MV proteome included plant cell wall degrading enzyme such as cellulase and xylosidase (Sidhu et al. 2008). T3S-effectors such as HopI1 (suppressors of T3E-triggered death in *Nicotiana benthamiana*) and avrA1 were also identified in the MV fraction of in vitro-grown *Pst* (Chowdhury and Jagannadham 2013). Additional virulence-related proteins found in *Pst* MVs included hydrolytic enzymes such as chitinase and phytase. Here too, these virulence-related proteins were detected in MVs although bacteria were grown in a rich medium, which does not simulate the plant environment.

When attempting to identify MV proteins with specific role in planta, it is important to consider that the MV secretion pathway also serves as a disposal machinery for the cell. Hence, MVs could be associated with a variety of disposed proteins that do not serve a specific function in MVs (McBroom and Kuehn 2007; Schwechheimer and Kuehn 2013). This, obviously, further complicates the task of identifying proteins with strict function in MVs. With this in mind, even a successful proteomic analysis of in planta produced bacterial MVs would still be difficult to interpret and would not necessarily allow us to distinguish MV-functional and nonfunctional proteins.

These two studies (Sidhu et al. 2008; Chowdhury and Jagannadham 2013) were the first proteomic analyses of plant pathogenic bacterial MVs. Interestingly, in both cases, a relatively low number of proteins was identified (30–40 proteins in Sidhu et al. 2008, 139 proteins in Chowdhury and Jagannadham 2013), compared with MV proteomic studies of other bacteria, such as *P. aeruginosa*, where several hundreds of proteins were identified by LC-MS/MS (Choi et al. 2011). The isolation of MVs is a step of the utmost importance in the study of MV function and cargo and should therefore be given careful attention. In a position statement by the International Society for Extracellular Vesicles, Lötvall et al. (2014) discusses important caveats and recommendations in eukaryotic extracellular vesicle (EV) isolation. These recommendations should also be adopted in MVs research in the plant–microbe interactions community, to standardize, where possible, the methodology and quality of MVs isolation.

Additional studies that describe the association of plant pathogens virulence factors with MVs include the study of Solé et al. (2015). Aiming to identify T2S-virulence factors from *X. euvesicatoria*, the authors found a predicted protease, a lipase, and two xylanases in *X. euvesicatoria* supernatants. Interestingly, while these enzymes were thought to be secreted by the T2SS; they were found in the extracellular milieu of *X. euvesicatoria* even in the absence of a functions T2SS (Solé et al. 2015). Immuno-gold labeling coupled with transmission electron microscopy observations revealed that these enzymes are present in MVs released by *X. euvesicatoria*. These results suggested that MVs could serve as an additional or alternative secretion pathway for T2S-enzymes by plant pathogens.

Interestingly, the secretion of cellulytic and hemicellulolytic enzymes via MVs was described nearly 40 years ago with the bacterium *Fibrobacter succinogenes* (Forsberg et al. 1981 (formerly *Bacteroides succinogenes*); Montgomery et al. 1988). The authors showed that this bacterium, which colonizes the rumen of cattle, when given cellulose as a carbohydrate source, releases much of its synthesized cellulytic enzymes to the medium. They further showed that over 50% of the cellulytic activity in the medium was associated with sedimentable subcellular MVs. The secreted MVs were observed adhering to cellulose and also free in the culture and exhibited endoglucanase, xylanase, and cellulase activities. Considering that plant pathogenic bacterial MVs are associated with cellulytic enzymes, these results may suggest that MV released during plant colonization may facilitate plant cell wall decomposition. Further examples for secretion of virulence factors in association with MVs from plant pathogens include the secretion of the lipase/esterase LesA by *Xylella fastidiosa*. This lipase esterase is a homolog of the *X. euvesicatoria* lipase (69% identities over 99% coverage) and was also shown to be associated with *Xylella* MVs and to contribute to Pierce's Disease symptoms and to *X. fastidiosa* virulence (Nascimento et al. 2016). In summary, virulence-related proteins are associated with bacterial plant pathogen MVs, as they are with mammalian bacteria pathogens MVs. Nevertheless, studying the function of these MVs-enclosed, or associated virulence factors during plant colonization will be a feat much more challenging to achieve and will most definitely be occupying this field in the near future.

## 6.3 Functions of MVs During Plant Colonization

Very few studies have been conducted thus far to examine the role bacterial MVs play during plant colonization. Interesting insights were gained from the work of Ionescu et al. (2014) that investigated the role of *Xylella fastidiosa* MVs during xylem vessel colonization. *X. fastidiosa* is a most serious crop-threatening pathogen (Mansfield et al. 2012) known to infect more than 300 different plant species (EFSA

Panel on Plant Health 2015). It is well-known for causing Pierce's Disease in grapevines, citrus variegated chlorosis, and more recently, olive quick decline syndrome in southern Italy (Almeida and Nunney 2015). *X. fastidiosa* is a Gram-negative bacterium transmitted to plants by insect vectors. In the plant, *X. fastidiosa* resides exclusively in the water conducting elements of plants, the xylem, hence the name *Xylella*.

Beautiful scanning electron microscopy images show that *X. fastidiosa* is a potent producer of MVs in vitro (Ionescu et al. 2014). The authors further showed that a cell–cell signaling system, mediated by a diffusible signal factor (DSF), significantly influences the levels of MV release. Interestingly, knocking out the gene responsible for DSF synthesis (regulation of pathogenicity factors, *rpfF*) and abolishing its cell–cell signaling function, resulted in enhanced release of MVs compared with the wild-type strain. To monitor MV production in grapevines, Ionescu et al. (Ionescu et al. 2014) first collected xylem fluids from *X. fastidiosa*-infected and healthy grapevine plants. They then used the XadA as an MV protein marker and nanoparticle tracking analysis to monitor the presence of *X. fastidiosa* MVs. With both approaches the authors were able to show that MVs are released by *X. fastidiosa* during xylem colonization. Moreover, similarly to in vitro conditions, the mutation in the *rpfF* gene led to higher levels of MVs in the xylem sap. As for the function of MVs, Ionescu et al. (2014) found that *X. fastidiosa* MVs act as an anti-adhesive extracellular factors, limiting the adherence of *X. fastidiosa* cells to a glass slide. The adherence of *X. fastidiosa* was also impaired in the presence of MVs when tested in a microfluidic device and in grapevine stems. These results led the authors to suggest a model whereby MVs regulate the transition of *X. fastidiosa* from an aggregated and sticky form to a free-swimming form, which supports bacterial spread through the plant and virulence (Ionescu et al. 2014). One mechanism suggested to explain these results was that binding of MVs to xylem cell walls could restrict the binding of *X. fastidiosa* cells, thereby limiting the number of attachment sites and driving the bacteria more into the free-swimming form over the surface adherent and aggregated form.

How exactly MVs bind to surfaces and block *X. fastidiosa* binding is not clear. Cell surface appendages, like type I pili in *X. fastidiosa*, and other surface molecules like lipopolysaccharides (LPS) were shown to facilitate bacterial cell adherence and biofilm formation (Abu-Lail and Camesano 2003; De La Fuente et al. 2007). Since MVs are basically a microcosm of the bacterial cell wall and carry many of these surface molecules, it is possible that their presence on the MV surface can facilitate MV binding despite being disconnected from the bacterial cell.

Another interesting aspect in relation to *X. fastidiosa* MVs is the theoretical ability to use MVs as a vehicle to transport the hydrophobic cell–cell signaling molecule DSF. Previous studies have shown that *P. aeruginosa* exploits MVs to carry and deliver cell–cell communication molecules of a hydrophobic nature, such as the *Pseudomonas* quinolone signal (PQS) (Mashburn and Whiteley 2005). Hence, it is tempting to speculate that *X. fastidiosa*, and other Xanthomonas in general, may utilize MVs to mediate DSF cell-to-cell signaling. This speculation was recently supported by the work of Feitosa-Junior et al. (2019), which showed that DSF molecules are associated with purified MVs from *X. fastidiosa* cultures.

## 6.4 Bacterial MVs and the Plant Immune System

Bacterial MVs are known immune modulators in mammalian systems. The presence of endotoxin and other bacterial surface molecules in MVs render them carriers of immunogenic material, which interacts with the host immune system (Kaparakis-Liaskos and Ferrero 2015). In mammals, both LPS and the protein cargo of MVs have been shown to induce immune responses (Ellis et al. 2010).

Plants possesses a similar innate immune system to that of mammals, composed of surface receptors that interact with conserved microbial determinants and mediate the elicitation of an immune response (Ronald and Beutler 2010). These receptors are termed pattern recognition receptors (PRRs) and their respective microbial elicitors are customarily termed microbe- or pathogen-associated molecular patterns (MAMPs/PAMPs).

Therefore, it was not surprising that MVs purified from plant pathogenic bacteria interacted with the plant immune system (Bahar et al. 2016). Challenging *Arabidopsis* seedlings with plant pathogenic bacterial MVs resulted in the production of a reactive oxygen species (ROS) burst and elevation in the expression of immune marker genes, both represent typical outputs of innate immune system activation (Bahar et al. 2016). These responses were shown to be partially mediated by membrane-bound co-receptors, such as brassinosteroid-insensitive 1-associated kinase (BAK1). BAK1 is a leucine-rich repeat (LRR)-receptor-like kinase (RLK), which forms a complex with multiple primary immune receptors immediately after ligand binding. These interactions stabilizes this immune complex and lead to auto- and transphosphorylation of the intracellular kinase domains of the primary and co-receptors, initiating downstream signaling (Schwessinger et al. 2011; Schwessinger and Rathjen 2015). The absence of a functional BAK1 co-receptor would therefore lead to the impairment of multiple primary immune receptors. Together with proteomic and biochemical studies of MVs, these results suggested that MVs carry multiple immune elicitors, which may interact simultaneously with multiple plant immune receptors. This hypothesis may also explain why single primary immune receptor knockouts in *Arabidopsis* plants do not show any reduction in the response to MVs, while knocking out co-receptors, which are important for the functionality of multiple PRRs, leads to a reduced immune response to MVs (Bahar et al. 2016).

Boiling and protease treatments applied to MVs did not appear to alter their immunogenic properties suggesting that similar to the mammalian immune system, both the protein and the nonprotein cargo of MVs elicit the plant immune response. Interestingly, one of the well-known bacterial MAMPs in plants, elongation factor—thermo unstable (EF-Tu) (Kunze et al. 2004) was found in association with MVs, hinting of its possible involvement in immune induction via MVs delivery (Bahar et al. 2016). While EF-Tu itself is heat unstable, its immunogenic properties in plants appeared not to be affected by boiling (Kunze et al. 2004). Since the immunogenic activity of EF-Tu does not depend on a functional protein, but rather on the conserved 18 amino acid epitope, elf18, it is not surprising that EF-Tu remained immunogenic despite the heat treatment. Immune marker activation assays, conducted with the *Arabidopsis* EF-Tu receptor (EFR) (Zipfel et al. 2006) knockout line, further supported the abovementioned notion that MVs induce plant immunity via multiple receptors, as the EFR knockout line and its wild parent responded similarly to MVs (Bahar et al. 2016).

Differently from mammalian hosts, plant cells possess a cell wall composed primarily of oligosaccharides such as cellulose, hemicellulose, and pectin. An intriguing question is how exactly do bacterial MAMPs come into contact with membrane-bound receptors? PRRs most often possess a LRR domain projecting from the plasma membrane into the cell wall. This protein domain is far shorter than the depth of the cell wall (0.2–1.0 μm) in which it is embedded. Hence, the LRR domain is not exposed to the extracellular space but rather engulfed by the cell wall matrix. The pore size of the sugar-based mesh of the cell wall was evaluated to be ~50 angstroms (Carpita et al. 1979). This is 1000-fold smaller than the diameter of a small bacterial MV, and ~10,000-fold smaller than the size of a bacterial cell. This indicates that intact bacterial cells or MVs most probably do not come into direct contact with the plant cell plasma membrane or the membrane-bound receptors as long as the plant cell wall is intact. One possible explanation of how bacterial immune elicitors come into contact with immune receptors is the lysis or breakdown of intact pathogen cells or of MVs into smaller pieces, which can diffuse through the cell wall pores. Another possibility is that plant cell wall degrading enzymes, such as those seen in association with MVs and discussed earlier in this chapter (i.e., cellulases and pectinases) facilitate the breaking down of the cell wall envelope, thereby exposing the plant cell plasma membrane and its embedded immune receptors to direct contact with bacterial cells or MVs. Further research will be needed in order to better understand the mechanisms by which bacterial MVs interact and activate the plant immune system.

## 6.5 Future Prospects and Major Questions

One of the most intriguing questions regarding the functions of MVs during plant colonization is whether MVs deliver specific cargo into host cells to facilitate infection. MV-mediated delivery of signal molecules, toxins, etc. is known to occur between bacterial cells (Li et al. 1998; Mashburn and Whiteley 2005) and mammalian host cells. These include the delivery of DNA, sensed by the mammalian toll-like receptor 9 (Laura et al. 2010), or the delivery of toxins to mammalian host cells to facilitate infection (Kesty et al. 2004). A critical question in this regard is whether and how MVs overcome the plant cell wall to interact with the plasma membrane. If indeed they do so, perhaps by the use of cell wall degrading enzymes as discussed above, is the interaction with the plant cell mediated by membrane receptors? Are the MVs endocytosed by plant cells, or are the MVs integrated into the plant cell plasma membrane?

Bacterial secretion systems have always drawn a lot of attention from the scientific community. Yet, the MVs secretion pathway in plant-associated microbes has barely been studied and most certainty holds many interesting discoveries yet to be made. Bacterial MVs have already been suggested to function as a complementary secretion system for T2S virulence factors. It would be very interesting to see whether MVs can also complement the functions of the T3SS, whose contribution to pathogenicity is paramount. First evidences for T3S-effectors in MV have already been given; however, the question of whether they serve a functional role in this form remains to be answered.

Other possible functions of MVs during plant colonization include the facilitation of long-distance cell–cell communication, quenching of antimicrobial molecules produced by the plant defense system, promotion of cell surface adherence and/or biofilm formation (or limiting it as seen with *X. fastidiosa)*, acquisition of nutrients, competition with other plant-associated microbes and more. Future research on this unique and scarcely explored secretion system will most likely help to shed more light on its functions during plant–microbe interactions.

**Acknowledgments** The author would like to thank Dr. Victor Gaba from ARO for critically reading the draft of this chapter and to Mr. Amit Lezmy-Reif from the Bahar laboratory at ARO for taking the TEM image presented in Fig. 6.1.

## References

Abu-Lail NI, Camesano TA (2003) Role of lipopolysaccharides in the adhesion, retention, and transport of *Escherichia coli* JM109. Environ Sci Technol 37:2173–2183

Almeida RPP, Nunney L (2015) How do plant diseases caused by *Xylella fastidiosa* emerge? Plant Dis 99:1457–1467

Andrade AE, Silva LP, Pereira JL, Noronha EF, Reis FB, Bloch C, dos Santos MF, Domont GB, Franco OL, Mehta A (2008) In vivo proteome analysis of *Xanthomonas campestris* pv. *campestris* in the interaction with the host plant *Brassica oleracea*. FEMS Microbiol Lett 281:167–174

Bahar O, Mordukhovich G, Luu DD, Schwessinger B, Daudi A, Jehle AK, Felix G, Ronald PC (2016) Bacterial outer membrane vesicles induce plant immune responses. Mol Plant-Microbe Interact 29:374–384

Baldrich P, Rutter BD, Karimi HZ, Podicheti R, Meyers BC, Innes RW (2019) Plant extracellular vesicles contain diverse small RNA species and are enriched in 10 to 17 nucleotide "tiny" RNAs. Plant Cell 31:315. https://doi.org/10.1105/tpc.18.00872

Carpita N, Sabularse D, Montezinos D, Delmer DP (1979) Determination of the pore size of cell walls of living plant cells. Science 205:1144–1147

Choi D-S, Kim D-K, Choi SJ, Lee J, Choi J-P, Rho S, Park S-H, Kim Y-K, Hwang D, Gho YS (2011) Proteomic analysis of outer membrane vesicles derived from *Pseudomonas aeruginosa*. Proteomics 11:3424–3429

Chowdhury C, Jagannadham MV (2013) Virulence factors are released in association with outer membrane vesicles of *Pseudomonas syringae* pv. *tomato* T1 during normal growth. Biochim Biophys Acta 1834:231–239

De La Fuente L, Burr T, Hoch H (2007) Mutations in type I and type IV pilus biosynthetic genes affect twitching motility rates in *Xylella fastidiosa*. J Bacteriol 189:7507–7510

Ellis TN, Leiman SA, Kuehn MJ (2010) Naturally produced outer membrane vesicles from *Pseudomonas aeruginosa* elicit a potent innate immune response via combined sensing of both lipopolysaccharide and protein components. Infect Immun 78:3822–3831

European Panel on Plant Health (PLH) (2015) Scientific opinion on the risk to plant health posed by *Xylella fastidiosa* in the EU territory, with the identification and evaluation of risk reduction options 1. EFSA J 13:3989. https://doi.org/10.2903/j.efsa.2004.50

Feitosa-Junior OR, Stefanello E, Zaini PA, Nascimento R, Pierry PM, Dandekar AM, Lindow SE, da Silva AM (2019) Proteomic and metabolomic analyses of OMV Enriched fractions reveal association with virulence factors and signaling molecules of the DSF family. Phytopathology 109(8):1344–1353

Forsberg CW, Beveridge TJ, Hellstrom A (1981) Cellulase and xylanase release from *Bacteroides succinogenes* and its importance in the rumen environment. Appl Environ Microbiol 42:886–896

Haurat MF, Aduse-Opoku J, Rangarajan M, Dorobantu L, Gray MR, Curtis M a, Feldman MF (2011) Selective sorting of cargo proteins into bacterial membrane vesicles. J Biol Chem 286:1269–1276

Haurat MF, Elhenawy W, Feldman MF (2014) Prokaryotic membrane vesicles: new insights on biogenesis and biological roles. Biol Chem 396:95–109

Ionescu M, Zaini PA, Baccari C, Tran S, da Silva AM, Lindow SE (2014) *Xylella fastidiosa* outer membrane vesicles modulate plant colonization by blocking attachment to surfaces. Proc Natl Acad Sci U S A 111:E3910–E3918

Jacobs JM, Babujee L, Meng F, Milling A, Allen C (2012) The in planta transcriptome of *Ralstonia solanacearum*: conserved physiological and virulence strategies during bacterial wilt of tomato. MBio 3:1–11

Kaparakis-Liaskos M, Ferrero RL (2015) Immune modulation by bacterial outer membrane vesicles. Nat Rev Immunol 15(6):375–387

Katsir L, Bahar O (2017) Bacterial outer membrane vesicles at the plant–pathogen interface. PLoS Pathog 13(6):e1006306. https://doi.org/10.1371/journal.ppat.1006306

Kesty NC, Mason KM, Reedy M, Miller SE, Kuehn MJ (2004) Enterotoxigenic *Escherichia coli* vesicles target toxin delivery into mammalian cells. EMBO J 23:4538–4549

Knox KW, Vesk M, Work E (1966) Relation between excreted lipopolysaccharide complexes and surface structures of a lysine-limited culture of *Escherichia coli*. J Bacteriol 92:1206–1217

Kunze G, Zipfel C, Robatzek S, Niehaus K, Boller T, Felix G (2004) The N terminus of bacterial elongation factor Tu elicits innate immunity in *Arabidopsis* plants. Plant Cell 16:3496–3507

Laura M, Vidakovics AP, Jendholm J, Mörgelin M, Månsson A, Larsson C, Cardell L-O, Riesbeck K (2010) B cell activation by outer membrane vesicles—a novel virulence mechanism. PLoS Pathog 6(1):e1000724. https://doi.org/10.1371/journal.ppat.1000724

Li Z, Clarke AJ, Beveridge TJ (1998) Gram-negative bacteria produce membrane vesicles which are capable of killing other bacteria. J Bacteriol 180:5478–5483

Lötvall J, Hill AF, Hochberg F, Buzás EI, Di Vizio D, Gardiner C, Gho YS, Kurochkin IV, Mathivanan S, Quesenberry P, Sahoo S, Tahara H, Wauben MH, Witwer KW, Théry C (2014) Minimal experimental requirements for definition of extracellular vesicles and their functions: a position statement from the International Society for Extracellular Vesicles. J Extracell Vesicles 3 (1):26913. https://doi.org/10.3402/jev.v3.26913

Mansfield J, Genin S, Magori S, Citovsky V, Sriariyanum M, Ronald P, Dow M, Verdier V, Beer SV, Machado MA, Toth I, Salmond G, Foster GD (2012) Top 10 plant pathogenic bacteria in molecular plant pathology. Mol Plant Pathol 13:614–629

Mashburn LM, Whiteley M (2005) Membrane vesicles traffic signals and facilitate group activities in a prokaryote. Nature 437:422–425

McBroom AJ, Kuehn MJ (2007) Release of outer membrane vesicles by Gram-negative bacteria is a novel envelope stress response. Mol Microbiol 63:545–558

Mendes JS, Santiago AS, Toledo MAS, Horta MAC, de Souza AA, Tasic L, de Souza AP (2016) In vitro determination of extracellular proteins from *Xylella fastidiosa*. Front Microbiol 7:2090. https://doi.org/10.3389/fmicb.2016.02090

Montgomery L, Flesher B, Stahl D (1988) Transfer of *Bacteroides succinogenes* (Hungate) to *Fibrobacter* gen. nov. as *Fibrobacter succinogenes* comb. nov. and description of *Fibrobacter intestinalis* sp. nov. Int J Syst Bacteriol 38:430–435

Nascimento R, Gouran H, Chakraborty S, Gillespie HW, Almeida-Souza HO, Tu A, Rao BJ, Feldstein PA, Bruening G, Goulart LR, Dandekar AM (2016) The type II secreted lipase/esterase LesA is a key virulence factor required for *Xylella fastidiosa* pathogenesis in grapevines. Sci Rep 6:18598. https://doi.org/10.1038/srep18598

Regente M, Pinedo M, San Clemente H, Balliau T, Jamet E, de la Canal L (2017) Plant extracellular vesicles are incorporated by a fungal pathogen and inhibit its growth. J Exp Bot 68:5485–5495

Ronald PC, Beutler B (2010) Plant and animal sensors of conserved microbial signatures. Science 330:1061–1064

Rutter BD, Innes RW (2017) Extracellular vesicles isolated from the leaf apoplast carry stress-response proteins. Plant Physiol 173:728–741

Rutter BD, Innes RW (2018) Extracellular vesicles as key mediators of plant–microbe interactions. Curr Opin Plant Biol 44:16–22

Ryan RP, Vorhölter F-J, Potnis N, Jones JB, Van Sluys M-A, Bogdanove AJ, Dow JM (2011) Pathogenomics of Xanthomonas: understanding bacterium–plant interactions. Nat Rev Microbiol 9(5):344–355

Rybak K, Robatzek S (2019) Functions of extracellular vesicles in immunity and virulence. Plant Physiol 179:1236. https://doi.org/10.1104/pp.18.01557

Schwechheimer C, Kuehn MJ (2013) Synthetic effect between envelope stress and lack of outer membrane vesicle production in *Escherichia coli*. J Bacteriol 195:4161–4173

Schwechheimer C, Kuehn MJ (2015) Outer-membrane vesicles from Gram-negative bacteria: biogenesis and functions. Nat Rev Microbiol 13:605–619

Schwessinger B, Rathjen JP (2015) Changing SERKs and priorities during plant life. Trends Plant Sci 20(9):531–533

Schwessinger B, Roux M, Kadota Y, Ntoukakis V, Sklenar J, Jones A, Zipfel C, Guttman DS (2011) Phosphorylation-dependent differential regulation of plant growth, cell death, and innate immunity by the regulatory receptor-like kinase BAK1. PLoS Genet 7(4):e1002046

Sidhu VK, Vorhölter F-J, Niehaus K, Watt SA (2008) Analysis of outer membrane vesicle associated proteins isolated from the plant pathogenic bacterium *Xanthomonas campestris* pv. *campestris*. BMC Microbiol 8:87. https://doi.org/10.1186/1471-2180-8-87

Solé M, Scheibner F, Hoffmeister A-K, Hartmann N, Hause G, Rother A, Jordan M, Lautier M, Arlat M, Büttner D (2015) *Xanthomonas campestris* pv. *vesicatoria* secretes proteases and xylanases via the Xps-type II secretion system and outer membrane vesicles. J Bacteriol 197:2879–2893

Wengelnik K, Bonas U (1996) HrpXv, an AraC-type regulator, activates expression of five of the six loci in the hrp cluster of *Xanthomonas campestris* pv. *vesicatoria*. J Bacteriol 178:3462–3469

Zipfel C, Kunze G, Chinchilla D, Caniard A, Jones JDG, Boller T, Felix G (2006) Perception of the bacterial PAMP EF-Tu by the receptor EFR restricts agrobacterium-mediated transformation. Cell 125(4):749–760

# Chapter 7
# Delivery of Virulence Factors by Bacterial Membrane Vesicles to Mammalian Host Cells

**Aftab Nadeem, Jan Oscarsson, and Sun Nyunt Wai**

**Abstract** Bacterial membrane vesicles represent a universal secretion mechanism enabling both Gram-negative and Gram-positive organisms to transfer cargo to eukaryotic cells, as well as to other bacterial cells. Bacterial vesicles can deliver to target cells an extremely wide range of virulence factors, including exotoxins, lipids, nucleic acids, and small molecules. Although there has been extensive research to decipher the mechanisms regulating cellular uptake of Gram-negative bacterial outer membrane vesicles (OMVs), much less is known about the cellular uptake of Gram-positive bacterial membrane vesicles (MVs). This chapter focuses on a selection of major bacterial pathogens and summarizes the present knowledge of OMV and MV-mediated virulence factor delivery, as well as mechanisms of bacterial vesicle–host cell interaction and uptake by mammalian cells.

## 7.1 Bacterial Membrane Vesicle-Mediated Protein Delivery

Bacterial pathogenicity is enhanced by secretion systems that export virulence factors, either by secretion or injection, into the environment or adjacent host cells. Once delivered, these virulence factors then interfere with or stimulate host cellular processes. Eight bacterial secretion systems designated types I–VIII have been characterized to date (Green and Mecsas 2016). Both Gram-negative and Gram-positive bacteria of several different bacterial species release membrane vesicles to augment their pathogenic potential. Release of bacterial membrane vesicles, a very basic and relevant mode of protein transport, presumably also occurs during infection. Compared to other secretion mechanisms, membrane vesicle release has special implications since vesicles can deliver cargo, e.g., virulence factors, over much

A. Nadeem · S. N. Wai (✉)
Department of Molecular Biology, Umeå University, Umeå, Sweden

The Umeå Centre for Microbial Research (UCMR), Umeå University, Umeå, Sweden
e-mail: sun.nyunt.wai@umu.se

J. Oscarsson
Oral Microbiology, Department of Odontology, Umeå University, Umeå, Sweden

 
M. Kaparakis-Liaskos, T. A. Kufer (eds.), *Bacterial Membrane Vesicles*,
https://doi.org/10.1007/978-3-030-36331-4_7

longer distances than secretory systems dependent upon direct bacterial contact. Effectively, these bacterial membrane vesicles become vehicles of multifunctional cargo, delivering a multitude of virulence factors, including metabolites, several protein toxins, nucleic acids, and immune modulators such as peptidoglycan (Berleman and Auer 2013).

### *7.1.1* Escherichia coli

Bacterial membrane vesicles were initially discovered as a product of Gram-negative bacterial outer membrane blebbing, and are therefore often referred to as outer membrane vesicles (OMVs). However, in 1976, Hoekstra et al. reported that membrane fragments, consisting of essentially unmodified outer membranes, were present in the culture supernatant of *E. coli* during normal growth (Hoekstra et al. 1976). Since then, a number of studies have described *E. coli* OMV biogenesis as well as the physiological cargo of these OMVs. When newly synthesized OMVs are released from *E. coli,* they contain active heat-labile enterotoxin (ELT) (Gankema et al. 1980; Wai et al. 2003). Because ELT is associated with lipopolysaccharide (LPS) on the OMV surface, host cell uptake of enterotoxigenic *E. coli* (ETEC) OMVs is allowed by binding to LT-receptor (GM1) (Horstman and Kuehn 2000, 2002; Kesty et al. 2004). The LeoA protein, a homolog of eukaryotic GTPase, secretes ELT from the periplasm of ETEC bacteria and it has been suggested that LeoA contributes to OMV formation and protein content (Brown and Hardwidge 2007).

Earlier studies demonstrating that a cytotoxic protein, cytolysin A (ClyA), in *E. coli* was exported through OMVs pointed to the potential physiological relevance of *E. coli* OMVs (Wai et al. 2003). In a process that involves redox-dependent oligomerization, ClyA is incorporated into OMVs, thus appearing to possess an intrinsic ability to translocate to the bacterial periplasm. ClyA incorporated into OMVs has considerably higher cytotoxicity toward mammalian cells compared to ClyA purified from the bacterial periplasm. Thus, protein localization in OMVs may play a direct role in activating and delivering virulence effector proteins (Wai et al. 2003).

Additional studies in *E. coli* examining vesicle-mediated export of bacterial virulence factors revealed the capacity of OMVs to deliver toxigenic cargo. For example, during infection, *E. coli* OMVs may represent an alternative pathway to deliver type I-secreted alpha-hemolysin from bacteria to host cells (Balsalobre et al. 2006). Another study reported that OMVs are a vehicle for bacteria to transfer cytotoxic necrotizing factor-1 (CNF1) to the environment and to infected tissue (Kouokam et al. 2006). Similarly, Enterohemorrhagic *E. coli* (EHEC) use OMVs to release hemolysin toxin (EHEC-Hly), a typical repeats-in-toxin protein (RTX) that lyses host cells through a mechanism of pore formation (Aldick et al. 2009). The toxin can exist as free EHEC-Hly and as EHEC-Hly associated with OMVs, and both forms are released during EHEC growth. Free EHEC-Hly is lytic toward human endothelial cells, whereas OMV-associated EHEC-Hly is not lytic toward microvascular endothelial cells (HBMEC) and the colon epithelial cell line Caco-2, although it can trigger apoptosis

(Bielaszewska et al. 2013). Research into whether an MV-associated genotoxin from intestinal *E. coli* can promote cancer development revealed that *E. coli*-derived OMVs are readily internalized into target cells. Within these target cells, OMVs have the potential to induce oxidative stress, which would lead to DNA damage, replication, and aneuploidy in susceptible cells (Tyrer et al. 2014).

During hyper-biofilm formation, the *kil* gene, located in a three-gene cluster on the *E. coli* ColE1 plasmid, induces release of proteinous materials and aberrant OMVs into the extracellular environment (Nakao et al. 2018). A variety of pathogen-associated molecular pattern molecules are enriched in OMVs isolated from *E. coli*. These include LPS, lipoproteins, CpG DNA, flagellin, and peptidoglycan, most of which are Toll-like receptor (TLR) and nucleotide-binding and oligomerization domain (NOD) ligands (Ellis et al. 2010). Therefore, bacterial OMVs are capable of activating epithelial cells, endothelial cells, macrophages, and dendritic cells to release TNF-α, IL-1β, IL-6, and IL-8 (Bauman and Kuehn 2006; Bielaszewska et al. 2018; Canas et al. 2018; Lee et al. 2018). OMVs from pathogenic *E. coli* have also been known to cause sepsis-induced cardiac dysfunction, demonstrated both in vitro and in vivo (Svennerholm et al. 2017).

OMVs secreted by clinical isolates of EHEC O157 cause cell death by delivering into host cells a cocktail of virulence factors, such as Shiga toxin 2a (Stx2a), cytolethal distending toxin V (CdtV), EHEC hemolysin, and flagellin (Bielaszewska et al. 2017). Interestingly, OMVs from the nonpathogenic *E. coli* strain Nissle 1917 (EcN) can cause anti-inflammatory responses by reinforcing epithelial barrier integrity, thus affecting intestinal homeostasis (Alvarez et al. 2016; Behrouzi et al. 2018; Fabrega et al. 2017). In contrast, OMVs from the Nissle 1917 strain can also cause eukaryotic DNA double-stranded breaks (Canas et al. 2016). This strain harbors a cluster of genes that encode for proteins involved in the biosynthesis of hybrid non-ribosomal peptide-polyketide(s). It has been suggested that polyketides may be involved in inducing these eukaryotic DNA double-stranded breaks (Olier et al. 2012). In addition to host inflammatory responses, in colon cancer cells, OMVs from nonpathogenic commensal *E. coli* can induce epigenetic modifications (Vdovikova et al. 2018). Furthermore, OMVs from nonpathogenic *E. coli* can suppress the growth of established tumors as well as prevent tumor metastasis. These activities occur via an interferon-γ-mediated antitumor response, whereby OMVs deliver trypsin-sensitive surface proteins to the target cancer cells (Kim et al. 2017). Taken together, these studies demonstrate the enormous potential of bacterial OMVs from nonpathogenic *E. coli* as novel therapeutic agents against various cancers.

### *7.1.2* Vibrio cholerae

The formation of OMVs by *Vibrio cholerae* and *Vibrio parahaemolyticus* was first observed by researchers analyzing the cell structure of *V. cholerae* and *V. parahaemolyticus* using electron microscopy with freeze-substitution (Kondo et al. 1993). *V. cholerae*, the causal agent of the diarrheal disease cholera, possesses cholera toxin (CT) as its major virulence factor. In addition to CT, many other secreted protein toxins and enzymes that are important to *V. cholerae* pathogenesis have been reported to be associated with OMVs. These include *Vibrio* cytolysin (VCC) (Olivier et al. 2007), metalloprotease of *Vibrio* (PrtV) (Vaitkevicius et al. 2006), Zn-dependent hemagglutinin protease (HAP) (Ghosh et al. 2006; Hase and Finkelstein 1991), accessory cholera enterotoxin (Ace) (Kaper et al. 1995), and trypsin-like serine protease (VesC) (Syngkon et al. 2010).

OMVs from the *V. cholerae* strain O395 secrete biologically active CT (Chatterjee and Chaudhuri 2011). Using a GM1-independent mechanism, CT-containing OMVs are trafficked to host cells. This GMl-independent mechanism represents a secondary mechanism for CT secretion, in addition to the well-studied type II secretion system (Chatterjee and Chaudhuri 2011). Future studies to elucidate the functional details of this secondary mechanism of CT delivery are important to fully understand *V. cholerae* pathogenesis (Rasti et al. 2018).

Non-O1 and non-O139 *V. cholerae* (NOVC) serogroups are the causal agents of gastroenteritis and extraintestinal infections in humans; however, the virulence of NOVC strains is not well understood. OMVs from NOVC strains elicit NOD1- and NOD2-mediated immune responses in mammalian hosts. Quorum-sensing machinery attenuates OMVs' inflammatory potential and thereby influences the immune responses (Bielig et al. 2011a, b).

Biologically active VCC, a pore-forming toxin, is released from the *V. cholerae* NOVC strain V:5/04 together with OMVs. OMV-associated VCC induces target cell autophagy, demonstrating that autophagy may play a role in cellular defense against an OMV-associated virulence factor (Elluri et al. 2014). The metalloprotease PrtV, a type II secretion system substrate protein, is also secreted from the *V. cholerae* strain C6706 together with OMVs (Rompikuntal et al. 2015). The biological activity of OMV-associated PrtV has been demonstrated in human colon carcinoma cells. Furthermore, the OMV-associated PrtV protease facilitates bacterial resistance toward the antimicrobial peptide LL-37 (Rompikuntal et al. 2015). HAP and VesC proteases are also released along with *V. cholerae* OMVs and the biologically active form of these proteases are delivered into human intestinal epithelial cells, causing cytotoxic and inflammatory responses (Mondal et al. 2016). Taken together, these studies demonstrate that the ability of *V. cholerae* to deliver virulence factors into host cells via OMV-mediated secretion is, therefore, a seemingly widespread feature among different *Vibrio* strains.

### *7.1.3* Pseudomonas aeruginosa

During normal growth, the Gram-negative bacterium *Pseudomonas aeruginosa* also releases OMVs. *Pseudomonas* OMVs carry and release several toxins and enzymes, including hemolysin, phospholipase C, alkaline phosphatase, protease, and elastase, which contribute to the organism's pathogenicity (Kadurugamuwa and Beveridge 1995). Peptidoglycan hydrolases associated with OMVs that are naturally released by several Gram-negative bacterial strains, including *Enterobacter*, *Citrobacter*, *Salmonella*, *Shigella*, *Escherichia*, *Klebsiella*, *Morganella*, *Pseudomonas*, and *Proteus*, enable the lysis of both Gram-negative and Gram-positive bacteria (Kadurugamuwa and Beveridge 1998; Li et al. 1998). This predatory interaction indicates that within biofilms, where bacteria compete for growth with other bacteria in the surrounding microflora, OMVs might play a fitness role, providing an increased survival benefit (Kadurugamuwa and Beveridge 1997). This hypothesis is supported by the finding that *P. aeruginosa* OMVs play an important role in the formation of biofilms (Beveridge et al. 1997; Murphy et al. 2014). Cif, the cystic fibrosis (CF) transmembrane conductance regulator (CFTR) inhibitory factor, is associated with *P. aeruginosa* PA14 OMVs (MacEachran et al. 2007). The zinc-dependent leucine aminopeptidase PaAP, an enzyme involved in bacterial association with host cells, is also found in OMVs from two *P. aeruginosa* clinical strains, PAO1 and CF2 (Bauman and Kuehn 2006). The major outer membrane proteins (OMPs) that are associated with OMVs have been identified by mass spectrometry to be OprE, OprF, OprG, OprH, OprI, PcoB, and PagL (Bauman and Kuehn 2006; Choi et al. 2017a; Tashiro et al. 2010). Previously, it was thought that OMVs contain only outer membrane and periplasmic proteins; however, proteomic analysis of *P. aeruginosa* revealed the possible presence of cytoplasmic proteins in naturally released OMVs (Choi et al. 2017a). *P. aeruginosa* primarily occupies the mucus layer of the lung epithelium in cystic fibrosis patients. Released OMVs can deliver virulence factors into the cytoplasm of host cells, resulting in modified innate immune responses. A recent study investigated whether the antibiotic tobramycin, which is commonly used to treat CF patient lung infections caused by *P. aeruginosa*, affects the abundance of virulence factors in OMVs. The study demonstrated that in CF patients, tobramycin may improve lung function by decreasing the abundance of several key virulence factors in OMVs, which restores chloride ion secretion necessary for bacterial clearance from the lungs (Koeppen et al. 2019). It is thus likely that OMVs carrying virulence factors are important contributors to the in vivo survival and adaptability of *P. aeruginosa* in CF lung infection.

### *7.1.4* Acinetobacter baumannii

*Acinetobacter baumannii* is an opportunistic pathogen responsible for a wide range of nosocomial infections. *A. baumannii* secretes OMVs that contain phospholipases and exhibit both hemolytic and leukocytic activities against target host cells (Jha et al. 2017). In a lung infection mouse model, surface proteins of *A. baumannii* OMVs can induce pro-inflammatory immune responses (Jun et al. 2013). Release of the elongation factor Tu (EF-Tu) from bacterial cells associated with OMVs from *A. baumannii* may contribute to fibronectin-mediated binding on the host cell (Dallo et al. 2012). Vaccinating a sepsis mouse model with OMVs purified from *A. baumannii* triggers high levels of IgM, IgG1, and IgG2c immunoglobulins, while levels of the pro-inflammatory cytokines IL-1β and IL-6 remain low (McConnell et al. 2011).

OMV vaccination protects mice against challenge with the *A. baumannii* ATCC 19606 strain (McConnell et al. 2011). Variation in OMV production is associated with a unique feature of *A. baumanii*, the reversible switching between formation of opaque and translucent colonies. In experiments assessing immune response in macrophages, OMVs from the *A. baumannii* opaque colony form appear to be more immunogenic than those from the translucent colony form (Ahmad et al. 2019). The rise in extensive antibiotic resistance to *A. baumannii* highlights the potential need for a vaccine against this organism (Li et al. 2006; Lei et al. 2019). Therefore, OMVs from *A. baumannii* represent a promising vaccine candidate due to its immunogenic properties.

### *7.1.5* Porphyromonas gingivalis

*Porphyromonas gingivalis* is a major pathogenic cause of adult periodontitis. The Gram-negative anaerobic bacterium *P. gingivalis* releases OMVs that contribute to pathogenesis due to their high proteolytic and hemagglutinating activities, as well as their ability to promote inter- and intra-bacterial species adherence (Olsen and Amano 2015). Multiple studies have shown that OMV-associated toxins and proteolytic enzymes have a major contribution to periodontal diseases (Bourgeau and Mayrand 1990; Duchesne et al. 1995; Ellen and Grove 1989; Kamaguchi et al. 2003; Patrick et al. 1996; Singh et al. 1989; Smalley et al. 1991). The specific OMV-associated virulence factor(s) involved in OMV-mediated pathogenesis of *P. gingivalis* are not well known despite the clinical importance of *P. gingivalis*. It is known that *P. gingivalis* produces gingipain proteinases that are preferentially packed into OMVs (Haurat et al. 2011; Veith et al. 2014). OMV-associated FimA, hemagglutinin A, and heat-stress protein (HtrA) are involved in the attachment of *P. gingivalis* to host cells and subsequent invasion (Belanger et al. 2012; Zhang et al. 2011). Also, major *P. gingivalis* outer membrane proteins are associated with OMVs, which are used to efficiently invade host cells (Ho et al. 2015; Mantri et al. 2015; Veith et al. 2014). Thus, it has been suggested that OMVs may be involved in the development of atherosclerosis and represent a "Trojan horse" strategy to cause an effect without employing intact bacterial cells (Xie 2015).

## 7.2 OMV-Mediated Virulence Factor Delivery by Other Gram-Negative Bacteria

OMVs from several different Gram-negative bacterial species effectively transport multifunctional cargo over long distances. For example, OMVs from *Aggregatibacter actinomycetemcomitans* carry proteins that function in antibiotic targeting, nutrient acquisition, and immune evasion, representing both offensive and defensive activities

(Kieselbach et al. 2015). This phenomenon was described for the first time in *Bordetella pertussis*, whose OMVs carry a virulence complex that includes an adhesin (Imagawa et al. 1979). A causal organism of acute respiratory tract infection, *B. pertussis* harbors a wide range of virulence factors, including pertussis toxin, filamentous hemagglutinin, adenylate cyclase hemolysin, and tracheal cytotoxin, which are secreted in association with OMVs (Hozbor et al. 1999). OMVs carrying adenylate cyclase toxin can induce murine macrophage and CHO-K1 cell death independent of the toxin's receptors (Donato et al. 2012). Immunization with *B. pertussis* OMVs may represent an effective next-generation pertussis vaccine strategy as evidenced by its ability to protect against bacterial colonization by eliciting antibody and Th1/Th17 type immune responses (Raeven et al. 2016).

One of the major virulence factors of *Campylobacter jejuni* and *A. actinomycetemcommitans*, cytolethal distending toxin (CDT), is secreted primarily from bacterial cells in association with OMVs, suggesting evolutionary conservation of this mode of CDT delivery (Berlanda Scorza et al. 2008; Lindmark et al. 2009; Rompikuntal et al. 2012). OMVs from *C. jejuni* carry three proteases, HtrA, Cj0511, and Cj1365c, and these OMV-associated proteases can cleave occludin and E-cadherin of T84 colon carcinoma cells (Elmi et al. 2016). In humans, *C. jejuni* can cause gastroenteritis, while in avian hosts, colonization is asymptomatic. The body temperature difference between human (37 °C) and avian (42 °C) hosts suggests that growth of *C. jejuni* at 37 °C potentially cues expression of bacterial virulence factors. Proteome analyses comparing OMVs from *C. jejuni* grown at 37 °C and at 42 °C revealed more virulence-related proteins associated with OMVs isolated from the bacteria grown at 37 °C (Taheri et al. 2018). The presence of bile in the growth medium also influences the selective packing of virulence factors in *C. jejuni* OMVs (Taheri et al. 2018), suggesting that the protein cargo of OMVs may also be regulated by the host environment.

The release and intracellular uptake of *Bacteroides fragilis* OMVs can activate caspase-11-dependent cell death and IL-1 responses to LPS (Vanaja et al. 2016). OMVs from *B. fragilis* carry polysaccharide A capsular antigen (PSA). PSA induces TLR2-mediated signaling in dendritic cells, which results in regulatory T cell maturation by production of the immunoregulatory cytokine IL-10 (Shen et al. 2012), thus implicating OMVs as an important mediator in establishing mutualism.

Bacteria employ various secretion systems to deliver virulence factors to target cells. The field of bacterial OMVs has become an exciting research area that is poised to improve our understanding of bacterial pathogenesis and provide alternative strategies to control infectious disease. Further investigation into the mechanisms and roles of OMV secretion systems may uncover novel targets and strategies for developing new antimicrobial therapies.

## 7.3 Delivery of Bacterial Nucleic Acids by OMVs

### 7.3.1 DNA

In 1989, *Neisseria gonorrhoeae* was shown to release RNA and DNA in association with OMVs (Dorward and Garon 1989). It was suggested that the RNA and linear DNA are associated with the exterior of the vesicles because nuclease treatment eliminated them from OMV preparations. However, circular DNA inside the OMVs, and thus resistant to nuclease treatment, was capable of transforming recipient cells. Hence, it was suggested that OMVs can act as a mechanism by which cells can exchange genetic information (Dorward et al. 1989). Since the first description in 1989, an increasing number of reports have described the OMV-associated release of plasmid DNA and/or chromosomal DNA (Biller et al. 2014; Lee et al. 2007; Perez-Cruz et al. 2015; Renelli et al. 2004; Yaron et al. 2000). DNA purified from *E. coli* O157:H7 OMVs contain the virulence genes *stx1*, *stx2*, *eae*, and *uidA* (Kolling and Matthews 1999). Among bacteria, OMVs can contribute to antibiotic resistance spread by two different mechanisms (Ciofu et al. 2000; Mashburn-Warren and Whiteley 2006). Spread of antibiotic resistance may occur by direct transfer of a resistance protein (e.g., β-lactamase) to neighboring cells or by lateral transfer of resistance genes following fusion of the OMV with the recipient cell membrane (Fulsundar et al. 2014; Renelli et al. 2004; Rumbo et al. 2011; Yaron et al. 2000). In the case of *P. aeruginosa*, it has been suggested that OMVs can also be formed after cell lysis, when membrane fragments and cytosolic contents including DNA are released from spontaneously lysed bacteria (Turnbull et al. 2016). DNA associated with OMVs contributes to establishing bacterial biofilms to facilitate bacterial host colonization (Liao et al. 2014). OMVs can deliver DNA into eukaryotic cells, suggesting a role for bacteria–host cell interactions and demonstrating potential for OMV-based DNA vaccines (Bitto et al. 2017). Interestingly, integration of bacterial DNA has been detected in the host genome, suggesting transfer of bacterial genetic material into human somatic cells (Riley et al. 2013). It remains to be determined, however, if OMV-associated DNA integrates into the host genome. Further studies are needed to investigate whether OMVs are capable of delivering DNA into the host nucleus as well as whether OMV-delivered DNA can integrate into the host genome or modulate the innate immune response via DNA sensors (Hornung 2014). Earlier studies indicated that among similar bacterial species, i.e., *N. gonorrheae*, it may be possible for genetic material to be transferred by OMVs (Dorward and Garon 1989). Moreover, OMVs can transfer carbapenem-resistance genes to surrounding *A. baumannii* bacterial isolates (Rumbo et al. 2011). Research on Gram-positive bacterial membrane vesicles (MVs) has shown that *Clostridium perfringens* releases MV-containing DNA and protein components that can be internalized by macrophages and induce dramatic pro-inflammatory cytokines both in vitro and in vivo (Jiang et al. 2014). In addition to providing a potential mechanism for exchange of genetic material between prokaryotes and eukaryotes, these studies reveal a new perspective on the immunogenic properties of bacterial OMV and MV-based DNA vaccines.

### *7.3.2 RNA*

Bacterial membrane vesicles share similarities with exomes, cell vesicles secreted by most mammalian cell types. Bacterial membrane vesicles and exosomes both carry payloads of proteins, lipids, and genetic material enclosed in membrane-bound spherical structures of similar size ranges. Both bacterial membrane vesicles and exosomes can deliver functional molecules to distant extracellular compartments and tissues. Exosomes are involved in the horizontal transfer of genetic material, such as mRNAs and miRNAs, from the donor cells to recipient cells (Valadi et al. 2007; Zhong et al. 2011; Zomer et al. 2010). Investigation of whether bacterial RNA associates with OMVs by us and other researchers revealed that RNA is indeed encapsulated inside the OMVs in the form of RNase-resistant secondary structures and/or is associated with proteins in RNase-stable complexes (Blenkiron et al. 2016; Choi et al. 2017a, b; Resch et al. 2016; Sjostrom et al. 2015). Emerging evidence indicates that OMVs contain short RNAs (sRNAs) that are differentially packaged and have the potential to target the function and/or stability of host mRNA. Interestingly, via a regulatory OMV-associated sRNA, a new mechanism of pathogen–host interaction attenuates the innate immune response in human airway epithelial cells as well as in mouse lung. A specific bacterial sRNA (sRNA52320) is transferred from *P. aeruginosa* OMVs to host cells, where in human airway epithelial cells it attenuates OMV-stimulated IL-8 secretion, and in the lungs of a mouse model it attenuates keratinocyte-derived cytokine secretion and neutrophil recruitment (Koeppen et al. 2016).

Different classes of RNA are present in OMV-associated fractions of *Salmonella enterica* serovar Typhimurium and are exported. These include rRNAs, mRNAs, tRNAs, and other ncRNAs (Malabirade et al. 2018). However, RNA associated with OMVs is clearly different when the bacteria are grown under host-mimic cultural conditions in comparison with ordinary laboratory culture media. At least a fraction of the extracellular RNA associated with OMVs is present as full-length transcripts, indicating that OMVs can protect RNA and that this RNA might be functionally active (Malabirade et al. 2018). Export of full-length transcripts via OMVs opens the possibility of numerous functional implications for bacteria–bacteria and bacteria–host communication.

#### 7.3.2.1 How Does RNA Associate with OMVs?

Several possibilities for RNA association with OMVs have been suggested (Blenkiron et al. 2016). First, extracellular RNA released by general bacterial cell lysis may be tightly reassociated to the OMV surface after secretion from bacterial cells. Second, RNA incorporation into OMVs could occur through an active and selective mechanism. Third, RNA association with OMVs may merely represent nonspecific envelopment of RNA in the cytoplasm within vesicle blebs. Finally, the phenomenon may be due to RNA riding as passengers on OMV-bound proteins, as bacterial mRNAs are frequently found at the sites of their future protein products (Nevo-Dinur et al. 2012). There are indeed many mRNAs that encode many membrane proteins present in OMVs. These include mRNAs for *ompA*, *lpp*, and *tonB* in OMVs from uropathogenic *E. coli* and mRNAs for *ompU*, *ompA*, and *tolC* in OMVs from *V. cholerae* (Blenkiron et al. 2016; Sjostrom et al. 2015). The ability of OMVs to deliver their associated RNA cargo into host cells poses the interesting question of whether these RNAs can function as novel signaling molecules in bacteria–host interactions.

## 7.4 Bacterial Lipid Release in Association with OMVs

OMVs contain bacterial phospholipids such as phosphatidylglycerol, phosphatidylethanolamine, and cardiolipin. Moreover, the phospholipid composition of OMVs generally resembles that of the outer membrane (OM) from which they are derived (Horstman and Kuehn 2000). Phospholipid and fatty acid compositions between OMVs and the cellular OM of *E. coli* do not differ significantly (Hoekstra et al. 1976). However, the phospholipid head groups and acyl chains compositions between the bacterial OM and OMVs from *P. aeruginosa* are quite different. Therefore, the OMV membrane is considered rigid compared to the cellular OM of *P. aeruginosa*. Thus, it has been suggested that the OMV blebbing mechanism may not be conserved among Gram-negative bacteria (Tashiro et al. 2011). OMVs contain, in addition to phospholipids, abundant LPS, which normally comprise the majority of the OM outer leaflet. *P. aeruginosa* can express both a common antigen (A-band) and serotype-specific antigen (B-band) in the O-antigen portion of LPS. Thus, these OMVs are highly enriched in B-band LPS, in contrast to the lipid composition of the OM (Kadurugamuwa and Beveridge 1995). Based on these differences, B-band LPS has been proposed to sort into OMVs, similar to the sorting of LPS and proteins seen in *Porphorymonas gingivalis* (Haurat et al. 2011).

## 7.5 Small Molecule Delivery Via OMVs

In addition to lipids, proteins, and nucleic acids, small molecules associate with OMVs as well. *Pseudomonas putida* strains that are resistant to toluene produce more OMVs upon exposure to toluene and release toluene-enriched OMVs as a detoxification system (Kobayashi et al. 2000). In *P. aeruginosa*, PQS (*Pseudomonas* quinolone signal), a quorum-sensing molecule, associates with OMVs (Mashburn and Whiteley 2005). Because PQS is more hydrophobic than the quorum-sensing signal acylhomoserine-lactone of *P. aeruginosa*, it is concentrated in the OMV membrane where it can interact specifically with LPS. It appears that such interactions contribute to physically stimulate the formation of vesicles into which the PQS is subsequently packaged (Mashburn and Whiteley 2005).

Gram-negative bacteria can employ OMVs to deliver peptidoglycan to cytosolic nucleotide-binding oligomerization domain-containing protein 1 (NOD1) in host cells. OMVs (containing peptidoglycan) purified from *P. aeruginosa*, *H. pylori*, and *Neisseria gonorrheae* can upregulate NF-κB and NOD1-dependent responses in vitro (Irving et al. 2014; Kaparakis et al. 2010). Moreover, when administered to mice intragastrically, *H. pylori* OMVs trigger NOD1-dependent but TLR-independent innate and adaptive immune responses (Kaparakis et al. 2010). In mammalian cells, *V. cholerae* OMVs induce NOD1- and NOD2-mediated immune responses. Quorum-sensing machinery attenuates the inflammatory potential of OMVs, playing an important role in regulating this process during infection (Bielig et al. 2011a, b). In human embryonic kidney cells, *A. actinomycetemcomitans* OMVs strongly induce NOD1- and NOD2-dependent NF-κB activation. Moreover, in myeloid THP1 cells, NOD1, the primary sensor of peptidoglycan delivered by MVs, contributes to the overall inflammatory responses induced by the vesicles (Thay et al. 2014).

OMVs from *P. aeruginosa*, *Shigella flexneri*, and *Myxococcus xanthus* contain molecules with bacteriolytic properties (Evans et al. 2012; Kadurugamuwa and Beveridge 1995, 1997, 1999). Thus, OMVs are critical to intra- and inter-species communication, although in bacterial cell–cell interactions occurring via OMVs, the selectivity of the interaction between MVs and bacterial cells is not fully understood. Recently, employing OMVs isolated from the Enterobacterium *Buttiauxella agrestis*, OMVs selectively interacted with target bacterial cells (Tashiro et al. 2017). These results offer a new avenue by which particular bacterial species can be controlled using bacterial OMVs in microbial communities.

## 7.6 Gram-Positive Bacteria Membrane Vesicles (MVs)

MVs are also released by Gram-positive bacteria such as *Staphylococcus aureus* (Lee et al. 2009), *Enterococcus faecium* (Wagner et al. 2018), *Streptococcus pneumoniae* (Codemo et al. 2018), *Streptococcus pyogenes* (Resch et al. 2016), *Mycobacterium ulcerans* (Marsollier et al. 2007), *Bacillus anthracis* (Rivera et al. 2010), *Listeria monocytogenes* (Vdovikova et al. 2017), and *Lactobacillus* (Dean et al. 2019) (see Chap. 3). These MVs are released both in vivo and in vitro as spherical, bilayered structures with a diameter of approximately 20–150 nm (Gurung et al. 2011; Rivera et al. 2010; Vdovikova et al. 2017).

## 7.6.1 *Staphylococcal Species*

### 7.6.1.1 *Staphylococcus aureus*

*S. aureus* is responsible for a wide spectrum of human infections that range from superficial cutaneous infections to life-threatening bacteremia (Lowy 1998). MVs from *S. aureus* have been isolated and analyzed by mass spectrometry. Proteins identified include the IgG-binding protein, ferritin, ferrichrome-binding lipoprotein precursor, ABC transporter extracellular binding protein, β-lactamase, and membrane protein OxaA (Gurung et al. 2011). *S. aureus* MVs deliver protein A to host cells by interacting with host cell plasma membranes through a cholesterol-rich microdomain in the membrane (Gurung et al. 2011; Rivera et al. 2010). *S. aureus* α-toxin (α-hemolysin), a 33-kDa pore-forming protein, is also associated with MVs. *S. aureus* α-toxin can lyse a wide range of human cells and induce apoptosis in T cells (Berube and Bubeck Wardenburg 2013). *S. aureus* MVs containing α-toxin are cytotoxic to HeLa cells and induce erythrocyte lysis (Thay et al. 2013). *S. aureus* MVs have also been reported to contain δ-hemolysin (Hld), γ-hemolysin, leukocidin D, exfoliative toxin C, and exfoliative toxin A, identified by proteomic analysis (Jeon et al. 2016). Comparative proteomics identified a total of 131 and 617 proteins in MVs from *S. aureus* grown in Luria-Bertani and brain-heart infusion broths, respectively, suggesting that culture media components can influence MV protein composition (Askarian et al. 2018). A study of the roles of MVs in bacteria–host interactions led to the suggestion that during systemic infection, *S. aureus* MVs can influence bacteria–host interactions and that they provide protective immunity in murine infection models (Askarian et al. 2018).

### 7.6.1.2 *Staphylococcus haemolyticus*

*S. haemolyticus* is a skin commensal microorganism. *S. haemolyticus* nosocomial isolates are the most antibiotic-resistant members of the coagulase-negative Staphylococci. However, little is known about *S. haemolyticus* virulence factors. Potential virulence proteins associated with MVs have been compared to the *S. haemolyticus* total secretome. This comparison revealed that the cargo carried by MVs is enriched in proteins involved in adhesion, acquisition of iron, and antimicrobial resistance (Cavanagh et al. 2018).

## 7.6.2 *Streptococcal Species*

### 7.6.2.1 *Streptococcus pyogenes*

Comprehensive studies have been performed on MVs produced by the Gram-positive human pathogen *S. pyogenes*, the etiological agent of necrotizing fasciitis and streptococcal toxic shock syndrome. These studies have provided an explanation for the MV-associated secretion of *S. pyogenes* macromolecules, including RNAs, lipids, and proteins, as well as described a two-component system that modulates *S. pyogenes* MV production (Resch et al. 2016; Biagini et al. 2015).

#### 7.6.2.2 *Streptococcus pneumoniae*

*S. pneumoniae*, a major Gram-positive respiratory pathogen, produces MVs that may serve as a vehicle for many bacterial proteins. Pneumolysin, a cytosolic pore-forming toxin, is significantly enriched in MVs (Codemo et al. 2018). Pneumococcal MVs are internalized into A549 lung epithelial cells and human monocyte-derived dendritic cells, where they trigger pro-inflammatory cytokine responses independent of pneumolysin content. It has been suggested that *S. pneumoniae* MVs act in an immunomodulatory manner by enabling transfer of vesicle-associated proteins and other macromolecules into host cells. In addition, MVs bind tightly to serum complement system components, sequestering complement factor C3 in human serum and decreasing pneumococcal opsonophagocytosis (Codemo et al. 2018).

### *7.6.3 Mycobacterial Species*

The etiologic agent of Buruli ulcers, the mycobacterium *Mycobacterium ulcerans* is slow-growing and infects the skin and subcutaneous tissues (George et al. 1999). Mycolactone, a poliketide-derived macrolide, is the only virulence factor known to be responsible for Buruli ulcers (George et al. 1999). MVs from *M. ulcerans* are cytotoxic to mouse macrophages because the vesicles contain mycolactone (Marsollier et al. 2007). MVs from *M. bovis* BCG and *M. tuberculosis* H37Rv are enriched in proteins associated with bacterial virulence, revealed by proteomic analysis. These proteins include a remarkable abundance of putative Toll-like receptor 2 (TLR2) ligands, such as 19 kDa *Mycobacterium* lipoproteins LpqH, LprA, and LprG. Interaction of MVs from either *M. bovis* BCG or *M. tuberculosis* H37Rv with murine macrophages induces TLR2-dependent cytokine and chemokine release. This evidence demonstrated that mycobacterial vesicles serve as a delivery mechanism for immunologically active molecules that contribute to the virulence of mycobacteria (Prados-Rosales et al. 2011). It was recently reported that the protein VirR (encoded by the gene rv0431) in *M. tuberculosis* (Mtb) regulates the amount of Mtb-derived MVs containing TLR2 ligands such as the lipoproteins LpqH and SodC, suggesting that VirR plays a role in immunomodulating properties of Mtb via MVs (Lee et al. 2013; Rath et al. 2013).

### *7.6.4* Enterococcus faecium

*E. faecium* is a commensal organism that is inherently resistant to several antimicrobial agents and can become a bacteremia-causing pathogen. Like other Gram-positive bacteria, *E. faecium* strains produce MVs (Gao et al. 2018). *E. faecium* MV-associated proteins include virulence factors, such as biofilm-promoting proteins, extracellular matrix-binding proteins, and antimicrobial resistance-related proteins, suggesting that *E. faecium* may utilize MVs to release proteins promoting virulence, pathogenicity, and antimicrobial resistance (Wagner et al. 2018).

### *7.6.5* Bacillus anthracis

*B. anthracis*, a spore-producing bacillus, causes anthrax in a range of vertebrates. *B. anthracis* releases vesicles that contain components of the anthrax toxins, the protective antigen (PA), lethal factor (LF), and edema toxin (ET), as well as anthrolysin (Rivera et al. 2010). Immunizing mice with *B. anthracis* MVs protects them against subsequent challenge with *B. anthracis* (Marsollier et al. 2007).

### *7.6.6* Listeria monocytogenes

*L. monocytogenes*, a Gram-positive pathogen, causes listeriosis, an illness transmitted through the consumption of contaminated food. Similar to other Gram-positive bacteria, MVs are released by *L. monocytogenes* in a process that is regulated by the general stress transcription factor σB (Lee et al. 2013). Internalin B (InlB), which is responsible for *L. monocytogenes* entry into target cells, and listeriolysin O (LLO), a pore-forming toxin, were identified in MVs from *L. monocytogenes* (Lee et al. 2013). In a detailed study, Vdovikova et al. demonstrated that *L. monocytogenes* produces MVs both in vitro and in vivo (Vdovikova et al. 2017). The pore-forming hemolysin LLO is a major virulence factor that is tightly associated with MVs in an oxidized, inactive form. Autophagy induced by pure LLO, by other bacterial pore-forming toxins or by Torin1-stimulated macroautophagy is effectively abrogated by MVs. Thus, it has been suggested that *L. monocytogenes* may survive inside host cells by controlling LLO activity and avoiding destruction from the autophagy system via intracellular release of MVs (Vdovikova et al. 2017).

### 7.6.7 Lactobacillus

MVs from three different *Lactobacillus* species have been characterized for their physiochemical properties and protein compositions. A recent study identified more than 80 protein components from *Lactobacillus*-derived MVs, including bacteriocin, which is enriched in MVs, suggesting that the vesicles serve as vehicles for delivery of the antimicrobial molecule (Dean et al. 2019). Collectively, these studies highlight the role of MVs in the pathogenesis of Gram-positive bacterial infections.

## 7.7 Entry and Trafficking of OMVs and MVs into Host Cells

Bacterial OMVs and MVs can enter host cells using various pathways, including clathrin- or caveolin-mediated pathways, or through fusion with plasma membranes (Bielaszewska et al. 2017; Mulcahy et al. 2014; Olofsson et al. 2014). Despite extensive research to understand the mechanisms that regulate cellular uptake of OMVs, little is known about the cellular uptake of MVs. Interestingly, Gram-negative (*V. cholerae*) and Gram-positive (*L. monocytogenes*) bacterial vesicles were efficiently internalized into the intracellular compartments of epithelial cells, which accumulated primarily in the lysosomal compartment of host epithelial cells (Fig. 7.1a and b). Importantly, a recent study showed that *E. coli* O157 (EHEC) OMVs are quickly internalized into intracellular compartments, where they deliver a cocktail of bacterial factors to different host cell compartments (Bielaszewska et al. 2017).

Endocytosis allows small molecules to traverse a cells membrane bilayer (Doherty and McMahon 2009). Host cells internalize OMVs and MVs from several bacteria mainly via various endocytic pathways. As described in recent reviews, endocytosis involves cell membrane invagination, and occurs through several different pathways that depend on the composition and cargo of the OMVs to be internalized (Bitto and Kaparakis-Liaskos 2017; Kaparakis-Liaskos and Ferrero 2015; Pathirana and Kaparakis-Liaskos 2016). Three primary cellular mechanisms regulating the cellular uptake of OMVs and MVs are: (i) clathrin-mediated endocytosis, (ii) cholesterol-enriched microdomains, also known as caveolae or lipid rafts, and (iii) F-actin-coated vacuoles, also known as macropinocytosis and phagocytosis (Table 7.1). These pathways produce endosomal compartments that allow cargo transfer to various subcellular sites in the host cell cytoplasm (Doherty and McMahon 2009). Furthermore, the size of OMVs has been recently shown to play an important role in the preferred mode of entry into host cells (Turner et al. 2018).

OMVs from several microorganisms, *H. pylori*, *A. actinomycetemcomitans*, *enterohemorrhagic E. coli* (EHEC), *enteroaggregative E. coli* (EAEC), and *Brucella abortus*, use clathrin-mediated endocytosis as their major mode of entry into host cells (Bielaszewska et al. 2013; Canas et al. 2016; O'Donoghue and Krachler 2016; Olofsson et al. 2014; Pollak et al. 2012; Thay et al. 2014). In addition to clathrin-

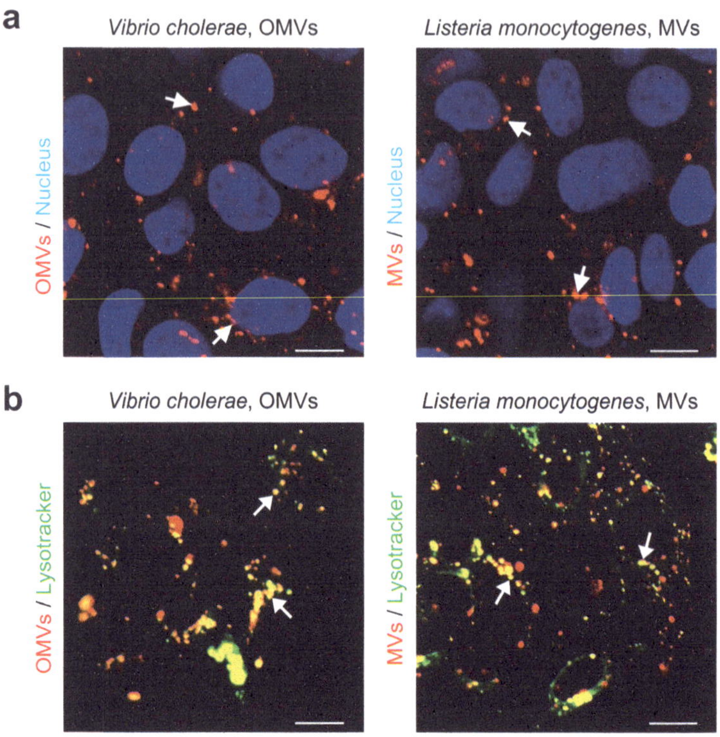


**Fig. 7.1** Lysosomal accumulation of bacterial OMVs and MVs. (**a**) Cellular uptake of PKH2-labeled OMVs (Red) isolated from *V. cholerae* or MVs from *L. monocytogenes*. Arrow head indicates their vesicular uptake into HCT8 cells. Nucleus is counter-stained with Hoechst 33342. Scale bars = 10 μm. (**b**) Co-localization of OMVs or MVs (red) with lysosomal marker, Lysotracker (green). Arrow head indicates co-localized spots, seen as yellow in the cytoplasm of HCT8 cells. Scale bars = 10 μm

mediated endocytosis, it has been proposed that dynamin also plays an important role in host cell-mediated uptake of OMVs (Bielaszewska et al. 2017; Kunsmann et al. 2015).

Membrane microdomains, called lipid rafts, are dynamic and abundant in several types of lipids, including cholesterol and sphingolipids, and proteins such as caveolin. The importance of membrane cholesterol for delivery of OMV cargo to the intracellular compartment of host cells has been reported by several investigators. OMVs from *V. cholerae* (Mondal et al. 2016), *V. vulnificus* (Kim et al. 2010), *A. actinomycetemcomitans* (Rompikuntal et al. 2012), *ETEC* (Johnson et al. 2009), *H. influenzae* (Sharpe et al. 2011), *P. gingivalis* (Furuta et al. 2009), *Moraxella catharralis* (Schaar et al. 2011), *H. pylori* (Kaparakis et al. 2010; Olofsson et al. 2014), and *C. jejuni* (Elmi et al. 2012) rely on lipid rafts to mediate internalization by host cells via endocytosis. Bacteria shed OMVs and MVs of different sizes, with the larger OMVs being engulfed by the host cell through ruffled cell membrane

**Table 7.1** Cellular mechanisms regulating host cell association and uptake of bacterial vesicles

| Mechanisms involved in vesicle uptake | Bacterial strain | References |
|---|---|---|
| Clathrin | | |
| | *H. pylori* | Olofsson et al. (2014), Turner et al. (2018) |
| | *A. actinomycetemcomitans* | Thay et al. (2014) |
| | EHEC | Bielaszewska et al. (2013) |
| | EAEC | Canas et al. (2016) |
| | *B. abortus* | Pollak et al. (2012) |
| Lipid raft | | |
| | *V. cholerae* | Mondal et al. (2016) |
| | *V. vulnificus* | Kim et al. (2010) |
| | *A. actinomycetemcomitans* | Rompikuntal et al. (2012) |
| | *P. aeruginosa* | Bauman and Kuehn (2009) |
| | *H. influenzae* | Sharpe et al. (2011) |
| | *M. catarrhalls* | Vidakovics et al. (2010), Schaar et al. (2011) |
| | ETEC | Johnson et al. (2009) |
| | *C. jejuni* | Elmi et al. (2012) |
| | *H. pylori* | Olofsson et al. (2014), Kaparakis et al. (2010) |
| | *P. gingivals* | Furuta et al. (2009) |
| | *S. aureus* | Gurung et al. (2011) |
| Membrane fusion | | |
| | *P. aeruginosa* | Bomberger et al. (2009) |
| | *A. actinomycetemcomitans* | Rompikuntal et al. (2012) |
| | *S. aureus* | Thay et al. (2013) |
| | *L. pneumophila* | Jager et al. (2015), Galka et al. (2008) |
| Macropinocytosis | | |
| | *H. pylori* | Turner et al. (2018) |
| | *P. aeruginosa* | Bomberger et al. (2009) |

protrusions driven by actin polymerization (Kaparakis-Liaskos and Ferrero 2015). Inhibiting actin polymerization using cytochalasin D or wiskostatin decreases entry of *P. aeruginosa* OMVs into the epithelial cells lining the airway tract (Bomberger et al. 2009).

The different pathways involved in entry of vesicles into host cells have been elucidated through experiments employing a large repertoire of inhibitors/binders specific for different components of each pathway, e.g., dynasore for dynamin, chlorpromazine for clathrin, filipin III, wortmannin, nystatin for lipid rafts, and cytochalasin D for pinocytosis (Amano et al. 2010; Canas et al. 2016; O'Donoghue and Krachler 2016; Rompikuntal et al. 2012). Vesicles make ideal delivery vehicles due to their ability to enter eukaryotic host cells and transfer their cargo to intracellular compartments. In addition, both Gram-positive and Gram-negative bacteria benefit from using vesicles for pathogenesis, intracellular communication and regulating host immunity (Bitto and Kaparakis-Liaskos 2017). Thus, discovering strategies to block vesicle entry into host cells may inhibit membrane vesicle-mediated pathogenesis of bacterial infections.

Membrane fusion is a mechanism by which Gram-negative and Gram-positive bacterial vesicles are internalized into host cells. Membrane fusion enables vesicles to deliver multiple virulence factors directly and simultaneously into the host cell cytoplasm in a coordinated fashion. This phenomenon was first reported by Bomberger et al. (2009) who demonstrated delivery of β-lactamase, alkaline phosphatase, hemolytic phospholipase C, and Cif by *P. aeruginosa* OMVs into human airway epithelial cells. Membrane fusion between *P. aeruginosa* vesicles and epithelial cells appear to occur preferentially at lipid raft domains on target host cells. Concomitantly, using filipin III, which sequesters cholesterol and disrupts lipid rafts, the membrane fusion events can be eliminated (Bomberger et al. 2009). *A. actinomycetemcomitans* OMVs deliver cytolethal distending toxin (CDT) in its biologically active form, and other proteins, including OmpA, into HeLa cells and human gingival fibroblasts, respectively. The OMV-mediated delivery of these proteins occurs in a cholesterol-dependent manner (Rompikuntal et al. 2012). Membrane fusion as a mechanism to deliver virulence factors into host cells has also been observed with MVs from Gram-positive bacteria, i.e., delivery of α-toxin (Hla) by *S. aureus* MVs into HeLa cells that occurs in a cholesterol-dependent manner and triggers death of the host cell (Thay et al. 2013). Despite the utility of filipin III in studying the dependence on lipid rafts for fusion of bacterial membrane vesicles with host cell vesicles, there is a limitation in its use. Filipin III affects a major component of eukaryotic cell plasma membranes, thus its inhibition of membrane fusion may also extend to processes beyond lipid rafts. In a study of the interaction of *L. pneumophila* OMVs with model membranes, the membrane material of the MV became incorporated into liposomes composed of different eukaryotic phospholipids, revealing that MVs have an inherent tendency to fuse with eukaryotic membranes (Jager et al. 2015).

## 7.8 Conclusions

Taken together, the present literature provides ample evidence that OMVs and MVs are capable of employing multiple routes to enter mammalian host cells. Due to their small size, adhesive properties, immunomodulating activity, and ability to carry and deliver specific effectors into mammalian cells, membrane vesicles of bacterial pathogens are well-suited to contribute significantly in the host interaction. Bacterial membrane vesicles allow the extracellular dispersal of particular proteins, as part of complexes of proteins, as well as lipids that can function synergistically to activate different pathways, either toxic or protective, in the host. Further investigations to dissect mechanisms of vesicle adhesion and entry, vesicle trafficking, and vesicle-associated contents will provide a critical foundation for future exploitation of OMVs and MVs for medical use. To date, innovative approaches based on engineered bacterial membrane vesicles have shown great clinical potential, and progress is being made to gain further insight and "know how" in using bacterial membrane vesicle-based technologies to enhance global human health.

## References

Ahmad I, Karah N, Nadeem A, Wai SN, Uhlin BE (2019) Analysis of colony phase variation switch in *Acinetobacter baumannii* clinical isolates. PLoS One 14:e0210082

Aldick T, Bielaszewska M, Uhlin BE, Humpf HU, Wai SN, Karch H (2009) Vesicular stabilization and activity augmentation of enterohaemorrhagic *Escherichia coli* haemolysin. Mol Microbiol 71:1496–1508

Alvarez CS, Badia J, Bosch M, Gimenez R, Baldoma L (2016) Outer membrane vesicles and soluble factors released by probiotic *Escherichia coli* Nissle 1917 and commensal ECOR63 enhance barrier function by regulating expression of tight junction proteins in intestinal epithelial cells. Front Microbiol 7:1981

Amano A, Takeuchi H, Furuta N (2010) Outer membrane vesicles function as offensive weapons in host-parasite interactions. Microbes Infect 12:791–798

Askarian F, Lapek JD Jr, Dongre M, Tsai CM, Kumaraswamy M, Kousha A, Valderrama JA, Ludviksen JA, Cavanagh JP, Uchiyama S, Mollnes TE, Gonzalez DJ, Wai SN, Nizet V, Johannessen M (2018) *Staphylococcus aureus* membrane-derived vesicles promote bacterial virulence and confer protective immunity in murine infection models. Front Microbiol 9:262

Balsalobre C, Silvan JM, Berglund S, Mizunoe Y, Uhlin BE, Wai SN (2006) Release of the type I secreted alpha-haemolysin via outer membrane vesicles from *Escherichia coli*. Mol Microbiol 59:99–112

Bauman SJ, Kuehn MJ (2006) Purification of outer membrane vesicles from *Pseudomonas aeruginosa* and their activation of an IL-8 response. Microbes Infect 8:2400–2408

Bauman SJ, Kuehn MJ (2009) *Pseudomonas aeruginosa* vesicles associate with and are internalized by human lung epithelial cells. BMC Microbiol 9:26

Behrouzi A, Vaziri F, Riazi Rad F, Amanzadeh A, Fateh A, Moshiri A, Khatami S, Siadat SD (2018) Comparative study of pathogenic and non-pathogenic *Escherichia coli* outer membrane vesicles and prediction of host-interactions with TLR signaling pathways. BMC Res Notes 11:539

Belanger M, Kozarov E, Song H, Whitlock J, Progulske-Fox A (2012) Both the unique and repeat regions of the *Porphyromonas gingivalis* hemagglutin A are involved in adhesion and invasion of host cells. Anaerobe 18:128–134

Berlanda Scorza F, Doro F, Rodriguez-Ortega MJ, Stella M, Liberatori S, Taddei AR, Serino L, Gomes Moriel D, Nesta B, Fontana MR, Spagnuolo A, Pizza M, Norais N, Grandi G (2008) Proteomics characterization of outer membrane vesicles from the extraintestinal pathogenic *Escherichia coli* DeltatolR IHE3034 mutant. Mol Cell Proteomics 7:473–485

Berleman J, Auer M (2013) The role of bacterial outer membrane vesicles for intra- and interspecies delivery. Environ Microbiol 15:347–354

Berube BJ, Bubeck Wardenburg J (2013) *Staphylococcus aureus* alpha-toxin: nearly a century of intrigue. Toxins (Basel) 5:1140–1166

Beveridge TJ, Makin SA, Kadurugamuwa JL, Li Z (1997) Interactions between biofilms and the environment. FEMS Microbiol Rev 20(3–4):291–303

Biagini M, Garibaldi M, Aprea S, Pezzicoli A, Doro F, Becherelli M, Taddei AR, Tani C, Tavarini S, Mora M, Teti G, D'Oro U, Nuti S, Soriani M, Margarit I, Rappuoli R, Grandi G,

Norais N (2015) The human pathogen *Streptococcus pyogenes* releases lipoproteins as lipoprotein-rich membrane vesicles. Mol Cell Proteomics 14:2138–2149

Bielaszewska M, Ruter C, Kunsmann L, Greune L, Bauwens A, Zhang W, Kuczius T, Kim KS, Mellmann A, Schmidt MA, Karch H (2013) Enterohemorrhagic *Escherichia coli* hemolysin employs outer membrane vesicles to target mitochondria and cause endothelial and epithelial apoptosis. PLoS Pathog 9:e1003797

Bielaszewska M, Ruter C, Bauwens A, Greune L, Jarosch KA, Steil D, Zhang W, He X, Lloubes R, Fruth A, Kim KS, Schmidt MA, Dobrindt U, Mellmann A, Karch H (2017) Host cell interactions of outer membrane vesicle-associated virulence factors of enterohemorrhagic *Escherichia coli* O157: intracellular delivery, trafficking and mechanisms of cell injury. PLoS Pathog 13: e1006159

Bielaszewska M, Marejkova M, Bauwens A, Kunsmann-Prokscha L, Mellmann A, Karch H (2018) Enterohemorrhagic *Escherichia coli* O157 outer membrane vesicles induce interleukin 8 production in human intestinal epithelial cells by signaling via toll-like receptors TLR4 and TLR5 and activation of the nuclear factor NF-kappaB. Int J Med Microbiol 308:882–889

Bielig H, Dongre M, Zurek B, Wai SN, Kufer TA (2011a) A role for quorum sensing in regulating innate immune responses mediated by *Vibrio cholerae* outer membrane vesicles (OMVs). Gut Microbes 2:274–279

Bielig H, Rompikuntal PK, Dongre M, Zurek B, Lindmark B, Ramstedt M, Wai SN, Kufer TA (2011b) NOD-like receptor activation by outer membrane vesicles from *Vibrio cholerae* non-O1 non-O139 strains is modulated by the quorum-sensing regulator HapR. Infect Immun 79:1418–1427

Biller SJ, Schubotz F, Roggensack SE, Thompson AW, Summons RE, Chisholm SW (2014) Bacterial vesicles in marine ecosystems. Science 343:183–186

Bitto NJ, Kaparakis-Liaskos M (2017) The therapeutic benefit of bacterial membrane vesicles. Int J Mol Sci 18:E1287

Bitto NJ, Chapman R, Pidot S, Costin A, Lo C, Choi J, D'Cruze T, Reynolds EC, Dashper SG, Turnbull L, Whitchurch CB, Stinear TP, Stacey KJ, Ferrero RL (2017) Bacterial membrane vesicles transport their DNA cargo into host cells. Sci Rep 7:7072

Blenkiron C, Simonov D, Muthukaruppan A, Tsai P, Dauros P, Green S, Hong J, Print CG, Swift S, Phillips AR (2016) Uropathogenic *Escherichia coli* releases extracellular vesicles that are associated with RNA. PLoS One 11:e0160440

Bomberger JM, Maceachran DP, Coutermarsh BA, Ye S, O'Toole GA, Stanton BA (2009) Long-distance delivery of bacterial virulence factors by *Pseudomonas aeruginosa* outer membrane vesicles. PLoS Pathog 5:e1000382

Bourgeau G, Mayrand D (1990) Aggregation of Actinomyces strains by extracellular vesicles produced by *Bacteroides gingivalis*. Can J Microbiol 36:362–365

Brown EA, Hardwidge PR (2007) Biochemical characterization of the enterotoxigenic *Escherichia coli* LeoA protein. Microbiology 153:3776–3784

Canas MA, Gimenez R, Fabrega MJ, Toloza L, Baldoma L, Badia J (2016) Outer membrane vesicles from the probiotic *Escherichia coli* Nissle 1917 and the commensal ECOR12 enter intestinal epithelial cells via clathrin-dependent endocytosis and elicit differential effects on DNA damage. PLoS One 11:e0160374

Canas MA, Fabrega MJ, Gimenez R, Badia J, Baldoma L (2018) Outer membrane vesicles from probiotic and commensal *Escherichia coli* activate NOD1-mediated immune responses in intestinal epithelial cells. Front Microbiol 9:498

Cavanagh JP, Pain M, Askarian F, Bruun JA, Urbarova I, Wai SN, Schmidt F, Johannessen M (2018) Comparative exoproteome profiling of an invasive and a commensal *Staphylococcus haemolyticus* isolate. J Proteomics 195:33–40

Chatterjee D, Chaudhuri K (2011) Association of cholera toxin with *Vibrio cholerae* outer membrane vesicles which are internalized by human intestinal epithelial cells. FEBS Lett 585:1357–1362

Choi JW, Kim SC, Hong SH, Lee HJ (2017a) Secretable Small RNAs via outer membrane vesicles in periodontal pathogens. J Dent Res 96:458–466

Choi JW, Um JH, Cho JH, Lee HJ (2017b) Tiny RNAs and their voyage via extracellular vesicles: secretion of bacterial small RNA and eukaryotic microRNA. Exp Biol Med (Maywood) 242:1475–1481

Ciofu O, Beveridge TJ, Kadurugamuwa J, Walther-Rasmussen J, Hoiby N (2000) Chromosomal beta-lactamase is packaged into membrane vesicles and secreted from *Pseudomonas aeruginosa*. J Antimicrob Chemother 45:9–13

Codemo M, Muschiol S, Iovino F, Nannapaneni P, Plant L, Wai SN, Henriques-Normark B (2018) Immunomodulatory effects of pneumococcal extracellular vesicles on cellular and humoral host defenses. MBio 9:e00559–e00518

Dallo SF, Zhang B, Denno J, Hong S, Tsai A, Haskins W, Ye JY, Weitao T (2012) Association of *Acinetobacter baumannii* EF-Tu with cell surface, outer membrane vesicles, and fibronectin. Sci World J 2012:128705

Dean SN, Leary DH, Sullivan CJ, Oh E, Walper SA (2019) Isolation and characterization of lactobacillus-derived membrane vesicles. Sci Rep 9:877

Doherty GJ, McMahon HT (2009) Mechanisms of endocytosis. Annu Rev Biochem 78:857–902

Donato GM, Goldsmith CS, Paddock CD, Eby JC, Gray MC, Hewlett EL (2012) Delivery of *Bordetella pertussis* adenylate cyclase toxin to target cells via outer membrane vesicles. FEBS Lett 586:459–465

Dorward DW, Garon CF (1989) DNA-binding proteins in cells and membrane blebs of *Neisseria gonorrhoeae*. J Bacteriol 171:4196–4201

Dorward DW, Garon CF, Judd RC (1989) Export and intercellular transfer of DNA via membrane blebs of *Neisseria gonorrhoeae*. J Bacteriol 171:2499–2505

Duchesne P, Grenier D, Mayrand D (1995) Demonstration of adherence properties of *Porphyromonas gingivalis* outer membrane vesicles using a new microassay. Oral Microbiol Immunol 10:76–80

Ellen RP, Grove DA (1989) Bacteroides gingivalis vesicles bind to and aggregate *Actinomyces viscosus*. Infect Immun 57:1618–1620

Ellis TN, Leiman SA, Kuehn MJ (2010) Naturally produced outer membrane vesicles from *Pseudomonas aeruginosa* elicit a potent innate immune response via combined sensing of both lipopolysaccharide and protein components. Infect Immun 78:3822–3831

Elluri S, Enow C, Vdovikova S, Rompikuntal PK, Dongre M, Carlsson S, Pal A, Uhlin BE, Wai SN (2014) Outer membrane vesicles mediate transport of biologically active *Vibrio cholerae* cytolysin (VCC) from V. cholerae strains. PLoS One 9:e106731

Elmi A, Watson E, Sandu P, Gundogdu O, Mills DC, Inglis NF, Manson E, Imrie L, Bajaj-Elliott M, Wren BW, Smith DG, Dorrell N (2012) *Campylobacter jejuni* outer membrane vesicles play an important role in bacterial interactions with human intestinal epithelial cells. Infect Immun 80:4089–4098

Elmi A, Nasher F, Jagatia H, Gundogdu O, Bajaj-Elliott M, Wren B, Dorrell N (2016) *Campylobacter jejuni* outer membrane vesicle-associated proteolytic activity promotes bacterial invasion by mediating cleavage of intestinal epithelial cell E-cadherin and occludin. Cell Microbiol 18:561–572

Evans AG, Davey HM, Cookson A, Currinn H, Cooke-Fox G, Stanczyk PJ, Whitworth DE (2012) Predatory activity of *Myxococcus xanthus* outer-membrane vesicles and properties of their hydrolase cargo. Microbiology 158:2742–2752

Fabrega MJ, Rodriguez-Nogales A, Garrido-Mesa J, Algieri F, Badia J, Gimenez R, Galvez J, Baldoma L (2017) Intestinal anti-inflammatory effects of outer membrane vesicles from *Escherichia coli* Nissle 1917 in DSS-experimental colitis in mice. Front Microbiol 8:1274

Fulsundar S, Harms K, Flaten GE, Johnsen PJ, Chopade BA, Nielsen KM (2014) Gene transfer potential of outer membrane vesicles of *Acinetobacter baylyi* and effects of stress on vesiculation. Appl Environ Microbiol 80:3469–3483

Furuta N, Tsuda K, Omori H, Yoshimori T, Yoshimura F, Amano A (2009) *Porphyromonas gingivalis* outer membrane vesicles enter human epithelial cells via an endocytic pathway and are sorted to lysosomal compartments. Infect Immun 77:4187–4196

Galka F, Wai SN, Kusch H, Engelmann S, Hecker M, Schmeck B, Hippenstiel S, Uhlin BE, Steinert M (2008) Proteomic characterization of the whole secretome of *Legionella pneumophila* and functional analysis of outer membrane vesicles. Infect Immun 76(5):1825–1836

Gankema H, Wensink J, Guinee PA, Jansen WH, Witholt B (1980) Some characteristics of the outer membrane material released by growing enterotoxigenic *Escherichia coli*. Infect Immun 29:704–713

Gao W, Howden BP, Stinear TP (2018) Evolution of virulence in *Enterococcus faecium*, a hospital-adapted opportunistic pathogen. Curr Opin Microbiol 41:76–82

George KM, Chatterjee D, Gunawardana G, Welty D, Hayman J, Lee R, Small PL (1999) Mycolactone: a polyketide toxin from *Mycobacterium ulcerans* required for virulence. Science 283:854–857

Ghosh A, Saha DR, Hoque KM, Asakuna M, Yamasaki S, Koley H, Das SS, Chakrabarti MK, Pal A (2006) Enterotoxigenicity of mature 45-kilodalton and processed 35-kilodalton forms of hemagglutinin protease purified from a cholera toxin gene-negative *Vibrio cholerae* non-O1, non-O139 strain. Infect Immun 74:2937–2946

Green ER, Mecsas J (2016) Bacterial secretion systems: an overview. Microbiol Spectr 4

Gurung M, Moon DC, Choi CW, Lee JH, Bae YC, Kim J, Lee YC, Seol SY, Cho DT, Kim SI, Lee JC (2011) *Staphylococcus aureus* produces membrane-derived vesicles that induce host cell death. PLoS One 6:e27958

Hase CC, Finkelstein RA (1991) Cloning and nucleotide sequence of the *Vibrio cholerae* hemagglutinin/protease (HA/protease) gene and construction of an HA/protease-negative strain. J Bacteriol 173:3311–3317

Haurat MF, Aduse-Opoku J, Rangarajan M, Dorobantu L, Gray MR, Curtis MA, Feldman MF (2011) Selective sorting of cargo proteins into bacterial membrane vesicles. J Biol Chem 286:1269–1276

Ho MH, Chen CH, Goodwin JS, Wang BY, Xie H (2015) Functional advantages of *Porphyromonas gingivalis* vesicles. PLoS One 10:e0123448

Hoekstra D, van der Laan JW, de Leij L, Witholt B (1976) Release of outer membrane fragments from normally growing *Escherichia coli*. Biochim Biophys Acta 455:889–899

Hornung V (2014) SnapShot: nucleic acid immune sensors, part 1. Immunity 41:868, 868 e1

Horstman AL, Kuehn MJ (2000) Enterotoxigenic *Escherichia coli* secretes active heat-labile enterotoxin via outer membrane vesicles. J Biol Chem 275:12489–12496

Horstman AL, Kuehn MJ (2002) Bacterial surface association of heat-labile enterotoxin through lipopolysaccharide after secretion via the general secretory pathway. J Biol Chem 277:32538–32545

Hozbor D, Rodriguez ME, Fernandez J, Lagares A, Guiso N, Yantorno O (1999) Release of outer membrane vesicles from *Bordetella pertussis*. Curr Microbiol 38:273–278

Imagawa T, Sonoda S, Kanoh M, Utsumi S (1979) Polymorphonuclear leukocyte-inhibitory factor of *Bordetella pertussis*. II. Localization in the outer membrane. Biken J 22:1–10

Irving AT, Mimuro H, Kufer TA, Lo C, Wheeler R, Turner LJ, Thomas BJ, Malosse C, Gantier MP, Casillas LN, Votta BJ, Bertin J, Boneca IG, Sasakawa C, Philpott DJ, Ferrero RL, Kaparakis-Liaskos M (2014) The immune receptor NOD1 and kinase RIP2 interact with bacterial peptidoglycan on early endosomes to promote autophagy and inflammatory signaling. Cell Host Microbe 15:623–635

Jager J, Keese S, Roessle M, Steinert M, Schromm AB (2015) Fusion of *Legionella pneumophila* outer membrane vesicles with eukaryotic membrane systems is a mechanism to deliver pathogen factors to host cell membranes. Cell Microbiol 17:607–620

Jeon H, Oh MH, Jun SH, Kim SI, Choi CW, Kwon HI, Na SH, Kim YJ, Nicholas A, Selasi GN, Lee JC (2016) Variation among *Staphylococcus aureus* membrane vesicle proteomes affects cytotoxicity of host cells. Microb Pathog 93:185–193

Jha C, Ghosh S, Gautam V, Malhotra P, Ray P (2017) In vitro study of virulence potential of *Acinetobacter baumannii* outer membrane vesicles. Microb Pathog 111:218–224

Jiang Y, Kong Q, Roland KL, Curtiss R 3rd (2014) Membrane vesicles of *Clostridium perfringens* type a strains induce innate and adaptive immunity. Int J Med Microbiol 304:431–443

Johnson AM, Kaushik RS, Francis DH, Fleckenstein JM, Hardwidge PR (2009) Heat-labile enterotoxin promotes *Escherichia coli* adherence to intestinal epithelial cells. J Bacteriol 191:178–186

Jun SH, Lee JH, Kim BR, Kim SI, Park TI, Lee JC, Lee YC (2013) *Acinetobacter baumannii* outer membrane vesicles elicit a potent innate immune response via membrane proteins. PLoS One 8: e71751

Kadurugamuwa JL, Beveridge TJ (1995) Virulence factors are released from *Pseudomonas aeruginosa* in association with membrane vesicles during normal growth and exposure to gentamicin: a novel mechanism of enzyme secretion. J Bacteriol 177:3998–4008

Kadurugamuwa JL, Beveridge TJ (1997) Natural release of virulence factors in membrane vesicles by *Pseudomonas aeruginosa* and the effect of aminoglycoside antibiotics on their release. J Antimicrob Chemother 40:615–621

Kadurugamuwa JL, Beveridge TJ (1998) Delivery of the non-membrane-permeative antibiotic gentamicin into mammalian cells by using *Shigella flexneri* membrane vesicles. Antimicrob Agents Chemother 42:1476–1483

Kadurugamuwa JL, Beveridge TJ (1999) Membrane vesicles derived from *Pseudomonas aeruginosa* and *Shigella flexneri* can be integrated into the surfaces of other gram-negative bacteria. Microbiology 145(Pt 8):2051–2060

Kamaguchi A, Nakayama K, Ichiyama S, Nakamura R, Watanabe T, Ohta M, Baba H, Ohyama T (2003) Effect of *Porphyromonas gingivalis* vesicles on coaggregation of *Staphylococcus aureus* to oral microorganisms. Curr Microbiol 47:485–491

Kaparakis M, Turnbull L, Carneiro L, Firth S, Coleman HA, Parkington HC, Le Bourhis L, Karrar A, Viala J, Mak J, Hutton ML, Davies JK, Crack PJ, Hertzog PJ, Philpott DJ, Girardin SE, Whitchurch CB, Ferrero RL (2010) Bacterial membrane vesicles deliver peptidoglycan to NOD1 in epithelial cells. Cell Microbiol 12:372–385

Kaparakis-Liaskos M, Ferrero RL (2015) Immune modulation by bacterial outer membrane vesicles. Nat Rev Immunol 15:375–387

Kaper JB, Morris JG Jr, Levine MM (1995) Cholera. Clin Microbiol Rev 8:48–86

Kesty NC, Mason KM, Reedy M, Miller SE, Kuehn MJ (2004) Enterotoxigenic *Escherichia coli* vesicles target toxin delivery into mammalian cells. EMBO J 23:4538–4549

Kieselbach T, Zijnge V, Granstrom E, Oscarsson J (2015) Proteomics of *Aggregatibacter actinomycetemcomitans* outer membrane vesicles. PLoS One 10:e0138591

Kim YR, Kim BU, Kim SY, Kim CM, Na HS, Koh JT, Choy HE, Rhee JH, Lee SE (2010) Outer membrane vesicles of *Vibrio vulnificus* deliver cytolysin-hemolysin VvhA into epithelial cells to induce cytotoxicity. Biochem Biophys Res Commun 399:607–612

Kim OY, Park HT, Dinh NTH, Choi SJ, Lee J, Kim JH, Lee SW, Gho YS (2017) Bacterial outer membrane vesicles suppress tumor by interferon-gamma-mediated antitumor response. Nat Commun 8:626

Kobayashi H, Uematsu K, Hirayama H, Horikoshi K (2000) Novel toluene elimination system in a toluene-tolerant microorganism. J Bacteriol 182:6451–6455

Koeppen K, Hampton TH, Jarek M, Scharfe M, Gerber SA, Mielcarz DW, Demers EG, Dolben EL, Hammond JH, Hogan DA, Stanton BA (2016) A novel mechanism of host-pathogen interaction through sRNA in bacterial outer membrane vesicles. PLoS Pathog 12:e1005672

Koeppen K, Barnaby R, Jackson AA, Gerber SA, Hogan DA, Stanton BA (2019) Tobramycin reduces key virulence determinants in the proteome of *Pseudomonas aeruginosa* outer membrane vesicles. PLoS One 14:e0211290

Kolling GL, Matthews KR (1999) Export of virulence genes and Shiga toxin by membrane vesicles of *Escherichia coli* O157:H7. Appl Environ Microbiol 65:1843–1848

Kondo K, Takade A, Amako K (1993) Release of the outer membrane vesicles from *Vibrio cholerae* and *Vibrio parahaemolyticus*. Microbiol Immunol 37:149–152

Kouokam JC, Wai SN, Fallman M, Dobrindt U, Hacker J, Uhlin BE (2006) Active cytotoxic necrotizing factor 1 associated with outer membrane vesicles from uropathogenic *Escherichia coli*. Infect Immun 74:2022–2030

Kunsmann L, Ruter C, Bauwens A, Greune L, Gluder M, Kemper B, Fruth A, Wai SN, He X, Lloubes R, Schmidt MA, Dobrindt U, Mellmann A, Karch H, Bielaszewska M (2015) Virulence from vesicles: novel mechanisms of host cell injury by *Escherichia coli* O104:H4 outbreak strain. Sci Rep 5:13252

Lee EY, Bang JY, Park GW, Choi DS, Kang JS, Kim HJ, Park KS, Lee JO, Kim YK, Kwon KH, Kim KP, Gho YS (2007) Global proteomic profiling of native outer membrane vesicles derived from *Escherichia coli*. Proteomics 7:3143–3153

Lee EY, Choi DY, Kim DK, Kim JW, Park JO, Kim S, Kim SH, Desiderio DM, Kim YK, Kim KP, Gho YS (2009) Gram-positive bacteria produce membrane vesicles: proteomics-based characterization of *Staphylococcus aureus*-derived membrane vesicles. Proteomics 9:5425–5436

Lee JH, Choi CW, Lee T, Kim SI, Lee JC, Shin JH (2013) Transcription factor sigmaB plays an important role in the production of extracellular membrane-derived vesicles in *Listeria monocytogenes*. PLoS One 8:e73196

Lee J, Yoon YJ, Kim JH, Dinh NTH, Go G, Tae S, Park KS, Park HT, Lee C, Roh TY, Di Vizio D, Gho YS (2018) Outer membrane vesicles derived from *Escherichia coli* regulate neutrophil migration by induction of endothelial IL-8. Front Microbiol 9:2268

Lei L, Yang F, Zou J, Jing H, Zhang J, Xu W, Zou Q, Zhang J (2019) Wang X (2019) DNA vaccine encoding OmpA and Pal from *Acinetobacter baumannii* efficiently protects mice against pulmonary infection. Mol Biol Rep 46(5):5397–5408

Li Z, Clarke AJ, Beveridge TJ (1998) Gram-negative bacteria produce membrane vesicles which are capable of killing other bacteria. J Bacteriol 180:5478–5483

Li J, Nation RL, Turnidge JD, Milne RW, Coulthard K, Rayner CR, Pater-son DL (2006) Colistin: the re-emerging antibiotic for multidrug-resistant Gram-negative bacterial infections. Lancet Infect Dis 6:589–601

Liao S, Klein MI, Heim KP, Fan Y, Bitoun JP, Ahn SJ, Burne RA, Koo H, Brady LJ, Wen ZT (2014) *Streptococcus mutans* extracellular DNA is upregulated during growth in biofilms, actively released via membrane vesicles, and influenced by components of the protein secretion machinery. J Bacteriol 196:2355–2366

Lindmark B, Rompikuntal PK, Vaitkevicius K, Song T, Mizunoe Y, Uhlin BE, Guerry P, Wai SN (2009) Outer membrane vesicle-mediated release of cytolethal distending toxin (CDT) from *Campylobacter jejuni*. BMC Microbiol 9:220

Lowy FD (1998) *Staphylococcus aureus* infections. N Engl J Med 339:520–532

MacEachran DP, Ye S, Bomberger JM, Hogan DA, Swiatecka-Urban A, Stanton BA, O'Toole GA (2007) The *Pseudomonas aeruginosa* secreted protein PA2934 decreases apical membrane expression of the cystic fibrosis transmembrane conductance regulator. Infect Immun 75:3902–3912

Malabirade A, Habier J, Heintz-Buschart A, May P, Godet J, Halder R, Etheridge A, Galas D, Wilmes P, Fritz JV (2018) The RNA complement of outer membrane vesicles from *Salmonella enterica* Serovar Typhimurium under distinct culture conditions. Front Microbiol 9:2015

Mantri CK, Chen CH, Dong X, Goodwin JS, Pratap S, Paromov V, Xie H (2015) Fimbriae-mediated outer membrane vesicle production and invasion of *Porphyromonas gingivalis*. Microbiology 4:53–65

Marsollier L, Brodin P, Jackson M, Kordulakova J, Tafelmeyer P, Carbonnelle E, Aubry J, Milon G, Legras P, Andre JP, Leroy C, Cottin J, Guillou ML, Reysset G, Cole ST (2007) Impact of *Mycobacterium ulcerans* biofilm on transmissibility to ecological niches and Buruli ulcer pathogenesis. PLoS Pathog 3:e62

Mashburn LM, Whiteley M (2005) Membrane vesicles traffic signals and facilitate group activities in a prokaryote. Nature 437:422–425

Mashburn-Warren LM, Whiteley M (2006) Special delivery: vesicle trafficking in prokaryotes. Mol Microbiol 61:839–846

McConnell MJ, Rumbo C, Bou G, Pachon J (2011) Outer membrane vesicles as an acellular vaccine against *Acinetobacter baumannii*. Vaccine 29:5705–5710

Mondal A, Tapader R, Chatterjee NS, Ghosh A, Sinha R, Koley H, Saha DR, Chakrabarti MK, Wai SN, Pal A (2016) Cytotoxic and inflammatory responses induced by outer membrane vesicle-associated biologically active proteases from *Vibrio cholerae*. Infect Immun 84:1478–1490

Mulcahy LA, Pink RC, Carter DR (2014) Routes and mechanisms of extracellular vesicle uptake. J Extracell Vesicles 3. https://doi.org/10.3402/jev.v3.24641.eCollection

Murphy K, Park AJ, Hao Y, Brewer D, Lam JS, Khursigara CM (2014) Influence of O polysaccharides on biofilm development and outer membrane vesicle biogenesis in *Pseudomonas aeruginosa* PAO1. J Bacteriol 196:1306–1317

Nakao R, Myint SL, Wai SN, Uhlin BE (2018) Enhanced biofilm formation and membrane vesicle release by *Escherichia coli* expressing a commonly occurring plasmid gene, *kil*. Front Microbiol 9:2605

Nevo-Dinur K, Govindarajan S, Amster-Choder O (2012) Subcellular localization of RNA and proteins in prokaryotes. Trends Genet 28:314–322

O'Donoghue EJ, Krachler AM (2016) Mechanisms of outer membrane vesicle entry into host cells. Cell Microbiol 18:1508–1517

Olier M, Marcq I, Salvador-Cartier C, Secher T, Dobrindt U, Boury M, Bacquie V, Penary M, Gaultier E, Nougayrede JP, Fioramonti J, Oswald E (2012) Genotoxicity of *Escherichia coli* Nissle 1917 strain cannot be dissociated from its probiotic activity. Gut Microbes 3:501–509

Olivier V, Haines GK 3rd, Tan Y, Satchell KJ (2007) Hemolysin and the multifunctional autoprocessing RTX toxin are virulence factors during intestinal infection of mice with *Vibrio cholerae* El Tor O1 strains. Infect Immun 75:5035–5042

Olofsson A, Nygard Skalman L, Obi I, Lundmark R, Arnqvist A (2014) Uptake of *Helicobacter pylori* vesicles is facilitated by clathrin-dependent and clathrin-independent endocytic pathways. MBio 5:e00979–e00914

Olsen I, Amano A (2015) Outer membrane vesicles - offensive weapons or good Samaritans? J Oral Microbiol 7:27468

Pathirana RD, Kaparakis-Liaskos M (2016) Bacterial membrane vesicles: biogenesis, immune regulation and pathogenesis. Cell Microbiol 18:1518–1524

Patrick S, McKenna JP, O'Hagan S, Dermott E (1996) A comparison of the haemagglutinating and enzymic activities of *Bacteroides fragilis* whole cells and outer membrane vesicles. Microb Pathog 20:191–202

Perez-Cruz C, Delgado L, Lopez-Iglesias C, Mercade E (2015) Outer-inner membrane vesicles naturally secreted by gram-negative pathogenic bacteria. PLoS One 10:e0116896

Pollak CN, Delpino MV, Fossati CA, Baldi PC (2012) Outer membrane vesicles from *Brucella abortus* promote bacterial internalization by human monocytes and modulate their innate immune response. PLoS One 7:e50214

Prados-Rosales R, Baena A, Martinez LR, Luque-Garcia J, Kalscheuer R, Veeraraghavan U, Camara C, Nosanchuk JD, Besra GS, Chen B, Jimenez J, Glatman-Freedman A, Jacobs WR Jr, Porcelli SA, Casadevall A (2011) Mycobacteria release active membrane vesicles that modulate immune responses in a TLR2-dependent manner in mice. J Clin Invest 121:1471–1483

Raeven RH, Brummelman J, Pennings JL, van der Maas L, Tilstra W, Helm K, van Riet E, Jiskoot W, van Els CA, Han WG, Kersten GF, Metz B (2016) *Bordetella pertussis* outer membrane vesicle vaccine confers equal efficacy in mice with milder inflammatory responses compared to a whole-cell vaccine. Sci Rep 6:38240

Rasti ES, Schappert ML, Brown AC (2018) Association of *Vibrio cholerae* 569B outer membrane vesicles with host cells occurs in a GM1-independent manner. Cell Microbiol 20:e12828

Rath P, Huang C, Wang T, Wang T, Li H, Prados-Rosales R, Elemento O, Casadevall A, Nathan CF (2013) Genetic regulation of vesiculogenesis and immunomodulation in *Mycobacterium tuberculosis*. Proc Natl Acad Sci U S A 110:E4790–E4797

Renelli M, Matias V, Lo RY, Beveridge TJ (2004) DNA-containing membrane vesicles of *Pseudomonas aeruginosa* PAO1 and their genetic transformation potential. Microbiology 150:2161–2169

Resch U, Tsatsaronis JA, Le Rhun A, Stubiger G, Rohde M, Kasvandik S, Holzmeister S, Tinnefeld P, Wai SN, Charpentier E (2016) A two-component regulatory system impacts extracellular membrane-derived vesicle production in group A Streptococcus. MBio 7: e00207–e00216

Riley DR, Sieber KB, Robinson KM, White JR, Ganesan A, Nourbakhsh S, Dunning Hotopp JC (2013) Bacteria-human somatic cell lateral gene transfer is enriched in cancer samples. PLoS Comput Biol 9:e1003107

Rivera J, Cordero RJ, Nakouzi AS, Frases S, Nicola A, Casadevall A (2010) *Bacillus anthracis* produces membrane-derived vesicles containing biologically active toxins. Proc Natl Acad Sci U S A 107:19002–19007

Rompikuntal PK, Thay B, Khan MK, Alanko J, Penttinen AM, Asikainen S, Wai SN, Oscarsson J (2012) Perinuclear localization of internalized outer membrane vesicles carrying active cytolethal distending toxin from *Aggregatibacter actinomycetemcomitans*. Infect Immun 80:31–42

Rompikuntal PK, Vdovikova S, Duperthuy M, Johnson TL, Ahlund M, Lundmark R, Oscarsson J, Sandkvist M, Uhlin BE, Wai SN (2015) Outer membrane vesicle-mediated export of processed PrtV protease from *Vibrio cholerae*. PLoS One 10:e0134098

Rumbo C, Fernandez-Moreira E, Merino M, Poza M, Mendez JA, Soares NC, Mosquera A, Chaves F, Bou G (2011) Horizontal transfer of the OXA-24 carbapenemase gene via outer membrane vesicles: a new mechanism of dissemination of carbapenem resistance genes in *Acinetobacter baumannii*. Antimicrob Agents Chemother 55:3084–3090

Schaar V, de Vries SP, Perez Vidakovics ML, Bootsma HJ, Larsson L, Hermans PW, Bjartell A, Morgelin M, Riesbeck K (2011) Multicomponent *Moraxella catarrhalis* outer membrane vesicles induce an inflammatory response and are internalized by human epithelial cells. Cell Microbiol 13:432–449

Sharpe SW, Kuehn MJ, Mason KM (2011) Elicitation of epithelial cell-derived immune effectors by outer membrane vesicles of nontypeable *Haemophilus influenzae*. Infect Immun 79:4361–4369

Shen Y, Giardino Torchia ML, Lawson GW, Karp CL, Ashwell JD, Mazmanian SK (2012) Outer membrane vesicles of a human commensal mediate immune regulation and disease protection. Cell Host Microbe 12:509–520

Singh U, Grenier D, McBride BC (1989) *Bacteroides gingivalis* vesicles mediate attachment of streptococci to serum-coated hydroxyapatite. Oral Microbiol Immunol 4:199–203

Sjostrom AE, Sandblad L, Uhlin BE, Wai SN (2015) Membrane vesicle-mediated release of bacterial RNA. Sci Rep 5:15329

Smalley JW, Birss AJ, McKee AS, Marsh PD (1991) Haemin-restriction influences haemin-binding, haemagglutination and protease activity of cells and extracellular membrane vesicles of *Porphyromonas gingivalis* W50. FEMS Microbiol Lett 69:63–67

Svennerholm K, Park KS, Wikstrom J, Lasser C, Crescitelli R, Shelke GV, Jang SC, Suzuki S, Bandeira E, Olofsson CS, Lotvall J (2017) *Escherichia coli* outer membrane vesicles can contribute to sepsis induced cardiac dysfunction. Sci Rep 7:17434

Syngkon A, Elluri S, Koley H, Rompikuntal PK, Saha DR, Chakrabarti MK, Bhadra RK, Wai SN, Pal A (2010) Studies on a novel serine protease of a DeltahapADeltaprtV *Vibrio cholerae* O1 strain and its role in hemorrhagic response in the rabbit ileal loop model. PLoS One 5:e13122

Taheri N, Mahmud A, Sandblad L, Fallman M, Wai SN, Fahlgren A (2018) *Campylobacter jejuni* bile exposure influences outer membrane vesicles protein content and bacterial interaction with epithelial cells. Sci Rep 8:16996

Tashiro Y, Ichikawa S, Shimizu M, Toyofuku M, Takaya N, Nakajima-Kambe T, Uchiyama H, Nomura N (2010) Variation of physiochemical properties and cell association activity of membrane vesicles with growth phase in *Pseudomonas aeruginosa*. Appl Environ Microbiol 76:3732–3739

Tashiro Y, Inagaki A, Shimizu M, Ichikawa S, Takaya N, Nakajima-Kambe T, Uchiyama H, Nomura N (2011) Characterization of phospholipids in membrane vesicles derived from *Pseudomonas aeruginosa*. Biosci Biotechnol Biochem 75:605–607

Tashiro Y, Hasegawa Y, Shintani M, Takaki K, Ohkuma M, Kimbara K, Futamata H (2017) Interaction of bacterial membrane vesicles with specific species and their potential for delivery to target cells. Front Microbiol 8:571

Thay B, Wai SN, Oscarsson J (2013) *Staphylococcus aureus* alpha-toxin-dependent induction of host cell death by membrane-derived vesicles. PLoS One 8:e54661

Thay B, Damm A, Kufer TA, Wai SN, Oscarsson J (2014) *Aggregatibacter actinomycetemcomitans* outer membrane vesicles are internalized in human host cells and trigger NOD1- and NOD2-dependent NF-kappaB activation. Infect Immun 82:4034–4046

Turnbull L, Toyofuku M, Hynen AL, Kurosawa M, Pessi G, Petty NK, Osvath SR, Carcamo-Oyarce G, Gloag ES, Shimoni R, Omasits U, Ito S, Yap X, Monahan LG, Cavaliere R, Ahrens CH, Charles IG, Nomura N, Eberl L, Whitchurch CB (2016) Explosive cell lysis as a mechanism for the biogenesis of bacterial membrane vesicles and biofilms. Nat Commun 7:11220

Turner L, Bitto NJ, Steer DL, Lo C, D'Costa K, Ramm G, Shambrook M, Hill AF, Ferrero RL, Kaparakis-Liaskos M (2018) *Helicobacter pylori* outer membrane vesicle size determines their mechanisms of host cell entry and protein content. Front Immunol 9:1466

Tyrer PC, Frizelle FA, Keenan JI (2014) *Escherichia coli*-derived outer membrane vesicles are genotoxic to human enterocyte-like cells. Infect Agent Cancer 9:2

Vaitkevicius K, Lindmark B, Ou G, Song T, Toma C, Iwanaga M, Zhu J, Andersson A, Hammarstrom ML, Tuck S, Wai SN (2006) A *Vibrio cholerae* protease needed for killing of *Caenorhabditis elegans* has a role in protection from natural predator grazing. Proc Natl Acad Sci U S A 103:9280–9285

Valadi H, Ekstrom K, Bossios A, Sjostrand M, Lee JJ, Lotvall JO (2007) Exosome-mediated transfer of mRNAs and microRNAs is a novel mechanism of genetic exchange between cells. Nat Cell Biol 9:654–659

Vanaja SK, Russo AJ, Behl B, Banerjee I, Yankova M, Deshmukh SD, Rathinam VAK (2016) Bacterial outer membrane vesicles mediate cytosolic localization of LPS and caspase-11 activation. Cell 165:1106–1119

Vdovikova S, Luhr M, Szalai P, Nygard Skalman L, Francis MK, Lundmark R, Engedal N, Johansson J, Wai SN (2017) A novel role of *Listeria monocytogenes* membrane vesicles in inhibition of autophagy and cell death. Front Cell Infect Microbiol 7:154

Vdovikova S, Gilfillan S, Wang S, Dongre M, Wai SN, Hurtado A (2018) Modulation of gene transcription and epigenetics of colon carcinoma cells by bacterial membrane vesicles. Sci Rep 8:7434

Veith PD, Chen YY, Gorasia DG, Chen D, Glew MD, O'Brien-Simpson NM, Cecil JD, Holden JA, Reynolds EC (2014) *Porphyromonas gingivalis* outer membrane vesicles exclusively contain outer membrane and periplasmic proteins and carry a cargo enriched with virulence factors. J Proteome Res 13:2420–2432

Vidakovics ML, Jendholm J, Morgelin M, Mansson A, Larsson C, Cardell LO, Riesbeck K (2010) B cell activation by outer membrane vesicles-a novel virulence mechanism. PLoS Pathog 6: e1000724

Wagner T, Joshi B, Janice J, Askarian F, Skalko-Basnet N, Hagestad OC, Mekhlif A, Wai SN, Hegstad K, Johannessen M (2018) *Enterococcus faecium* produces membrane vesicles containing virulence factors and antimicrobial resistance related proteins. J Proteome 187:28–38

Wai SN, Lindmark B, Soderblom T, Takade A, Westermark M, Oscarsson J, Jass J, Richter-Dahlfors A, Mizunoe Y, Uhlin BE (2003) Vesicle-mediated export and assembly of pore-forming oligomers of the enterobacterial ClyA cytotoxin. Cell 115:25–35

Xie H (2015) Biogenesis and function of *Porphyromonas gingivalis* outer membrane vesicles. Future Microbiol 10:1517–1527

Yaron S, Kolling GL, Simon L, Matthews KR (2000) Vesicle-mediated transfer of virulence genes from *Escherichia coli* O157:H7 to other enteric bacteria. Appl Environ Microbiol 66:4414–4420

Zhang W, Ju J, Rigney T, Tribble GD (2011) Fimbriae of *Porphyromonas gingivalis* are important for initial invasion of osteoblasts, but not for inhibition of their differentiation and mineralization. J Periodontol 82:909–916

Zhong H, Yang Y, Ma S, Xiu F, Cai Z, Zhao H, Du L (2011) Induction of a tumour-specific CTL response by exosomes isolated from heat-treated malignant ascites of gastric cancer patients. Int J Hyperth 27:604–611

Zomer A, Vendrig T, Hopmans ES, van Eijndhoven M, Middeldorp JM, Pegtel DM (2010) Exosomes: fit to deliver small RNA. Commun Integr Biol 3:447–450

# Chapter 8
# Immunodetection and Pathogenesis Mediated by Bacterial Membrane Vesicles

**Ella L. Johnston, Thomas A. Kufer, and Maria Kaparakis-Liaskos**

**Abstract** Animals have evolved sophisticated means to detect and defend themselves against colonization by microorganisms. Mammals differentiate self from nonself by the use of germline encoded pattern recognition receptors (PRRs) of the innate immune system in addition to pathogen-specific recognition by the adaptive immune system. Conserved microbial structures, termed microbe-associated molecular patterns (MAMPs) are recognized by the innate immune system. Pathogens succeed in subverting these immune mechanisms by the production of effector proteins and modification of MAMPs. While many PRRs and their cognate ligands have been extensively examined, most studies addressed the immune response toward the whole microorganism, isolated MAMPs, or effector proteins. Both Gram-negative and Gram-positive bacteria produce membrane vesicles that contain several MAMPs and bacterial effector proteins, and the study of the interplay of these with immune PRRs is a recently emerging field. Here we summarize key components of the innate and adaptive immune systems and discuss current knowledge regarding the immune recognition and immunomodulatory functions of bacterial membrane vesicles in mammalian hosts.

E. L. Johnston · M. Kaparakis-Liaskos (✉)
Department of Physiology, Anatomy and Microbiology, La Trobe University, Melbourne, VIC, Australia

Research Centre for Extracellular Vesicles, La Trobe Institute for Molecular Science, La Trobe University, Melbourne, VIC, Australia
e-mail: m.liaskos@latrobe.edu.au

T. A. Kufer
Department of Immunology, Institute of Nutritional Medicine, University of Hohenheim, Stuttgart, Baden-Württemberg, Germany



M. Kaparakis-Liaskos, T. A. Kufer (eds.), *Bacterial Membrane Vesicles*,
https://doi.org/10.1007/978-3-030-36331-4_8

## 8.1 Bacterial Membrane Vesicles

Both Gram-negative and Gram-positive bacteria produce vesicles known as outer membrane vesicles (OMVs) or membrane vesicles (MVs), respectively. They contain components derived from the parent bacterium, including proteins, lipids, and nucleic acids (reviewed in Schwechheimer and Kuehn 2015). Originally thought to be an artifact of bacterial growth, recent studies have demonstrated that OMVs and MVs may contribute to disease pathogenesis and progression due to their immunostimulatory cargo. OMVs are naturally released by bacteria as part of their normal growth. There are a number of mechanisms by which OMVs are produced and bacteria can actively increase their production (Kulp and Kuehn 2010). In contrast to OMVs, the mechanisms of MV biogenesis are not well characterized. This is because Gram-negative OMVs were discovered long before Gram-positive MVs, and consequently we have a greater understanding of OMV production, composition, and their effects on the host. Although research has been focused on examining the immunostimulatory effects of OMVs derived from pathogens, recent studies have begun investigating the role of OMVs from commensal bacteria in modulating the hosts immune response. In contrast, due to their more recent discovery, limited studies have examined the immunostimulatory effects of Gram-positive MVs on the host.

A number of studies have suggested that OMV production may be increased during the course of infection; however, the mechanisms behind OMV biogenesis during infection have not been completely elucidated. It has been identified that the disruption of genes encoding for outer membrane proteins and peptidoglycan synthesis can lead to changes in OMV production (McBroom et al. 2006). OMV production is also thought to be a response to stress stimuli such as antibiotic treatment, envelope stress, and increase in temperature (Baumgarten et al. 2012; Kadurugamuwa and Beveridge 1995; Macdonald and Kuehn 2013; McBroom and Kuehn 2007; Yun et al. 2018). Moreover, the immunogenicity of OMV cargo can be altered by certain bacteria, such as *Vibrio cholerae,* in response to changes in the expression of virulence factors upon quorum sensing (Bielig et al. 2011). Therefore, it is thought that during infection, factors such as environmental stress, bacterial regulators, and host immune molecules contribute to increased production of bacterial OMVs. Collectively, this increased production of OMVs and their composition of immunogenic cargo is thought to contribute to pathogenesis.

### *8.1.1 Bacterial Membrane Vesicles Contain Immunogenic Cargo*

Numerous studies have demonstrated the pro-inflammatory nature of Gram-negative OMVs, their ability to stimulate the immune system, and their use as vaccines for humans (Holst et al. 2009; Kaparakis-Liaskos and Ferrero 2015; Manning and

Kuehn 2011; Ellis and Kuehn 2010; Pathirana and Kaparakis-Liaskos 2016). OMVs have been observed in a number of samples from human patients and tissue biopsies. For example, OMVs from *Neisseria meningitidis* were observed using electron microscopy in the cerebrospinal fluid of an infected infant, and in samples from a patient with fatal meningococcal disease (Stephens et al. 1982; Namork and Brandtzaeg 2002). Additionally, OMVs from *Helicobacter pylori* have been found in gastric biopsies from infected individuals and in cultures from several clinical patient samples (Fiocca et al. 1999; Keenan et al. 2000). OMVs have also been observed in samples from the sinus of a child infected with *Moraxella catarrhalis* (Vidakovics et al. 2010), and in the lungs of patients with nontypeable *Haemophilus influenza* infections (Ren et al. 2012). In addition to their presence during infection, a number of studies have revealed that OMVs contain specific microbe-associated molecular patterns (MAMPs), known to stimulate the immune system, such as lipopolysaccharide (LPS), peptidoglycan (PG), nucleic acids, as well as other bacterial cell wall components (Vanaja et al. 2016; Chatterjee and Das 1967; Renelli et al. 2004; Choi et al. 2011).

In addition to their ability to carry immunostimulatory cargo, OMVs have been shown to be important in the structure and function of biofilms, in the environment as well as during the course of infection (Schooling and Beveridge 2006; Yonezawa et al. 2009). Additionally, they are able to act as bacteriophage and antibiotic decoys and can "trap" complement proteins produced by the host immune system (reviewed in Manning and Kuehn 2011; Tan et al. 2007). Because of this, it is thought that OMVs may be harnessed by bacteria to further establish colonization and pathogenesis within the host.

Conversely, the biogenesis, contents and immunostimulatory ability of Gram-positive MVs is not well characterized. Some studies have reported that Gram-positive MVs contain immunogenic cargo (Lee et al. 2009; Gurung et al. 2011; Peschel et al. 2001; Rivera et al. 2010); however, further studies are required to elucidate their immunogenic profiles and interactions with the host immune system. Chapter 3 contains a more detailed review of the biogenesis and functions of Gram-positive MVs.

The contribution of bacterial OMVs and MVs to pathogenesis and immunity is quickly becoming apparent as more studies uncover their roles as a novel transport system for bacterial derived molecules. Here we discuss the various mechanisms by which OMVs are able to interact with the host's innate and adaptive immune system, their contributions to pathogenesis, as well as the recent discovery of immunogenic Gram-positive MVs.

## 8.2 Bacterial Membrane Vesicles Interact with and Modulate the Hosts Innate Immune System

In order to survive within a human host, microorganisms must circumvent host immune defenses. All multicellular organisms have developed innate defense mechanisms which have the capacity to be activated by foreign microorganisms and that function to defend the host by destroying pathogens and neutralizing their virulence factors (Medzhitov and Janeway 1997). One of the major mechanisms by which bacteria are able to interact with host cells is by the secretion of immunostimulatory products (Young et al. 1999; Bruno et al. 2009). Bacterial membrane vesicles are released from both Gram-negative and Gram-positive bacteria and act as secretory transporters for proteins, lipids, and other bacterial products to host cells (reviewed in Ellis and Kuehn 2010). The innate immune system is the first line of defense against invading and resident bacteria, and it is able to rapidly detect and initiate an immune response to bacteria membrane vesicles. In humans, the immune system is comprised of innate and adaptive immunity. The innate immune system is the frontline defense against bacteria, whereas adaptive immunity functions to develop pathogenic-specific immune responses while generating long-lived immunity.

### 8.2.1 OMVs Protect Bacteria Against Host Innate Immune Molecules

The hosts innate immune system is multifaceted in its ability to respond to pathogens and is composed of a range of defense mechanisms which include innate immune receptors, innate immune cells, and the production of antimicrobials and inflammatory molecules. For example, pathogens can trigger the activation of the complement system, a group of small proteins made by hepatocytes which circulate in the bloodstream. The complement system enhances the ability of antibodies and phagocytic cells to clear microbes, in addition to damaging bacteria by attacking their cell membrane (Rus et al. 2005). Although part of the innate immune system, complement can also regulate aspects of the adaptive immune system by recruiting and activating immune cells using anaphylatoxins, small peptide mediators (Barrington et al. 2001).

Pathogens are often able to inhibit activation of the complement system; however, the mechanisms by which this can occur have not been entirely elucidated (Joiner 1988). It has been shown that *Porphyromonas gingivalis* OMVs can degrade complement components such as the complement 5a (C5a) cluster of differentiation 88 receptor (CD88) (Jagels et al. 1996) and CD14 from the surface of human macrophage cells. CD88 and CD14 are both involved in the recognition of bacteria (Duncan et al. 2004) and their targeting leads to reduced efficacy of immune responses and allows bacteria to evade the immune system. Similarly, *Haemophilus influenzae* can survive complement-mediated attacks when supplemented with *M. catarrhalis* OMVs, further potentiating a mutually beneficial role for invading pathogens and their OMVs (Tan et al. 2007).

Moreover, the innate immune system also employs antimicrobial peptides (AMPs), such as defensins, as another defense mechanism against invading pathogens. AMPs can directly attack invading microbes to limit pathogenesis; however, resistance to AMPs has been observed for a number of pathogens, including *Staphylococcus aureus* (Peschel et al. 2001), *Salmonella enterica* serovar

Typhimurium (*S.* Typhimurium) (Gunn et al. 2000), *Listeria monocytogenes* (Abachin et al. 2002), and *Pseudomonas aeruginosa* (Moskowitz et al. 2004). OMVs have been demonstrated to contribute to AMP resistance, as *V. cholerae* OMVs were shown to protect other bacteria from membrane attack by AMPs (Duperthuy et al. 2013). This was proposed to be a more general mechanism of immune evasion, where OMVs can function to sequester host immune molecules (Band and Weiss 2015). Moreover, the host limits the availability of transition metals by siderophores, such as calprotectin, to restrict pathogen growth, referred to as "nutritional immunity" (Hood and Skaar 2012).

### *8.2.2 Recognition of Bacterial Membrane Vesicles by Host Pattern Recognition Receptors Results in a Pro-Inflammatory Response*

An integral feature of the innate immune system involves a number of specific pattern recognition receptors (PRRs), which recognize MAMPs and subsequently results in the generation of a pro-inflammatory innate immune response (Janeway and Medzhitov 2002). MAMPs are conserved microbial structures such as bacterial nucleic acids and bacterial cell wall components including LPS and PG, which are distinguishable from host molecules (Boller and Felix 2009). PRRs are protein receptors expressed by a number of innate immune cells such as neutrophils, macrophages, and dendritic cells (DCs) (Thomas and Schroder 2013; Takeuchi and Akira 2010) as well as non-immune cells, such as epithelial cells (Kawai and Akira 2010). Once PRRs are activated by detection of their respective MAMP ligand, a signaling cascade initiates the release of inflammatory mediators such as cytokines and chemokines which are responsible for the recruitment and the activation of immune cells and for mediating an inflammatory response (Kawai and Akira 2010). PRRs are vital for the recognition of, and downstream effects initiated by invading bacteria.

PRRs include Toll-like receptors (TLRs) which are membrane bound (Fig. 8.1), intracellular nucleotide-binding oligomerization domain (NOD)-like receptors (NLRs) (Fig. 8.1), retinoic acid inducible gene-I (RIG-I)-like receptors (RLRs) and C-type lectins, which can be either soluble or function as membrane associated receptors (Kawai and Akira 2008; Kanneganti et al. 2007). TLRs and NLRs are activated upon recognition of MAMPs, such as components of the bacterial cell wall or microbial DNA (Fig. 8.1). RIG-I-like receptors are responsible for the recognition of virally derived dsRNA within the cytoplasm (Wells et al. 2011), and C-type lectins contribute to innate immune responses to many pathogens (reviewed in Robinson et al. 2006). Activation of PRRs leads to a number of outcomes, namely the induction of signaling pathways resulting in the transcription of pro-inflammatory cytokines that drive inflammation, in an effort to combat the invading pathogen (Fig. 8.1). Bacterial membrane vesicles can activate these signaling pathways in a number of ways, which will be discussed below.

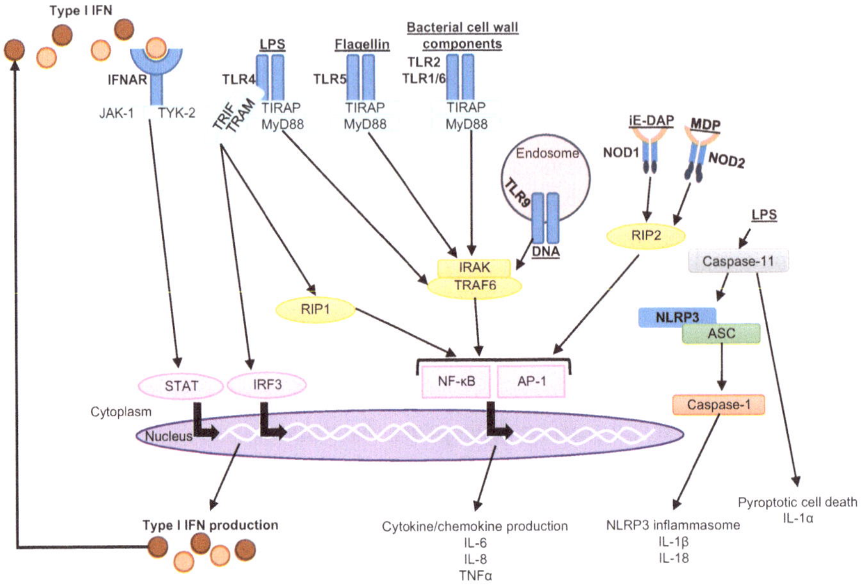


**Fig. 8.1** Bacterial products are detected by TLRs and NLRs by the host cell. On the surface of the host cell, TLR5 detects flagellin, TLR2 and TLR1/6 detect bacterial cell wall components and TLR4 detects LPS. Within the cytoplasm, caspase-11 detects LPS, and NOD1 and NOD2 detect peptidoglycan subunits iE-DAP and MDP, respectively. Endosome-bound TLR9 detects bacterial DNA. Detection of bacterial MAMPs leads to the recruitment of adaptor proteins, downstream activation of transcription factors and subsequent production of pro-inflammatory chemokines and cytokines or pyroptotic cell death in the case of caspase-11 activation. Type I IFN production can also occur in response to the detection of bacteria or bacterial products, via the JAK/STAT pathway

#### 8.2.2.1 Recognition of Bacteria by Toll-like Receptors

Toll-like receptors (TLRs) are a family of transmembrane proteins expressed by both leukocytes and non-immune cells, such as epithelial and endothelial cells. They contain an extracellular leucine-rich repeat (LRR) domain that is necessary for the detection of MAMPs, a transmembrane domain and a cytoplasmic tail (Toll/interleukin-1 receptor (TIR) domain). TLRs can be found on the cell surface or on intracellular endosomal membranes. To date, 10 members of the TLR family have been identified in humans (Akira and Takeda 2004). TLRs are capable of detecting a range of molecules derived from bacteria, viruses, and fungi. Importantly, TLRs detect molecules such as LPS, PG, and nucleic acids from both Gram-positive and Gram-negative bacteria (Fig. 8.1).

Upon detection of MAMPs, most TLRs form either homodimers or heterodimers (Fig. 8.1). Once activated, adaptor molecules are recruited to TLRs to initiate downstream signaling. TLR2, TLR4, TLR5, and TLR1/6 signal via the adaptor proteins Myeloid Differentiation Primary Response Gene 88 (MyD88) and TIR-Domain-Containing Adaptor Protein (TIRAP). This leads to the activation of IL-1

Receptor-Associated Kinases (IRAKs) and the adaptor molecules TNF Receptor-Associated Factor 6 (TRAF6) and eventually the nuclear translocation of transcription factors Nuclear Factor Kappa B (NF-κB) and Activator Protein 1 (AP-1). TLR9 also uses this pathway; however, it is located on the endosomal membrane (Fig. 8.1). Both NF-κB and AP-1 have functions in the expression of pro-inflammatory cytokines (Fig. 8.1). As a result of NF-κB and AP-1 activation, cytokines and chemokines are produced, for example interleukin-6 (IL-6), IL-8 (CXCL8) and Tumor Necrosis Factor Alpha (TNF/TNFα).

Additionally, TLR4 can signal independently of MyD88, via TIR-Domain-Containing Adaptor Inducing Interferon-β (TRIF) and TRIF-related adaptor molecule (TRAM) (Fig. 8.1). This leads to the activation of adaptor molecules, such as Receptor-Interacting Protein 1 (RIP1), and the translocation of NF-κB and AP-1 or Interferon Regulatory Factor-3 (IRF3) to the nucleus. IRF3 is a transcription factor which induces the production of type 1 interferons (Fig. 8.1, Kumar et al. 2009). Type I IFNs can signal in a paracrine or autocrine manner by interacting with Interferon Alpha/Beta Receptor (IFNAR) on the cell surface, and signal using the Janus Kinases and Signal Transducer and Activation of Transcription Proteins (JAK/STAT) pathway (Fig. 8.1, Akira and Takeda 2004). Type 1 interferons have been demonstrated to limit bacterial infections by upregulating the production of host defense molecules (reviewed in Boxx and Cheng 2016). Therefore, membrane-bound TLRs are collectively important receptors of bacteria and bacterial products. In addition, intracellular receptors from other protein families are also vital in the host response to internalized MAMPs, as discussed below.

#### 8.2.2.2 Recognition of Bacteria by Cytoplasmic Innate Immune Receptors

In the cytoplasm of the host cell, a large family of PRRs known as NOD-like receptors (NLRs) are expressed. They are characterized by the presence of a central conserved NTPase (NACHT) domain, required for nucleotide binding and self-oligomerization (Inohara et al. 2000). These proteins generally contain a variable N-terminal effector-binding domain, which is responsible for protein–protein interactions, as well as C-terminal LRRs to detect MAMPs (Inohara and Nunez 2001). NOD1 and NOD2 are cytoplasmic receptors that recognize bacterial peptidoglycan (PG). NOD1 recognizes γ-D-Glu-mDAP (iE-DAP), a dipeptide present in the PG of almost all Gram-negative and some Gram-positive bacteria. NOD2 recognizes muramyl dipeptide (MDP) which is common to PG of all bacteria (Girardin et al. 2003a, b).

Activation of either NOD1 or NOD2 leads to the recruitment of Receptor-Interacting Serine/Threonine-Protein Kinase 2 (RIP2, RIPK2) by their CARD domain, the activation of NF-κB (Inohara et al. 2000) and the production of pro-inflammatory cytokines and chemokines (Fig. 8.1). Other NLRs, such as those from the NLR-pyrin domain containing (NLRP) subfamily are namely responsible for the formation of "platforms," known as inflammasomes, in response to host signals, which contribute to inflammation (Elliott and Sutterwala 2015). NLRP3, for example, can be triggered by caspase-11 activation by cytoplasmic LPS (Shi et al. 2014), cell membrane damage, and extracellular ATP released by damaged cells (Fig. 8.1, Martinon et al. 2002). Activation of these inflammasome forming NLRs leads to the recruitment of the adaptor Apoptosis-Associated Speck-Like Protein Containing a CARD (ASC) that links NLRP3 to caspase-1 (Fig. 8.1). This complex, termed inflammasome, triggers caspase-1 activation, resulting in the production and release of the key inflammatory mediators IL-1β and IL-18, and leads to pyroptotic cell death (Schroder et al. 2010).

The activation of PRRs is a pivotal prerequisite to prime the adaptive immune system and to induce antigen-specific responses to bacterial pathogens (Fritz et al. 2007). As a result, TLRs and NLRs are key regulators of the innate immune response (Milward et al. 2007; Faure et al. 2000). Collectively, there are a range of mechanisms by which bacteria can interact with host PRRs to trigger TLR and NLR signaling to mediate a pro-inflammatory response.

#### 8.2.2.3 Bacterial Membrane Vesicles Are Detected by Host Pattern Recognition Receptors Resulting in the Induction of an Innate Immune Response

The mucosal epithelial cell surface is the first line of defense in the host, and often the location where bacterial vesicles first come into contact with the host. The gut epithelium and the underlying lamina propria contains a diverse subset of innate and adaptive immune cells within the gut-associated lymphoid tissue (GALT). Long seen as a classical barrier, we now know that the intestinal epithelium is actively involved in immune regulation (Rhee et al. 2004). Epithelial cells express PRRs and can detect pathogens and respond to cytokines, such as IL-22, to produce antimicrobial peptides to control the microbiota composition and to secrete cytokines that recruit immune cells (Peterson and Artis 2014). The mucosal epithelial cell surface has evolved to cope with the constant presence of microbial products and foreign antigens without inducing inflammation. Indeed, the presence of microbiota and their derived MAMPs are needed for the development of the GALT and to maintain tissue homeostasis (Stappenbeck et al. 2002). A wealth of current work aims to delineate how the host can mediate tolerance of commensals at the mucosal surface, while still being able to detect and respond toward bacterial pathogens. We still know little about the microbial substances and molecular details that mediate these responses, and the role of OMVs and MVs in these processes remains largely elusive. It has been demonstrated that the interaction of OMVs with host epithelial cells results in the production of cytokines and chemokines. Therefore, OMVs and MVs are thought to be a mechanism whereby bacteria can transport their products to non-phagocytic cells at the mucosal surface in order to modulate innate and adaptive immune responses which will be discussed below.

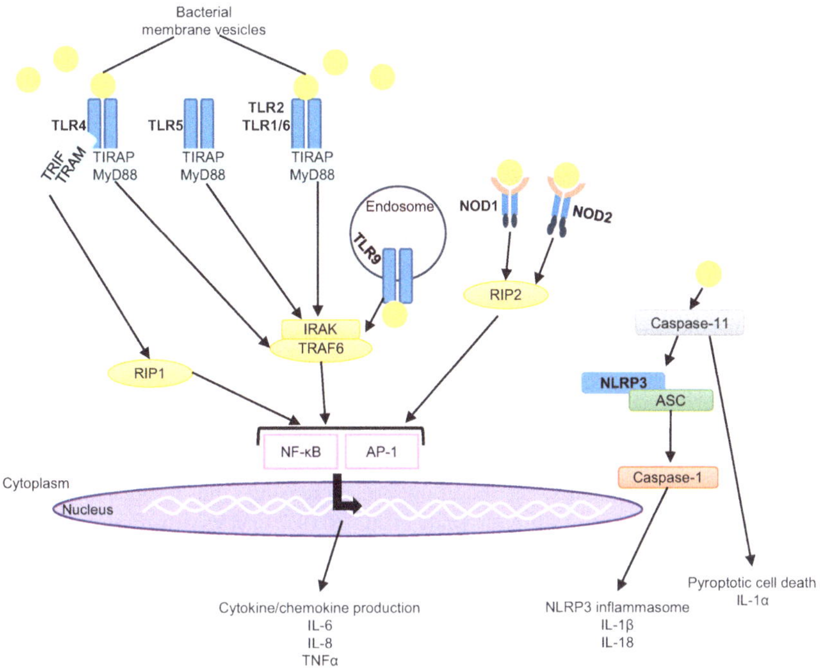


**Fig. 8.2.** Bacterial membrane vesicles interact with PRRs on the surface and cytoplasm of host cells to initiate an innate immune response. OMVs produced by *E. coli, V. cholerae*, and *H. pylori*, for example, can interact with TLR4, TLR2, and TLR1/6 on the surface of the cell. OMVs produced by *H. pylori* and *A. actinomycetemcomitans* can interact with NOD1 and NOD2, respectively. The activation of these PRRs results in the induction of an intracellular signaling cascade that ultimately results in the production of pro-inflammatory cytokines and chemokines

#### 8.2.2.4 Gram-Negative OMVs Are Detected by PRRs Expressed by Epithelial Cells

A number of in vitro studies using human epithelial cells have reported the production of chemokines and cytokines in response to stimulation by bacterial OMVs. *H. pylori* is a gastric pathogen, which produces OMVs demonstrated to modulate the proliferation of, and IL-8 production by gastric adenocarcinoma cells (Ismail et al. 2003). This was the first study to indicate that OMVs could induce pro-inflammatory cytokine production by human epithelial cells. Further investigation revealed that the production of IL-8, stimulated by *H. pylori* OMVs was dependent on detection of their peptidoglycan cargo by the intracellular NOD1 receptor (Fig. 8.2, Kaparakis et al. 2010). Similarly, OMVs from the pathogen *V. cholerae* were found to induce the production of IL-8 and granulocyte-macrophage colony-stimulating factor (GM-CSF) in a NOD1-dependent manner (Fig. 8.2, Chatterjee and Chaudhuri 2013; Bielig et al. 2011). NOD2 is able to detect PG from both Gram-positive and Gram-negative membrane vesicles, and has been shown to induce a pro-inflammatory immune response upon the detection of OMVs, such is the case for the pathogen *Aggregatibacter actinomycetemcomitans* (Fig. 8.2, Thay et al. 2014). In addition, OMVs from the Gram-negative pathogen *Legionella pneumophila* were found to induce the production of a number of chemokines and cytokines by human alveolar epithelial cells (Galka et al. 2008). OMVs from periodontal pathogens have also been implicated in the progression of disease by interaction with epithelial cells, similarly to the studies discussed above (Cecil et al. 2019).

Opportunistic pathogens can utilize OMVs as a virulence factor when establishing disease. *P. aeruginosa* produces OMVs which can stimulate IL-8 production by primary human bronchial epithelial cells (Bauman and Kuehn 2006). OMVs from the opportunistic pathogen, *A. baumannii* were also found to upregulate the production of a number of cytokines and chemokines by human epithelial cells (Jun et al. 2013).

Lastly, OMVs derived from the commensal Gram-negative bacterium *Bacteroides fragilis* have been shown to modify the expression of TLR2 and TLR4 in epithelial cells, in addition to increasing the expression of anti-inflammatory cytokines (Ahmadi Badi et al. 2019). OMVs from commensal and probiotic *E. coli* can interact with epithelial cells and activate NOD1, leading to the release of the pro-inflammatory cytokines IL-6 and IL-8 (Fig. 8.2, Canas et al. 2016). Collectively, these studies demonstrate the complex role Gram-negative OMVs play at the epithelium in immune tolerance, and conversely, in the initiation of an innate immune response by pathogens.

#### 8.2.2.5 Gram-Positive MVs Interact with Host Epithelial Cells

Conversely, a limited number of studies have demonstrated the immunogenic ability of Gram-positive MVs. A proteomic study examining MVs produced by *S. aureus* found that they are enriched in penicillin-binding proteins which contribute to antibiotic resistance as well as proteins which have pathological functions during infection (Lee et al. 2009). Later, it was demonstrated that *S. aureus* MVs can induce inflammation in dermal fibroblasts, but not human epithelial cells, suggesting a potential role for MVs in mediating inflammation during atopic dermatitis (Jun et al. 2017). Another study found that MVs from *S. aureus* were able to enter and induce host cell death in the human epithelial cell line (HEp-2) (Gurung et al. 2011).

In addition to epithelial cells, MVs are thought to interact with various other cell types throughout the host, both immune and non-immune. Overall, bacterial membrane vesicles are known to interact with epithelial cells and are thought to disseminate throughout the host to interact with cells, such as macrophages, neutrophils, and DCs, distal to the site of infection.

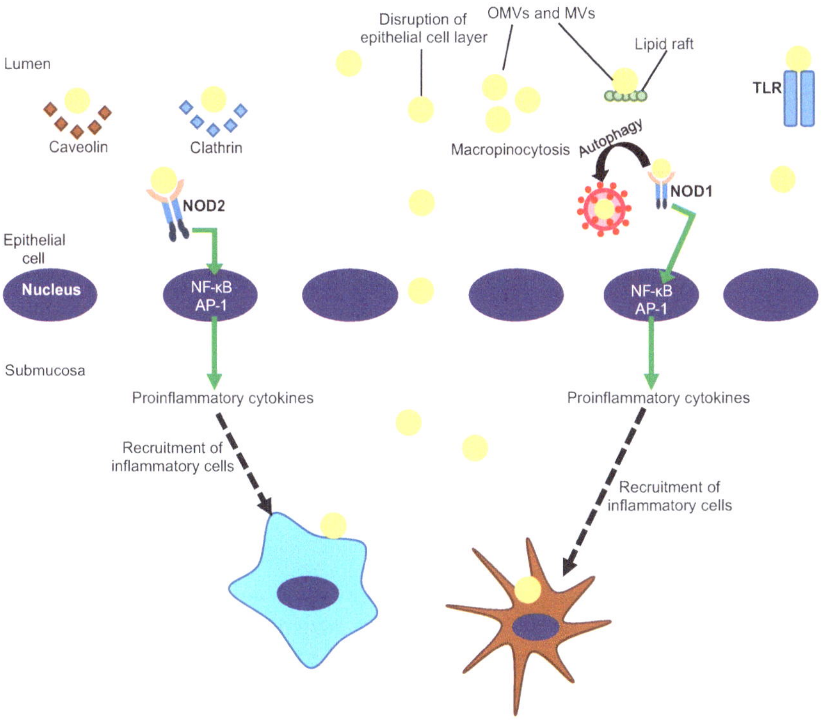


**Fig. 8.3** Bacterial membrane vesicles interact with epithelial cells via a range of different mechanisms to mediate pathogenesis in the host. OMVs and MVs enter host cells via lipid rafts. OMVs can enter cells in a lipid-raft independent manner, via caveolin- or clathrin-mediated endocytosis, macropinocytosis, or by TLR-mediated endocytosis. Adherence and entry into host cells results in the production of pro-inflammatory cytokines and the recruitment of inflammatory innate immune cells. Within epithelial cells, OMVs can be degraded by autophagy. These processes, as well as the delivery of cytotoxic molecules by OMVs to the host cell can initiate the disruption of the epithelial cell layer, allowing OMVs to enter the submucosa

## 8.3 Bacterial Membrane Vesicles Adhere to, Fuse and Enter Host Cells to Mediate a Pro-Inflammatory Response

OMVs have been identified in a diverse range of host tissues, signifying their ability to travel and exist in a range of environments in the host. In addition to interacting with host surface PRRs, OMVs have been shown to adhere, fuse with, and enter host cells in order to contribute to inflammation, pro-inflammatory immune response, and mediate pathogenesis (Fig. 8.3, Bomberger et al. 2009; Furuta et al. 2009; Kaparakis et al. 2010). Early studies demonstrated the ability of OMVs to fuse with host cells and subsequently release their contents into the cytoplasm (Kadurugamuwa and Beveridge 1998). More recent studies confirmed these findings, and additional examples of OMV uptake into host cells were identified.

### 8.3.1 Lipid Raft-Dependent Entry of OMVs and MVs into Epithelial Cells

The most commonly observed mechanism of entry for bacterial OMVs into non-phagocytic host cells involves cholesterol-rich lipid rafts (Fig. 8.3). This has been demonstrated for entry of OMVs produced by Gram-negative pathogens such as enterotoxigenic *E. coli*, *Campylobacter jejuni*, *H. pylori*, *H. influenzae*, *P. aeruginosa*, and *P. gingivalis* (Bauman and Kuehn 2009; Sharpe et al. 2011; Kaparakis et al. 2010; Kesty et al. 2004; Elmi et al. 2012; Furuta et al. 2009; Bomberger et al. 2009). Although not well-characterized, MVs are also thought to enter host cells via cholesterol-rich lipid rafts (Gurung et al. 2011; Thay et al. 2013).

### 8.3.2 Lipid Raft-Independent Mechanisms of OMV Entry into Epithelial Cells

OMVs have also been shown to enter host cells by mechanisms which are independent of lipid rafts (Parker et al. 2010). Clathrin-mediated endocytosis of bacterial OMVs into host cells can be triggered by receptor-mediated binding of ligands at the cell surface and subsequent internalization (Fig. 8.3) (reviewed in O'Donoghue and Krachler 2016). The internalization of clathrin-coated pits leads to degradation of the material contained within these endocytic vacuoles (Rewatkar et al. 2015). Several studies have identified OMV entry into host epithelial cells via clathrin-mediated endocytosis, including OMVs produced by *H. pylori* (Olofsson et al. 2014). However, other studies have demonstrated that inhibition of clathrin-mediated endocytosis did not abolish, but only reduced the entry of *H. pylori* OMVs into host epithelial cells implying there are several mechanisms of entry into host cells (Parker et al. 2010). Caveolin-mediated endocytosis has also been implicated in the uptake of bacterial OMVs into host cells (Fig. 8.3, reviewed in O'Donoghue and Krachler 2016). Another suggested mechanism of OMV entry into host cells is by macropinocytosis; however, this is not thought to be a deliberate process (Fig. 8.3, Lim and Gleeson 2011). OMVs produced by *P. aeruginosa* have been shown to enter airway epithelial cells by macropinocytosis (Bomberger et al. 2009). Furthermore, it has recently been shown that OMV size can regulate their mechanism of entry into host cells, as *H. pylori* OMVs utilized distinct mechanisms of endocytosis or micropinocytosis to enter host cells based on their size (Turner et al. 2018). Collectively, these studies show that OMVs can utilize a range of mechanisms to enter host cells, and their mechanism of entry may be regulated based on receptor-mediated entry, OMV size or their content.

### *8.3.3 Detection and Degradation of Intracellular OMVs by Epithelial Cells*

Until recently, the intracellular fate of OMVs following their entry into host cells remained unknown. It was identified that upon entry into host epithelial cells, *H. pylori* and *P. aeruginosa* OMVs were detected by the innate immune receptor NOD1, resulting in the induction of an inflammatory response (Kaparakis et al. 2010). This finding was further supported by subsequent studies showing the ability of *V. cholerae* OMVs to be detected by NOD1 and NOD2 which resulted in NF-κB activation and the production of pro-inflammatory cytokines (Chatterjee and Chaudhuri 2013; Bielig et al. 2011). It was later shown that NOD1 was essential for the degradation of intracellular OMVs in host epithelial cells, as internalized OMVs were degraded via the host cellular degradation pathway of autophagy in a NOD1-dependent manner (Fig. 8.3, Irving et al. 2014). Specifically, the detection of PG contained within *H. pylori* and *P. aeruginosa* OMVs by the immune receptor NOD1, resulted in their degradation via autophagy (Fig. 8.3, Irving et al. 2014). Therefore, despite OMVs having a range of mechanisms to enter host cells, they can rapidly be degraded once intracellular via the host degradation pathway of autophagy.

Alternatively, bacterial OMVs are not only restricted to interacting with the mucosal cell surface or being internalized by host epithelial cells, as they are capable of disrupting and passing the epithelial cell layer entirely (Fig. 8.3). This has been demonstrated in studies using OMVs from *Treponema denticola* which can disrupt intact epithelium as a means to further mediate pathogenesis in the host (Chi et al. 2003), discussed further below.

## 8.4 Bacterial Membrane Vesicles Are Cytotoxic and Disrupt the Epithelial Cell Barrier

A significant aspect of pathogenesis is the invasion of bacteria or their products into the host tissue. Bacterial membrane vesicles have a range of mechanisms whereby they can function to destroy the epithelial cell barrier integrity and promote cell death to ultimately create damage and promote disease in the host. These mechanisms involve disruption of the epithelial cell barrier and having cytotoxic activities.

### *8.4.1 OMVs Are Cytotoxic to Host Cells*

OMVs have been shown to disrupt the epithelial cell layer, enabling further penetration of bacteria and their products (Fig. 8.3). This has been observed for OMVs including those derived from enterobacteria such as *E. coli* (Wai et al. 2003; Kouokam et al. 2006), and the dental pathogens *T. denticola* and *P. gingivalis* (Chi et al. 2003;

Nakao et al. 2014). Once OMVs disrupt the integrity of the epithelial cell layer and gain access to the submucosa, they can interact directly with immune cells and stimulate an inflammatory response (Fig. 8.3, Nakao et al. 2014).

In addition to the disruption of the epithelium, OMVs can interrupt division of epithelial cells and induce death of innate immune cells, such as macrophages, neutrophils, and DCs. For example, *Acinetobacter baumannii* was shown to deliver cytotoxic OMVs to macrophages (Jin et al. 2011), and OMVs derived from *Actinobacillus actinomycetemcomitans* contained leukotoxins which could kill neutrophil-like cells in vitro (Kato et al. 2002). Moreover, *C. jejuni* OMVs contained cytolethal distending toxin, which disrupted the epithelial cell layer by interrupting cell division (Fig. 8.3, Lindmark et al. 2009).

Antibiotic treatment and stress of bacterial cells has widely been demonstrated to affect the production of OMVs; however, only a few studies have addressed the potential downstream immunostimulatory functions of these OMVs. One study demonstrated that *A. baumannii* produced twofold more OMVs when treated with antibiotics, and these OMVs had an altered proteome compared to control OMVs. When stimulated with OMVs obtained from cells treated with antibiotic, human lung carcinoma cells underwent severe apoptotic cell death, which is suggested to be attributable to the differential expression of OMV proteins in the presence of antibiotic stress (Yun et al. 2018). Collectively, these studies highlight the cytotoxic potential of OMVs on host cells, and the ability of bacteria to modify the cytotoxic abilities of OMVs during conditions of antibiotic stress, to ultimately mediate pathogenesis in the host.

### 8.4.2 MVs Are Cytotoxic to Host Cells

Gram-positive MVs have been shown to also have cytotoxic effects on host cells. MVs produced by *S. aureus* have been shown to deliver the α-haemolysin toxin to host cells, inducing cell death (Thay et al. 2013). However, not all MVs of the same genus are cytotoxic or pathogenic, highlighting that these MVs may have different functions in vivo. For example, analysis of MVs from the Gram-positive resident skin bacterium, *Staphylococcus haemolyticus,* revealed an enrichment of proteins involved in the acquisition of iron by MVs from both commensal and pathogenic *S. haemolyticus* (Cavanagh et al. 2018). However, MVs from a clinical isolate of a coagulase-negative staphylococci strain were enriched in proteins involved in antimicrobial resistance, such as β-lactamase and other virulence factors when compared to MVs isolated from a commensal *S. haemolyticus* strain (Cavanagh et al. 2018). This suggests that bacteria from the same genus can produce MVs which differ in functions within the host. Therefore, it should be noted that MVs produced by a number of commensal and probiotic strains are not cytotoxic to host epithelial cells (Canas et al. 2016; Shen et al. 2012). This further demonstrates the dichotomy between MVs produced by commensal and pathogenic microbes and their role in host immunity and pathogenesis, and suggests that bacterial vesicles have a range of functions in the host.

## 8.5 Bacterial Membrane Vesicles Interact with Host Innate Immune Cells

In addition to their interactions with epithelial cells and cytotoxic abilities, bacterial OMVs have been shown to induce inflammatory responses in innate immune cells such as macrophages, DCs, and neutrophils, which are phagocytic cells responsible for detecting and responding to pathogens. These cells express PRRs, including TLRs and NLRs at high levels (Kawai and Akira 2010). Due to their immunostimulatory abilities, OMVs are thought to interact directly with immune cells within the host and the mechanisms by which they interact with host immune cells are discussed below.

### *8.5.1 OMVs Interact with Macrophages*

A number of studies have examined the effects of OMVs on innate immune cells such as macrophages. Analysis of macrophages from the lung tissue of patients infected with *L. pneumophila* revealed OMVs within the cytoplasm, suggesting that OMVs directly interact with macrophages in vivo (Jager et al. 2014). Another intracellular pathogen, *S.* Typhimurium, is able to penetrate and divide within innate immune cells. One study investigated the release of OMVs from *S.* Typhimurium within macrophages, and showed that OMVs containing virulence proteins were transported to the macrophage cytoplasm from within the *Salmonella*-containing vacuole (Yoon et al. 2011). Additionally, OMVs from *S.* Typhimurium could activate DCs and macrophages and stimulate their expression of a number of inflammatory molecules, including TNFα and IL-12 (Alaniz et al. 2007).

OMVs produced by the oral pathogen *P. gingivalis* are thought to contribute to atherosclerosis by inducing lipid droplet accumulation in macrophages (Qi et al. 2003). *P. gingivalis* OMVs have also been shown to induce the production of nitric oxide and the expression of inducible nitric oxide synthase by mouse macrophages to exacerbate disease (Imayoshi et al. 2011). Additionally, following treatment with *P. gingivalis* OMVs, macrophages were shown to have reduced CD14 on their surface (Duncan et al. 2004). CD14 is a co-receptor involved in TLR4-mediated recognition of LPS, and therefore loss of CD14 results in a reduced immune response to secondary *E. coli* LPS stimulation. These OMVs also suppressed TNFα production, further illustrating the effects of OMVs on immune cells (Duncan et al. 2004). Although detected by membrane-bound TLR4, cytoplasmic LPS can also activate the intracellular sensor caspase-11 in mice and caspase-4/5 in human, resulting in the release of IL-1β and pyroptotic cell death. A recent study reported the ability of *E. coli* OMV-associated LPS to enter the cytosol of host macrophages resulting in caspase-11 activation (Fig. 8.2, Gu et al. 2019). The transport of OMV-associated LPS into the cytosol of macrophages required TLR4 activation and occurred in a TIR domain-containing adaptor-inducing interferon-β (TRIF)-dependent manner (Fig. 8.2, Gu et al. 2019).

Similarly, *P. aeruginosa* OMVs have been shown to upregulate the production of inflammatory cytokines such as macrophage inflammatory protein-2 (MIP-2), TNFα and IL-6 in mouse macrophages (Ellis et al. 2010). *H. pylori* OMVs have also been shown to promote IL-6 production by human peripheral blood mononuclear cells (Winter et al. 2014) and *Salmonella* spp. OMVs induced production of TNF and nitric oxide by mouse macrophages (Alaniz et al. 2007).

*N. meningitidis* OMVs have also been shown to induce cytokines and chemokines such as CC-chemokine ligand 2 (CCL2), CCL3, CCL5, CXCL8 (IL-8), IL-1β, IL-6, IL-10, IL-12p40, IL-12p70, and TNF production in monocytes and macrophages (Tavano et al. 2009). These OMVs also induced the production of monocyte-derived thrombotic factors such as tissue factor and plasminogen activator inhibitor 2. These molecules facilitate intravascular coagulation, microthrombosis, and organ dysfunction which is observed during septicemia caused by this bacterium (Tavano et al. 2009). Additionally, this study revealed that *N. meningitidis* OMVs can activate macrophages to induce adaptive immune responses by the upregulation of co-stimulatory molecules HLA-DR, CD80, and CD86, and intracellular adhesion molecule 1 (ICAM1) (Tavano et al. 2009).

#### 8.5.1.1 Anti-Inflammatory Effects of OMVs on Host Macrophages

Despite a number of pro-inflammatory properties, OMVs can also have anti-inflammatory effects on macrophage precursor cells, known as monocytes. *Neisseria gonorrhoeae*, the causative agent of gonorrhea in humans, has been shown to release OMVs within the mucosa that contain PorB. PorB is a protein which targets the macrophage mitochondrial membrane resulting in host cell death, thus down-regulating the innate immune response to the pathogen (Deo et al. 2018). Moreover, *H. pylori* OMVs induce the production of anti-inflammatory IL-10 by human peripheral blood mononuclear cells to limit inflammation and bacterial clearance, facilitating persistence of the pathogen (Winter et al. 2014). Similarly, *Brucella abortus* OMVs inhibit TLR2, TLR4, and TLR5 responses, limit IFNγ-induced major histocompatibility complex (MHC) class II expression, and promote the internalization of *B. abortus* bacteria by monocytes in an effort to enhance pathogenesis (Pollak et al. 2012). Overall, OMVs have been shown to interact with different cell types, such as macrophages and monocytes to result in pro- or anti-inflammatory responses; however, this has not been demonstrated as widely for Gram-positive MVs.

### *8.5.2 OMVs and MVs Interact with Neutrophils*

Similar to macrophages, neutrophil responses can be modulated by bacterial OMVs in different ways. When stimulated with OMVs derived from *N. meningitidis*, human neutrophils produced TNFα and IL-1β and exhibited upregulation of IL-8,

CCL3, and CCL4 expression (Lapinet et al. 2000). Likewise, OMVs from uropathogenic *E. coli* which contained a toxin, cytotoxic necrotizing factor type 1, impaired the phagocytic and chemotactic abilities of neutrophils (Davis et al. 2006). Additionally, *E. coli* OMVs were recently demonstrated to effectively recruit neutrophils into the lung by stimulating the release of IL-8 in TLR4-dependent and NF-κB manners (Fig. 8.3, Lee et al. 2018).

During cell death, neutrophils produce neutrophil extracellular traps (NETs) which contain antimicrobial peptides, DNA, and histones. NETs are fibrous in structure to assist in trapping and killing pathogens (Brinkmann et al. 2004). OMVs produced by *N. meningitidis* and *Histophilus somni* induced neutrophil cell death and NET formation resulting in the killing of these pathogens. Although *N. meningitidis* OMVs induced NET formation, they also facilitated pathogen colonization and may be a contributing factor to the substantial amount of OMVs in patients with meningococcal disease (Lappann et al. 2013).

More recently, Gram-positive *S. aureus* MVs were also shown to activate neutrophils and induce an immune response. These MVs induced NET formation and neutrophil cell death. Moreover, mice immunized with MVs had increased production of antibodies such as immunoglobulin M (IgM) and total IgG, and were protected against subsequent *S. aureus* challenge (Askarian et al. 2018). Collectively, these studies reveal the multiple mechanisms whereby bacterial membrane vesicles interact with neutrophils to ultimately facilitate the onset of pathogenesis in the host.

### *8.5.3 OMVs Induce DC Maturation*

In addition to their interactions with other immune cells, OMVs from pathogens induce the maturation of DCs, which is important for antigen presentation and adaptive immunity. DC maturation is characterized by their upregulation of MHC class II and production of pro-inflammatory cytokines and chemokines (Inaba et al. 1990). *N. meningitidis* OMVs facilitate their own delivery and internalization into DCs (Schultz et al. 2007), and are known to induce DC maturation, as characterized by the upregulation of MHC class II and co-stimulatory molecules, and their production of pro-inflammatory cytokines and chemokines (Durand et al. 2009). Similarly, *Salmonella* spp. OMVs induced the expression of CD86 and MHC class II on DCs, and their production of TNF and IL-12, as well as promoting protective T and B cell responses (Alaniz et al. 2007). These studies demonstrate the abilities of OMVs to induce DC maturation, leading to antigen presentation to further facilitate an adaptive immune response.

### 8.5.4 *The Effects of Commensal OMVs on Innate Immune Cells*

The effects of commensal OMVs on various innate immune cells is of interest due to their ability to traverse the epithelial cell layer (Chi et al. 2003). Commensal bacteria have been long thought to shape the immune system, and maintain homeostasis between tolerance and immunity (Macpherson and Harris 2004). The contributions of bacterial OMVs and MVs to the host-commensal homeostatic balance is of interest due to the versatility of bacterial vesicles. The Gram-negative commensal *B. fragilis* was found to produce OMVs containing LPS, a well-characterized MAMP. These OMVs were shown to have a protective effect against experimental colitis in mice, by stimulating DCs through TLR2 to adopt an anti-inflammatory profile (Shen et al. 2012). Additionally, OMVs from the commensal *Bacteroides thetaiotaomicron* have been demonstrated to package and deliver BtMinpp (the bacterial homolog of eukaryotic multiple inositol polyphosphate phosphatase 1) into host cells. This is thought to have beneficial effects for both the host and the bacterial community by providing the anticarcinogenic properties of BtMinpp and nutrients, respectively (Stentz et al. 2014).

Conversely, *B. thetaiotaomicron* OMVs were demonstrated to enter and activate mouse macrophages inducing an innate immune response, which may contribute to pathogenesis involved in inflammatory bowel disease (Hickey et al. 2015). Furthermore, OMVs from *Bacteroides vulgatus* were found to induce semi-maturation of bone marrow-derived cells (BMDCs). OMV-treated BMDCs had increased expression of MHC class II, CD40, CD80, and CD86 on their surface. After stimulation with colitogenic *E. coli*, BMDCs treated with OMVs from *B. vulgatus* had unchanged expression of MHC class II and secretion of TNF in the supernatant, indicating the induction of tolerance to bacterial challenge (Maerz et al. 2018). This study highlighted the role of commensal OMVs in immune cell priming. Collectively, these studies highlight the broad and opposing roles of commensal OMVs in host immunity, and their abilities to interact with innate immune cells to ultimately shape the host immune response.

## 8.6 Interactions of OMVs with Endothelial Cells

In addition to interacting with epithelial cells and innate immune cells, OMVs have been shown to interact with endothelial cells in a number of ways to initiate an immune response. OMVs produced by *E. coli* induce the production of IL-6 (Soult et al. 2013), tissue factor, thrombomodulin and adhesion molecules P- and E-selectin (Soult et al. 2014) by human endothelial cells. This leads to the recruitment of pro-inflammatory immune cells, platelet aggregation, and coagulation (Soult et al. 2013). *E. coli*-derived OMVs also upregulate the expression of ICAM1 and vascular cell adhesion molecule 1 on the surface of human microvascular endothelial cells

(Kim et al. 2013a). Mice injected intraperitoneally with these OMVs showed neutrophil aggregation in the lung endothelium in an ICAM1- and TLR4-dependent manner (Kim et al. 2013a).

As well as pro-inflammatory responses in the endothelium, OMVs from *P. gingivalis* were demonstrated to downregulate inflammation in IFNγ-stimulated endothelial cells. OMVs did this by inhibiting the upregulation of MHC class II molecules on the host cell surface, thus limiting antigen presentation (Srisatjaluk et al. 2002). Recently, another study revealed that *P. gingivalis* OMVs suppress endothelial nitric oxide synthase (eNOS) expression in human umbilical vein endothelial cells in vitro, which may contribute to cardiovascular disease (Jia et al. 2015).

The interaction of bacterial vesicles with host cells, including innate immune cells, can lead to activation of the adaptive immune system. Bacterial vesicles have also been described to interact with the adaptive immune system directly, which is discussed below.

## 8.7 Adaptive Immune Responses to Bacterial Membrane Vesicles

The adaptive immune system is responsible for generating a pathogen-specific immune response that is capable of clearing the pathogen, in addition to generating immune memory that can be reactivated when the host reencounters the pathogen. Specifically, the adaptive immune system is responsible for the production of antigen-specific T and B cells that can target the pathogen directly, or produce pathogen-specific antibodies (Iwasaki and Medzhitov 2015). The hallmarks of this system are the random generation of receptor specificities (T cell receptor and immunoglobulins) by somatic recombination. This results in the possibility of generating a nearly unlimited spectrum of antigen specificity. Populations of cells with cognate receptors for the pathogen are selected and enriched by clonal selection, which means that specific activation of the cell bearing the receptor with the highest affinity leads to its proliferation.

Phagocytic cells such as macrophages, DCs and B cells act as antigen-presenting cells (APCs) in this process. These cells can present peptide antigens via surface receptors (MHC class I and II), which can be recognized by naïve T lymphocytes, leading to their activation, proliferation, and differentiation to effector cell subsets (reviewed in Unanue 1984; Guermonprez et al. 2002; Rodriguez-Pinto 2005). Helper T lymphocytes (CD4 positive cells) are also essential for the activation of B cells and their differentiation into antigen producing plasma cells. Collectively, the activation of the adaptive immune response results in the generation of memory T and B cells which can confer long lived immunity, and is the basis for the idea of bacterial membrane vesicles functioning as vaccines.

### 8.7.1 Bacterial Membrane Vesicles Shape Adaptive Immunity

Due to their immune stimulating properties and vast evidence of their interactions with the innate immune system, OMVs and MVs are thought to influence the hosts adaptive immune response. OMVs from *H. pylori* were able to adsorb antibodies in sera from infected patients, thus providing protection for the bacteria (Hynes et al. 2005). Following this, a number of in vivo studies have further revealed the complexity of OMVs as vehicles for antigens and their initiation of adaptive immune responses, which will be discussed here briefly.

### 8.7.2 Generation of OMV-Specific Protective Immune Responses

Many in vivo studies have demonstrated the generation of specific antibody responses to bacterial OMVs. For example, OMVs produced by *N. meningitidis* induce the upregulation of MHC class II receptors as well as costimulatory molecules such as CD80 (B7-1) and CD86 (B7-2) on macrophages (Tavano et al. 2009). Furthermore, *S.* Typhimurium and *V. cholerae* OMVs were also described as an abundant source of antigens, which could activate $CD4^+$ T cells in order to generate antibody-specific B cell responses (Alaniz et al. 2007; Chatterjee and Chaudhuri 2013). Specifically, in a mouse model of infection and sepsis, *Burkholderia pseudomallei* OMVs administered subcutaneously were shown to induce serum IgG and T helper 1 (Th1) cell responses (Nieves et al. 2011). A similar finding was observed in mice immunized intraperitoneally with *E. coli* OMVs. Exposure to OMVs from *E. coli* resulted in stimulation of T cell immunity primarily driven by Th1 and Th17 cells (Kim et al. 2013b).

Mice vaccinated intramuscularly with OMVs produced by *A. baumannii* produced high levels of OMV-specific IgG antibodies, and subsequent challenge with *A. baumannii* intranasally resulted in mucosal IgA and IgG responses in a pneumonia model (Huang et al. 2014). These mice were also challenged with *A. baumannii* intraperitoneally as a model for bacterial sepsis, and were found to have increased survival when compared to the control group. In this study, mice passively immunized intravenously with antiserum from vaccinated mice also had protection from *A. baumannii* intranasal challenge. The antisera from immunized mice was also shown to have opsonophagocytic properties and resulted in killing of numerous clinical isolates of *A. baumannii* by macrophages in vitro (Huang et al. 2014).

Furthermore, OMVs produced by streptomycin resistant clinical strains of nontypeable *H. influenzae* used to immunize mice both intranasally and intraperitoneally resulted in robust IgG1 antibody production, and IgA production following intranasal immunization. However, when intraperitoneally immunized mice were challenged with *H. influenzae* strains, they were found to have a higher rate of bacterial colonization when compared to intranasally immunized mice (Roier et al.

2012). These reports demonstrate the shared capacity of OMVs from different bacteria to induce OMV-specific immune responses in vivo.

### *8.7.3 OMVs Can Modulate B Cell Responses*

OMVs have also been demonstrated to modulate B cell responses directly. OMVs produced by *M. catarrhalis* were shown to contain an outer membrane protein known as *Moraxella* immunoglobulin D binding protein (MID). MID is an antigen which potently activates human tonsillar B cells via clustering of B cell receptors which mediates endocytosis. TLR9 was also involved in detection of DNA associated with these OMVs, which induced full B cell activation (Vidakovics et al. 2010). *H. influenzae* also uses a similar approach to modulate B cell responses, by producing OMVs which sensitize B cells resulting in an increase in IgG and IgM production; however, these antibodies do not recognize *H. influenzae* (Deknuydt et al. 2014). In a manner independent of T cells, *Neisseria lactamica* OMVs induce B cell proliferation (Vaughan et al. 2010) and polyclonal antibody production (Vaughan et al. 2009) highlighting the versatile interactions of OMVs with adaptive immune cells.

### *8.7.4 Gram-Positive MVs Mediate Adaptive Immune Responses*

Only a handful of studies have analyzed the effects of Gram-positive MVs on mediating adaptive immunity. MVs produced by the pathogen *Clostridium perfringens* have been shown to induce the production of high titer IgG1 in mice (Jiang et al. 2014). Additionally, *B. anthracis* MVs have been demonstrated to induce a robust IgM response in mice (Rivera et al. 2010). Mice immunized intramuscularly with *Streptococcus pneumoniae* MVs were protected from intranasal *S. pneumoniae* challenge. These mice had increased IgG levels, and increased survival (Olaya-Abril et al. 2014).

*S. aureus* MVs administered intramuscularly or intravenously have been demonstrated to induce protective immune responses in mice using various models of infection including oropharyngeal or intravenous challenge, respectively, the latter inducing a Th1-mediated response (Choi et al. 2015; Wang et al. 2018). It is because of the antigenic properties of OMVs and MVs that bacterial membrane vesicles are currently being considered as future vesicle vaccines.

### 8.7.5 Development of OMVs and MVs for Use as Vaccines

Due to their ability to mediate innate and adaptive immune responses in vivo, OMVs have been examined for their ability to confer protection against bacterial challenge and refine their use as vaccine candidates. There are numerous studies that have demonstrated the effect of OMVs on the adaptive immune system, with studies demonstrating the ability of OMVs and MVs to mediate protection from infection, sepsis, and death (Fantappie et al. 2014; Huang et al. 2014; Camacho et al. 2013; Kim et al. 2013b). OMVs have also been proven to be effective at generating a robust immune response and have been found to be more protective against live challenge than vaccination using bacterial membrane fractions (Tavano et al. 2009). In addition, vaccination with OMVs is more effective at generating a bactericidal serum response in mice when compared to serum responses generated as a result of infection alone (Zhu et al. 2005). There are many examples of the effective use of OMVs and MVs as vaccine candidates, in addition to their future therapeutic applications, and these are discussed in further detail in Chap. 10.

## 8.8 Conclusions and Perspectives

The innate and adaptive immune systems are complex, as are their interactions with invading and commensal organisms. Bacterial membrane vesicles and their immunostimulatory functions have not been fully characterized; however, studies have examined their effects on host epithelial, innate and adaptive immune cells (reviewed in Kaparakis-Liaskos and Ferrero 2015; Manning and Kuehn 2011). OMVs from various bacterial species have been shown to induce a constant interplay between pro-inflammatory and anti-inflammatory signals, and the overall response can vary depending on the amount of OMVs, as well as their content (reviewed in Kaparakis-Liaskos and Ferrero 2015). Bacterial OMVs are able to stimulate TLRs and NLRs of host cells. Furthermore, OMVs and MVs have a wide variety of inherent properties which can be beneficial to the pathogenic capabilities of their parent bacteria, including their contributions to biofilm formation, and antibiotic and host immune molecule defense (Tan et al. 2007; Jagels et al. 1996; Baumgarten et al. 2012; Schooling and Beveridge 2006; Yonezawa et al. 2009). Overall, the vast mechanisms by which OMVs and MVs can interact with and contribute to immunity and pathogenesis are only emerging at present, and further studies should aim to understand the complex nature of membrane vesicle–immune interactions. This knowledge may contribute to forthcoming nanoparticle technologies, whereby OMVs and MVs can be utilized for therapeutics in addition to novel targets to limit bacterial vesicle-mediated pathology.

# References

Abachin E, Poyart C, Pellegrini E, Milohanic E, Fiedler F, Berche P, Trieu-Cuot P (2002) Formation of D-alanyl-lipoteichoic acid is required for adhesion and virulence of *Listeria monocytogenes*. Mol Microbiol 43(1):1–14

Ahmadi Badi S, Khatami SH, Irani SH, Siadat SD (2019) Induction effects of *Bacteroides fragilis* derived outer membrane vesicles on toll like receptor 2, toll like receptor 4 genes expression and cytokines concentration in human intestinal epithelial cells. Cell J 21(1):57–61

Akira S, Takeda K (2004) Toll-like receptor signaling. Nat Rev Immunol 4(7):499–511

Alaniz RC, Deatherage BL, Lara JC, Cookson BT (2007) Membrane vesicles are immunogenic facsimiles of *Salmonella* Typhimurium that potently activate dendritic cells, prime B and T cell responses, and stimulate protective immunity in vivo. J Immunol 179(11):7692–7701

Askarian F, Lapek JD Jr, Dongre M, Tsai CM, Kumaraswamy M, Kousha A, Valderrama JA, Ludviksen JA, Cavanagh JP, Uchiyama S, Mollnes TE, Gonzalez DJ, Wai SN, Nizet V, Johannessen M (2018) *Staphylococcus aureus* membrane-derived vesicles promote bacterial virulence and confer protective immunity in murine infection models. Front Microbiol 9:262

Band VI, Weiss DS (2015) Mechanisms of antimicrobial peptide resistance in Gram-negative bacteria. Antibiotics (Basel) 4(1):18–41

Barrington R, Zhang M, Fischer M, Carroll MC (2001) The role of complement in inflammation and adaptive immunity. Immunol Rev 180:5–15

Bauman SJ, Kuehn MJ (2006) Purification of outer membrane vesicles from *Pseudomonas aeruginosa* and their activation of an IL-8 response. Microbes Infect 8(9-10):2400–2408

Bauman SJ, Kuehn MJ (2009) *Pseudomonas aeruginosa* vesicles associate with and are internalized by human lung epithelial cells. BMC Microbiol 9:26

Baumgarten T, Sperling S, Seifert J, von Bergen M, Steiniger F, Wick LY, Heipieper HJ (2012) Membrane vesicle formation as a multiple-stress response mechanism enhances *Pseudomonas putida* DOT-T1E cell surface hydrophobicity and biofilm formation. Appl Environ Microbiol 78(17):6217–6224

Bielig H, Rompikuntal PK, Dongre M, Zurek B, Lindmark B, Ramstedt M, Wai SN, Kufer TA (2011) NOD-like receptor activation by outer membrane vesicles from *Vibrio cholerae* non-O1 non-O139 strains is modulated by the quorum-sensing regulator HapR. Infect Immun 79(4): 1418–1427

Boller T, Felix G (2009) A renaissance of elicitors: perception of microbe-associated molecular patterns and danger signals by pattern-recognition receptors. Annu Rev Plant Biol 60:379–406

Bomberger JM, Maceachran DP, Coutermarsh BA, Ye S, O'Toole GA, Stanton BA (2009) Long-distance delivery of bacterial virulence factors by *Pseudomonas aeruginosa* outer membrane vesicles. PLoS Pathog 5(4):e1000382

Boxx GM, Cheng G (2016) The roles of type I interferon in bacterial infection. Cell Host Microbe 19(6):760–769

Brinkmann V, Reichard U, Goosmann C, Fauler B, Uhlemann Y, Weiss DS, Weinrauch Y, Zychlinsky A (2004) Neutrophil extracellular traps kill bacteria. Science 303(5663):1532–1535

Bruno VM, Hannemann S, Lara-Tejero M, Flavell RA, Kleinstein SH, Galan JE (2009) *Salmonella* Typhimurium type III secretion effectors stimulate innate immune responses in cultured epithelial cells. PLoS Pathog 5(8):e1000538

Camacho AI, Irache JM, de Souza J, Sanchez-Gomez S, Gamazo C (2013) Nanoparticle-based vaccine for mucosal protection against *Shigella flexneri* in mice. Vaccine 31(32):3288–3294

Canas MA, Gimenez R, Fabrega MJ, Toloza L, Baldoma L, Badia J (2016) Outer membrane vesicles from the probiotic *Escherichia coli* Nissle 1917 and the commensal ECOR12 enter intestinal epithelial cells via clathrin-dependent endocytosis and elicit differential effects on DNA damage. PLoS One 11(8):e0160374

Cavanagh JP, Pain M, Askarian F, Bruun JA, Urbarova I, Wai SN, Schmidt F, Johannessen M (2018) Comparative exoproteome profiling of an invasive and a commensal *Staphylococcus haemolyticus* isolate. J Proteomics 197:106–114

Cecil JD, Sirisaengtaksin N, O'Brien-Simpson NM, Krachler AM (2019) Outer membrane vesicle-host cell interactions. Microbiol Spectr 7(1)
Chatterjee D, Chaudhuri K (2013) *Vibrio cholerae* O395 outer membrane vesicles modulate intestinal epithelial cells in a NOD1 protein-dependent manner and induce dendritic cell-mediated Th2/Th17 cell responses. J Biol Chem 288(6):4299–4309
Chatterjee SN, Das J (1967) Electron microscopic observations on the excretion of cell-wall material by *Vibrio cholerae*. J Gen Microbiol 49(1):1–11
Chi B, Qi M, Kuramitsu HK (2003) Role of dentilisin in *Treponema denticola* epithelial cell layer penetration. Res Microbiol 154(9):637–643
Choi DS, Kim DK, Choi SJ, Lee J, Choi JP, Rho S, Park SH, Kim YK, Hwang D, Gho YS (2011) Proteomic analysis of outer membrane vesicles derived from *Pseudomonas aeruginosa*. Proteomics 11(16):3424–3429
Choi SJ, Kim MH, Jeon J, Kim OY, Choi Y, Seo J, Hong SW, Lee WH, Jeon SG, Gho YS, Jee YK, Kim YK (2015) Active immunization with extracellular vesicles derived from *Staphylococcus aureus* effectively protects against staphylococcal lung infections, mainly via Th1 cell-mediated immunity. PLoS One 10(9):e0136021
Davis JM, Carvalho HM, Rasmussen SB, O'Brien AD (2006) Cytotoxic necrotizing factor type 1 delivered by outer membrane vesicles of uropathogenic *Escherichia coli* attenuates polymorphonuclear leukocyte antimicrobial activity and chemotaxis. Infect Immun 74(8): 4401–4408
Deknuydt F, Nordstrom T, Riesbeck K (2014) Diversion of the host humoral response: a novel virulence mechanism of *Haemophilus influenzae* mediated via outer membrane vesicles. J Leukoc Biol 95(6):983–991
Deo P, Chow SH, Hay ID, Kleifeld O, Costin A, Elgass KD, Jiang JH, Ramm G, Gabriel K, Dougan G, Lithgow T, Heinz E, Naderer T (2018) Outer membrane vesicles from *Neisseria gonorrhoeae* target PorB to mitochondria and induce apoptosis. PLoS Pathog 14(3):e1006945
Duncan L, Yoshioka M, Chandad F, Grenier D (2004) Loss of lipopolysaccharide receptor CD14 from the surface of human macrophage-like cells mediated by *Porphyromonas gingivalis* outer membrane vesicles. Microb Pathog 36(6):319–325
Duperthuy M, Sjostrom AE, Sabharwal D, Damghani F, Uhlin BE, Wai SN (2013) Role of the *Vibrio cholerae* matrix protein Bap1 in cross-resistance to antimicrobial peptides. PLoS Pathog 9(10):e1003620
Durand V, Mackenzie J, de Leon J, Mesa C, Quesniaux V, Montoya M, Le Bon A, Wong SY (2009) Role of lipopolysaccharide in the induction of type I interferon-dependent cross-priming and IL-10 production in mice by meningococcal outer membrane vesicles. Vaccine 27 (13):1912–1922
Elliott EI, Sutterwala FS (2015) Initiation and perpetuation of NLRP3 inflammasome activation and assembly. Immunol Rev 265(1):35–52
Ellis TN, Kuehn MJ (2010) Virulence and immunomodulatory roles of bacterial outer membrane vesicles. Microbiol Mol Biol Rev 74(1):81–94
Ellis TN, Leiman SA, Kuehn MJ (2010) Naturally produced outer membrane vesicles from *Pseudomonas aeruginosa* elicit a potent innate immune response via combined sensing of both lipopolysaccharide and protein components. Infect Immun 78(9):3822–3831
Elmi A, Watson E, Sandu P, Gundogdu O, Mills DC, Inglis NF, Manson E, Imrie L, Bajaj-Elliott M, Wren BW, Smith DG, Dorrell N (2012) *Campylobacter jejuni* outer membrane vesicles play an important role in bacterial interactions with human intestinal epithelial cells. Infect Immun 80(12):4089–4098
Fantappie L, de Santis M, Chiarot E, Carboni F, Bensi G, Jousson O, Margarit I, Grandi G (2014) Antibody-mediated immunity induced by engineered *Escherichia coli* OMVs carrying heterologous antigens in their lumen. J Extracell Vesicles 3
Faure E, Equils O, Sieling PA, Thomas L, Zhang FX, Kirschning CJ, Polentarutti N, Muzio M, Arditi M (2000) Bacterial lipopolysaccharide activates NF-kappaB through toll-like receptor 4 (TLR-4) in cultured human dermal endothelial cells. Differential expression of TLR-4 and TLR-2 in endothelial cells. J Biol Chem 275(15):11058–11063

Fiocca R, Necchi V, Sommi P, Ricci V, Telford J, Cover TL, Solcia E (1999) Release of *Helicobacter pylori* vacuolating cytotoxin by both a specific secretion pathway and budding of outer membrane vesicles. Uptake of released toxin and vesicles by gastric epithelium. J Pathol 188(2):220–226

Fritz JH, Le Bourhis L, Sellge G, Magalhaes JG, Fsihi H, Kufer TA, Collins C, Viala J, Ferrero RL, Girardin SE, Philpott DJ (2007) Nod1-mediated innate immune recognition of peptidoglycan contributes to the onset of adaptive immunity. Immunity 26(4):445–459

Furuta N, Takeuchi H, Amano A (2009) Entry of *Porphyromonas gingivalis* outer membrane vesicles into epithelial cells causes cellular functional impairment. Infect Immun 77(11): 4761–4770

Galka F, Wai SN, Kusch H, Engelmann S, Hecker M, Schmeck B, Hippenstiel S, Uhlin BE, Steinert M (2008) Proteomic characterization of the whole secretome of *Legionella pneumophila* and functional analysis of outer membrane vesicles. Infect Immun 76(5):1825–1836

Girardin SE, Boneca IG, Carneiro LA, Antignac A, Jehanno M, Viala J, Tedin K, Taha MK, Labigne A, Zahringer U, Coyle AJ, DiStefano PS, Bertin J, Sansonetti PJ, Philpott DJ (2003a) Nod1 detects a unique muropeptide from gram-negative bacterial peptidoglycan. Science 300(5625):1584–1587

Girardin SE, Boneca IG, Viala J, Chamaillard M, Labigne A, Thomas G, Philpott DJ, Sansonetti PJ (2003b) Nod2 is a general sensor of peptidoglycan through muramyl dipeptide (MDP) detection. J Biol Chem 278(11):8869–8872

Gu L, Meng R, Tang Y, Zhao K, Liang F, Zhang R, Xue Q, Chen F, Xiao X, Wang H, Wang H, Billiar TR, Lu B (2019) Toll-like receptor 4 signaling licenses the cytosolic transport of lipopolysaccharide from bacterial outer membrane vesicles. Shock 51(2):256–265

Guermonprez P, Valladeau J, Zitvogel L, Thery C, Amigorena S (2002) Antigen presentation and T cell stimulation by dendritic cells. Annu Rev Immunol 20:621–667

Gunn JS, Ryan SS, Van Velkinburgh JC, Ernst RK, Miller SI (2000) Genetic and functional analysis of a PmrA-PmrB-regulated locus necessary for lipopolysaccharide modification, antimicrobial peptide resistance, and oral virulence of *Salmonella enterica serovar* Typhimurium. Infect Immun 68(11):6139–6146

Gurung M, Moon DC, Choi CW, Lee JH, Bae YC, Kim J, Lee YC, Seol SY, Cho DT, Kim SI, Lee JC (2011) *Staphylococcus aureus* produces membrane-derived vesicles that induce host cell death. PLoS One 6(11):e27958

Hickey CA, Kuhn KA, Donermeyer DL, Porter NT, Jin C, Cameron EA, Jung H, Kaiko GE, Wegorzewska M, Malvin NP, Glowacki RW, Hansson GC, Allen PM, Martens EC, Stappenbeck TS (2015) Colitogenic *Bacteroides thetaiotaomicron* antigens access host immune cells in a sulfatase-dependent manner via outer membrane vesicles. Cell Host Microbe 17(5): 672–680

Holst J, Martin D, Arnold R, Huergo CC, Oster P, O'Hallahan J, Rosenqvist E (2009) Properties and clinical performance of vaccines containing outer membrane vesicles from *Neisseria meningitidis*. Vaccine 27(Suppl 2):B3–B12

Hood MI, Skaar EP (2012) Nutritional immunity: transition metals at the pathogen-host interface. Nat Rev Microbiol 10(8):525–537

Huang W, Yao Y, Long Q, Yang X, Sun W, Liu C, Jin X, Li Y, Chu X, Chen B, Ma Y (2014) Immunization against multidrug-resistant *Acinetobacter baumannii* effectively protects mice in both pneumonia and sepsis models. PLoS One 9(6):e100727

Hynes SO, Keenan JI, Ferris JA, Annuk H, Moran AP (2005) Lewis epitopes on outer membrane vesicles of relevance to *Helicobacter pylori* pathogenesis. Helicobacter 10(2):146–156

Imayoshi R, Cho T, Kaminishi H (2011) NO production in RAW264 cells stimulated with *Porphyromonas gingivalis* extracellular vesicles. Oral Dis 17(1):83–89

Inaba K, Metlay JP, Crowley MT, Steinman RM (1990) Dendritic cells pulsed with protein antigens in vitro can prime antigen-specific, MHC-restricted T cells in situ. J Exp Med 172(2):631–640

Inohara N, Nunez G (2001) The NOD: a signaling module that regulates apoptosis and host defense against pathogens. Oncogene 20(44):6473–6481

Inohara N, Koseki T, Lin J, del Peso L, Lucas PC, Chen FF, Ogura Y, Nunez G (2000) An induced proximity model for NF-kappa B activation in the Nod1/RICK and RIP signaling pathways. J Biol Chem 275(36):27823–27831

Irving AT, Mimuro H, Kufer TA, Lo C, Wheeler R, Turner LJ, Thomas BJ, Malosse C, Gantier MP, Casillas LN, Votta BJ, Bertin J, Boneca IG, Sasakawa C, Philpott DJ, Ferrero RL, Kaparakis-Liaskos M (2014) The immune receptor NOD1 and kinase RIP2 interact with bacterial peptidoglycan on early endosomes to promote autophagy and inflammatory signaling. Cell Host Microbe 15(5):623–635

Ismail S, Hampton MB, Keenan JI (2003) *Helicobacter pylori* outer membrane vesicles modulate proliferation and interleukin-8 production by gastric epithelial cells. Infect Immun 71(10): 5670–5675

Iwasaki A, Medzhitov R (2015) Control of adaptive immunity by the innate immune system. Nat Immunol 16(4):343–353

Jagels MA, Travis J, Potempa J, Pike R, Hugli TE (1996) Proteolytic inactivation of the leukocyte C5a receptor by proteinases derived from *Porphyromonas gingivalis*. Infect Immun 64(6): 1984–1991

Jager J, Marwitz S, Tiefenau J, Rasch J, Shevchuk O, Kugler C, Goldmann T, Steinert M (2014) Human lung tissue explants reveal novel interactions during *Legionella pneumophila* infections. Infect Immun 82(1):275–285

Janeway CA Jr, Medzhitov R (2002) Innate immune recognition. Annu Rev Immunol 20:197–216

Jia Y, Guo B, Yang W, Zhao Q, Jia W, Wu Y (2015) Rho kinase mediates *Porphyromonas gingivalis* outer membrane vesicle-induced suppression of endothelial nitric oxide synthase through ERK1/2 and p38 MAPK. Arch Oral Biol 60(3):488–495

Jiang Y, Kong Q, Roland KL, Curtiss R 3rd (2014) Membrane vesicles of *Clostridium perfringens* type A strains induce innate and adaptive immunity. Int J Med Microbiol 304(3-4):431–443

Jin JS, Kwon SO, Moon DC, Gurung M, Lee JH, Kim SI, Lee JC (2011) *Acinetobacter baumannii* secretes cytotoxic outer membrane protein A via outer membrane vesicles. PLoS One 6(2): e17027

Joiner KA (1988) Complement evasion by bacteria and parasites. Annu Rev Microbiol 42:201–230

Jun SH, Lee JH, Kim BR, Kim SI, Park TI, Lee JC, Lee YC (2013) *Acinetobacter baumannii* outer membrane vesicles elicit a potent innate immune response via membrane proteins. PLoS One 8(8):e71751

Jun SH, Lee JH, Kim SI, Choi CW, Park TI, Jung HR, Cho JW, Kim SH, Lee JC (2017) *Staphylococcus aureus*-derived membrane vesicles exacerbate skin inflammation in atopic dermatitis. Clin Exp Allergy 47(1):85–96

Kadurugamuwa JL, Beveridge TJ (1995) Virulence factors are released from *Pseudomonas aeruginosa* in association with membrane vesicles during normal growth and exposure to gentamicin: a novel mechanism of enzyme secretion. J Bacteriol 177(14):3998–4008

Kadurugamuwa JL, Beveridge TJ (1998) Delivery of the non-membrane-permeative antibiotic gentamicin into mammalian cells by using *Shigella flexneri* membrane vesicles. Antimicrob Agents Chemother 42(6):1476–1483

Kanneganti TD, Lamkanfi M, Nunez G (2007) Intracellular NOD-like receptors in host defense and disease. Immunity 27(4):549–559

Kaparakis M, Turnbull L, Carneiro L, Firth S, Coleman HA, Parkington HC, Le Bourhis L, Karrar A, Viala J, Mak J, Hutton ML, Davies JK, Crack PJ, Hertzog PJ, Philpott DJ, Girardin SE, Whitchurch CB, Ferrero RL (2010) Bacterial membrane vesicles deliver peptidoglycan to NOD1 in epithelial cells. Cell Microbiol 12(3):372–385

Kaparakis-Liaskos M, Ferrero RL (2015) Immune modulation by bacterial outer membrane vesicles. Nat Rev Immunol 15(6):375–387

Kato S, Kowashi Y, Demuth DR (2002) Outer membrane-like vesicles secreted by *Actinobacillus actinomycetemcomitans* are enriched in leukotoxin. Microb Pathog 32(1):1–13

Kawai T, Akira S (2008) Toll-like receptor and RIG-I-like receptor signaling. Ann N Y Acad Sci 1143:1–20

Kawai T, Akira S (2010) The role of pattern-recognition receptors in innate immunity: update on Toll-like receptors. Nat Immunol 11(5):373–384

Keenan J, Day T, Neal S, Cook B, Perez-Perez G, Allardyce R, Bagshaw P (2000) A role for the bacterial outer membrane in the pathogenesis of *Helicobacter pylori* infection. FEMS Microbiol Lett 182(2):259–264

Kesty NC, Mason KM, Reedy M, Miller SE, Kuehn MJ (2004) Enterotoxigenic *Escherichia coli* vesicles target toxin delivery into mammalian cells. EMBO J 23(23):4538–4549

Kim JH, Yoon YJ, Lee J, Choi EJ, Yi N, Park KS, Park J, Lotvall J, Kim YK, Gho YS (2013a) Outer membrane vesicles derived from *Escherichia coli* up-regulate expression of endothelial cell adhesion molecules in vitro and in vivo. PLoS One 8(3):e59276

Kim OY, Hong BS, Park KS, Yoon YJ, Choi SJ, Lee WH, Roh TY, Lotvall J, Kim YK, Gho YS (2013b) Immunization with *Escherichia coli* outer membrane vesicles protects bacteria-induced lethality via Th1 and Th17 cell responses. J Immunol 190(8):4092–4102

Kouokam JC, Wai SN, Fallman M, Dobrindt U, Hacker J, Uhlin BE (2006) Active cytotoxic necrotizing factor 1 associated with outer membrane vesicles from uropathogenic *Escherichia coli*. Infect Immun 74(4):2022–2030

Kulp A, Kuehn MJ (2010) Biological functions and biogenesis of secreted bacterial outer membrane vesicles. Annu Rev Microbiol 64:163–184

Kumar H, Kawai T, Akira S (2009) Toll-like receptors and innate immunity. Biochem Biophys Res Commun 388(4):621–625

Lapinet JA, Scapini P, Calzetti F, Perez O, Cassatella MA (2000) Gene expression and production of tumor necrosis factor alpha, interleukin-1beta (IL-1beta), IL-8, macrophage inflammatory protein 1alpha (MIP-1alpha), MIP-1beta, and gamma interferon-inducible protein 10 by human neutrophils stimulated with group B meningococcal outer membrane vesicles. Infect Immun 68(12):6917–6923

Lappann M, Danhof S, Guenther F, Olivares-Florez S, Mordhorst IL, Vogel U (2013) In vitro resistance mechanisms of *Neisseria meningitidis* against neutrophil extracellular traps. Mol Microbiol 89(3):433–449

Lee EY, Choi DY, Kim DK, Kim JW, Park JO, Kim S, Kim SH, Desiderio DM, Kim YK, Kim KP, Gho YS (2009) Gram-positive bacteria produce membrane vesicles: proteomics-based characterization of *Staphylococcus aureus*-derived membrane vesicles. Proteomics 9(24):5425–5436

Lee J, Yoon YJ, Kim JH, Dinh NTH, Go G, Tae S, Park KS, Park HT, Lee C, Roh TY, Di Vizio D, Gho YS (2018) Outer membrane vesicles derived from *Escherichia coli* regulate neutrophil migration by induction of endothelial IL-8. Front Microbiol 9:2268

Lim JP, Gleeson PA (2011) Macropinocytosis: an endocytic pathway for internalising large gulps. Immunol Cell Biol 89(8):836–843

Lindmark B, Rompikuntal PK, Vaitkevicius K, Song T, Mizunoe Y, Uhlin BE, Guerry P, Wai SN (2009) Outer membrane vesicle-mediated release of cytolethal distending toxin (CDT) from *Campylobacter jejuni*. BMC Microbiol 9:220

Macdonald IA, Kuehn MJ (2013) Stress-induced outer membrane vesicle production by *Pseudomonas aeruginosa*. J Bacteriol 195(13):2971–2981

Macpherson AJ, Harris NL (2004) Interactions between commensal intestinal bacteria and the immune system. Nat Rev Immunol 4(6):478–485

Maerz JK, Steimle A, Lange A, Bender A, Fehrenbacher B, Frick JS (2018) Outer membrane vesicles blebbing contributes to *B. vulgatus* mpk-mediated immune response silencing. Gut Microbes 9(1):1–12

Manning AJ, Kuehn MJ (2011) Contribution of bacterial outer membrane vesicles to innate bacterial defense. BMC Microbiol 11:258

Martinon F, Burns K, Tschopp J (2002) The inflammasome: a molecular platform triggering activation of inflammatory caspases and processing of proIL-beta. Mol Cell 10(2):417–426

McBroom AJ, Kuehn MJ (2007) Release of outer membrane vesicles by Gram-negative bacteria is a novel envelope stress response. Mol Microbiol 63(2):545–558

McBroom AJ, Johnson AP, Vemulapalli S, Kuehn MJ (2006) Outer membrane vesicle production by *Escherichia coli* is independent of membrane instability. J Bacteriol 188(15):5385–5392

Medzhitov R, Janeway CA Jr (1997) Innate immunity: the virtues of a nonclonal system of recognition. Cell 91(3):295–298

Milward MR, Chapple IL, Wright HJ, Millard JL, Matthews JB, Cooper PR (2007) Differential activation of NF-kappaB and gene expression in oral epithelial cells by periodontal pathogens. Clin Exp Immunol 148(2):307–324
Moskowitz SM, Ernst RK, Miller SI (2004) PmrAB, a two-component regulatory system of *Pseudomonas aeruginosa* that modulates resistance to cationic antimicrobial peptides and addition of aminoarabinose to lipid A. J Bacteriol 186(2):575–579
Nakao R, Takashiba S, Kosono S, Yoshida M, Watanabe H, Ohnishi M, Senpuku H (2014) Effect of *Porphyromonas gingivalis* outer membrane vesicles on gingipain-mediated detachment of cultured oral epithelial cells and immune responses. Microbes Infect 16(1):6–16
Namork E, Brandtzaeg P (2002) Fatal meningococcal septicaemia with "blebbing" meningococcus. Lancet 360(9347):1741
Nieves W, Asakrah S, Qazi O, Brown KA, Kurtz J, Aucoin DP, McLachlan JB, Roy CJ, Morici LA (2011) A naturally derived outer-membrane vesicle vaccine protects against lethal pulmonary *Burkholderia pseudomallei* infection. Vaccine 29(46):8381–8389
O'Donoghue EJ, Krachler AM (2016) Mechanisms of outer membrane vesicle entry into host cells. Cell Microbiol 18(11):1508–1517
Olaya-Abril A, Prados-Rosales R, McConnell MJ, Martin-Pena R, Gonzalez-Reyes JA, Jimenez-Munguia I, Gomez-Gascon L, Fernandez J, Luque-Garcia JL, Garcia-Lidon C, Estevez H, Pachon J, Obando I, Casadevall A, Pirofski LA, Rodriguez-Ortega MJ (2014) Characterization of protective extracellular membrane-derived vesicles produced by *Streptococcus pneumoniae*. J Proteome 106:46–60
Olofsson A, Nygard Skalman L, Obi I, Lundmark R, Arnqvist A (2014) Uptake of *Helicobacter pylori* vesicles is facilitated by clathrin-dependent and clathrin-independent endocytic pathways. MBio 5(3):e00979–e00914
Parker H, Chitcholtan K, Hampton MB, Keenan JI (2010) Uptake of *Helicobacter pylori* outer membrane vesicles by gastric epithelial cells. Infect Immun 78(12):5054–5061
Pathirana RD, Kaparakis-Liaskos M (2016) Bacterial membrane vesicles: biogenesis, immune regulation and pathogenesis. Cell Microbiol 18(11):1518–1524
Peschel A, Jack RW, Otto M, Collins LV, Staubitz P, Nicholson G, Kalbacher H, Nieuwenhuizen WF, Jung G, Tarkowski A, van Kessel KP, van Strijp JA (2001) *Staphylococcus aureus* resistance to human defensins and evasion of neutrophil killing via the novel virulence factor MprF is based on modification of membrane lipids with l-lysine. J Exp Med 193(9):1067–1076
Peterson LW, Artis D (2014) Intestinal epithelial cells: regulators of barrier function and immune homeostasis. Nat Rev Immunol 14(3):141–153
Pollak CN, Delpino MV, Fossati CA, Baldi PC (2012) Outer membrane vesicles from *Brucella abortus* promote bacterial internalization by human monocytes and modulate their innate immune response. PLoS One 7(11):e50214
Qi M, Miyakawa H, Kuramitsu HK (2003) *Porphyromonas gingivalis* induces murine macrophage foam cell formation. Microb Pathog 35(6):259–267
Ren D, Walker AN, Daines DA (2012) Toxin-antitoxin loci vapBC-1 and vapXD contribute to survival and virulence in nontypeable *Haemophilus influenzae*. BMC Microbiol 12:263
Renelli M, Matias V, Lo RY, Beveridge TJ (2004) DNA-containing membrane vesicles of *Pseudomonas aeruginosa* PAO1 and their genetic transformation potential. Microbiology 150(Pt 7): 2161–2169
Rewatkar PV, Parton RG, Parekh HS, Parat MO (2015) Are caveolae a cellular entry route for non-viral therapeutic delivery systems? Adv Drug Deliv Rev 91:92–108
Rhee KJ, Sethupathi P, Driks A, Lanning DK, Knight KL (2004) Role of commensal bacteria in development of gut-associated lymphoid tissues and preimmune antibody repertoire. J Immunol 172(2):1118–1124
Rivera J, Cordero RJ, Nakouzi AS, Frases S, Nicola A, Casadevall A (2010) *Bacillus anthracis* produces membrane-derived vesicles containing biologically active toxins. Proc Natl Acad Sci U S A 107(44):19002–19007
Robinson MJ, Sancho D, Slack EC, LeibundGut-Landmann S, Reis e Sousa C (2006) Myeloid C-type lectins in innate immunity. Nat Immunol 7(12):1258–1265
Rodriguez-Pinto D (2005) B cells as antigen presenting cells. Cell Immunol 238(2):67–75

Roier S, Leitner DR, Iwashkiw J, Schild-Prufert K, Feldman MF, Krohne G, Reidl J, Schild S (2012) Intranasal immunization with nontypeable *Haemophilus influenzae* outer membrane vesicles induces cross-protective immunity in mice. PLoS One 7(8):e42664

Rus H, Cudrici C, Niculescu F (2005) The role of the complement system in innate immunity. Immunol Res 33(2):103–112

Schooling SR, Beveridge TJ (2006) Membrane vesicles: an overlooked component of the matrices of biofilms. J Bacteriol 188(16):5945–5957

Schroder K, Zhou R, Tschopp J (2010) The NLRP3 inflammasome: a sensor for metabolic danger? Science 327(5963):296–300

Schultz H, Hume J, Zhang DS, Gioannini TL, Weiss JP (2007) A novel role for the bactericidal/permeability increasing protein in interactions of gram-negative bacterial outer membrane blebs with dendritic cells. J Immunol 179(4):2477–2484

Schwechheimer C, Kuehn MJ (2015) Outer-membrane vesicles from Gram-negative bacteria: biogenesis and functions. Nat Rev Microbiol 13(10):605–619

Sharpe SW, Kuehn MJ, Mason KM (2011) Elicitation of epithelial cell-derived immune effectors by outer membrane vesicles of nontypeable *Haemophilus influenzae*. Infect Immun 79(11): 4361–4369

Shen Y, Giardino Torchia ML, Lawson GW, Karp CL, Ashwell JD, Mazmanian SK (2012) Outer membrane vesicles of a human commensal mediate immune regulation and disease protection. Cell Host Microbe 12(4):509–520

Shi J, Zhao Y, Wang Y, Gao W, Ding J, Li P, Hu L, Shao F (2014) Inflammatory caspases are innate immune receptors for intracellular LPS. Nature 514(7521):187–192

Soult MC, Lonergan NE, Shah B, Kim WK, Britt LD, Sullivan CJ (2013) Outer membrane vesicles from pathogenic bacteria initiate an inflammatory response in human endothelial cells. J Surg Res 184(1):458–466

Soult MC, Dobrydneva Y, Wahab KH, Britt LD, Sullivan CJ (2014) Outer membrane vesicles alter inflammation and coagulation mediators. J Surg Res 192(1):134–142

Srisatjaluk R, Kotwal GJ, Hunt LA, Justus DE (2002) Modulation of gamma interferon-induced major histocompatibility complex class II gene expression by *Porphyromonas gingivalis* membrane vesicles. Infect Immun 70(3):1185–1192

Stappenbeck TS, Hooper LV, Gordon JI (2002) Developmental regulation of intestinal angiogenesis by indigenous microbes via Paneth cells. Proc Natl Acad Sci U S A 99(24): 15451–15455

Stentz R, Osborne S, Horn N, Li AW, Hautefort I, Bongaerts R, Rouyer M, Bailey P, Shears SB, Hemmings AM, Brearley CA, Carding SR (2014) A bacterial homolog of a eukaryotic inositol phosphate signaling enzyme mediates cross-kingdom dialog in the mammalian gut. Cell Rep 6(4):646–656

Stephens DS, Edwards KM, Morris F, McGee ZA (1982) Pili and outer membrane appendages on *Neisseria meningitidis* in the cerebrospinal fluid of an infant. J Infect Dis 146(4):568

Takeuchi O, Akira S (2010) Pattern recognition receptors and inflammation. Cell 140(6):805–820

Tan TT, Morgelin M, Forsgren A, Riesbeck K (2007) *Haemophilus influenzae* survival during complement-mediated attacks is promoted by Moraxella catarrhalis outer membrane vesicles. J Infect Dis 195(11):1661–1670

Tavano R, Franzoso S, Cecchini P, Cartocci E, Oriente F, Arico B, Papini E (2009) The membrane expression of *Neisseria meningitidis* adhesin A (NadA) increases the proimmune effects of MenB OMVs on human macrophages, compared with NadA-OMVs, without further stimulating their proinflammatory activity on circulating monocytes. J Leukoc Biol 86(1):143–153

Thay B, Wai SN, Oscarsson J (2013) *Staphylococcus aureus* alpha-toxin-dependent induction of host cell death by membrane-derived vesicles. PLoS One 8(1):e54661

Thay B, Damm A, Kufer TA, Wai SN, Oscarsson J (2014) *Aggregatibacter actinomycetemcomitans* outer membrane vesicles are internalized in human host cells and trigger NOD1- and NOD2-dependent NF-kappaB activation. Infect Immun 82(10):4034–4046

Thomas CJ, Schroder K (2013) Pattern recognition receptor function in neutrophils. Trends Immunol 34(7):317–328

Turner L, Bitto NJ, Steer DL, Lo C, D'Costa K, Ramm G, Shambrook M, Hill AF, Ferrero RL, Kaparakis-Liaskos M (2018) *Helicobacter pylori* outer membrane vesicle size determines their mechanisms of host cell entry and protein content. Front Immunol 9:1466

Unanue ER (1984) Antigen-presenting function of the macrophage. Annu Rev Immunol 2:395–428

Vanaja SK, Russo AJ, Behl B, Banerjee I, Yankova M, Deshmukh SD, Rathinam VAK (2016) Bacterial outer membrane vesicles mediate cytosolic localization of LPS and caspase-11 activation. Cell 165(5):1106–1119

Vaughan AT, Gorringe A, Davenport V, Williams NA, Heyderman RS (2009) Absence of mucosal immunity in the human upper respiratory tract to the commensal bacteria *Neisseria lactamica* but not pathogenic *Neisseria meningitidis* during the peak age of nasopharyngeal carriage. J Immunol 182(4):2231–2240

Vaughan AT, Brackenbury LS, Massari P, Davenport V, Gorringe A, Heyderman RS, Williams NA (2010) *Neisseria lactamica* selectively induces mitogenic proliferation of the naive B cell pool via cell surface Ig. J Immunol 185(6):3652–3660

Vidakovics ML, Jendholm J, Morgelin M, Mansson A, Larsson C, Cardell LO, Riesbeck K (2010) B cell activation by outer membrane vesicles—a novel virulence mechanism. PLoS Pathog 6(1): e1000724

Wai SN, Lindmark B, Soderblom T, Takade A, Westermark M, Oscarsson J, Jass J, Richter-Dahlfors A, Mizunoe Y, Uhlin BE (2003) Vesicle-mediated export and assembly of pore-forming oligomers of the enterobacterial ClyA cytotoxin. Cell 115(1):25–35

Wang X, Thompson CD, Weidenmaier C, Lee JC (2018) Release of *Staphylococcus aureus* extracellular vesicles and their application as a vaccine platform. Nat Commun 9(1):1379

Wells JM, Rossi O, Meijerink M, van Baarlen P (2011) Epithelial crosstalk at the microbiota-mucosal interface. Proc Natl Acad Sci U S A 108(Suppl 1):4607–4614

Winter J, Letley D, Rhead J, Atherton J, Robinson K (2014) *Helicobacter pylori* membrane vesicles stimulate innate pro- and anti-inflammatory responses and induce apoptosis in Jurkat T cells. Infect Immun 82(4):1372–1381

Yonezawa H, Osaki T, Kurata S, Fukuda M, Kawakami H, Ochiai K, Hanawa T, Kamiya S (2009) Outer membrane vesicles of *Helicobacter pylori* TK1402 are involved in biofilm formation. BMC Microbiol 9:197

Yoon H, Ansong C, Adkins JN, Heffron F (2011) Discovery of *Salmonella* virulence factors translocated via outer membrane vesicles to murine macrophages. Infect Immun 79(6): 2182–2192

Young GM, Schmiel DH, Miller VL (1999) A new pathway for the secretion of virulence factors by bacteria: the flagellar export apparatus functions as a protein-secretion system. Proc Natl Acad Sci U S A 96(11):6456–6461

Yun SH, Park EC, Lee SY, Lee H, Choi CW, Yi YS, Ro HJ, Lee JC, Jun S, Kim HY, Kim GH, Kim SI (2018) Antibiotic treatment modulates protein components of cytotoxic outer membrane vesicles of multidrug-resistant clinical strain, *Acinetobacter baumannii* DU202. Clin Proteomics 15:28

Zhu W, Thomas CE, Chen CJ, Van Dam CN, Johnston RE, Davis NL, Sparling PF (2005) Comparison of immune responses to gonococcal PorB delivered as outer membrane vesicles, recombinant protein, or Venezuelan equine encephalitis virus replicon particles. Infect Immun 73(11):7558–7568

# Chapter 9
# Membrane Vesicles from the Gut Microbiota and Their Interactions with the Host

**Josefa Badia and Laura Baldomà**

**Abstract** Gut microbiota plays an essential role in maintaining intestinal homeostasis and human health. Microbiota establishes a complex network of dynamic and reciprocal interactions with the intestinal epithelium and immune system. The mucin layer that covers the epithelium prevents luminal bacteria from accessing host cells. Thus, microbiota–host communication mainly relies on secreted factors and membrane vesicles (MVs), which can cross the inner mucus layer and reach the epithelium. This chapter focuses on the role of microbiota-secreted MVs as key players in signaling processes in the intestinal mucosa. This is an emerging research topic, with the first reports dating from 2012. Microbiota-derived MVs are involved in interspecies communication in the gut, between bacteria and between microbiota and host. Here we present current knowledge on the mechanisms used by microbiota MVs to assist and control the gut microbial community and to modulate host immune and defense responses. Constant stimulation of immune receptors by microbiota MVs results in tightly controlled inflammation that contributes to tolerogenic responses essential to maintaining intestinal homeostasis. Moreover, gut microbiota MVs are emerging as physical vehicles for distribution and delivery of bacterial effectors to distal tissues in human health and disease.

## Abbreviations

| | |
|---|---|
| BMCDs | bone marrow-derived dendritic cells |
| CME | clathrin-mediated endocytosis |
| DCs | dendritic cells |
| DSS | dextran sodium sulfate |
| Gadd45α | growth arrest and DNA-damage-inducible protein, |

J. Badia · L. Baldomà (✉)
Secció de Bioquímica i Biología Molecular, Departament de Bioquímica i Fisiologia, Facultat de Farmàcia i Ciències de l'Alimentació, Universitat de Barcelona, Barcelona, Spain

Institut de Biomedicina de la Universitat de Barcelona (IBUB), Institut de Recerca Sant Joan de Déu (IRSJD), Barcelona, Spain
e-mail: lbaldoma@ub.edu



M. Kaparakis-Liaskos, T. A. Kufer (eds.), *Bacterial Membrane Vesicles*,
https://doi.org/10.1007/978-3-030-36331-4_9

| | |
|---|---|
| GI | gastrointestinal |
| HFD | high-fat diet |
| IBD | inflammatory bowel disease |
| IECs | intestinal epithelial cells |
| iNOS | inducible nitric oxide synthase |
| LPS | lipopolysaccharide |
| MAMPs | microbial-associated molecular patterns |
| miRNAs | microRNAs |
| MMP | matrix metalloprotease |
| MUC | mucin |
| MVs | membrane vesicles |
| ncRNAs | non coding RNAs |
| NOD | nucleotide-binding oligomerization domain protein |
| PGN | peptidoglycan |
| PRRs | pattern-recognition receptors |
| PSA | polysaccharide A |
| RIP2 | receptor-interacting protein 2 |
| TFF-3 | trefoil factor 3 |
| Th | T helper |
| TLRs | tol-like receptors |
| TNBS | 2,4,6-trinitrobenzene sulfonic acid |
| Treg | regulatory T cells |
| ZO | zonula occludens |

## 9.1 Gut Microbiota

Gut microbiota refers to the entire set of microbial communities that colonize the human gastrointestinal (GI) tract. This community is mainly composed of bacteria, but other groups such as archaea, fungi, protozoa, and viruses are also represented. Recent advances in omics technologies such as metagenomics, transcriptomics, and proteomics are currently being applied to study intestinal microbial ecology at a molecular level. From these studies, we have learned that the gut microbiome comprises more than three million genes, which greatly exceed and complement the genetic information encoded by the human genome (Qin et al. 2010). Microbiota-encoded products provide trophic, metabolic, and protective signals that are beneficial to the host. In fact, the gut microbiota is considered as a "hidden organ" as it plays fundamental roles in intestinal homeostasis and human health (Jandhyala et al. 2015). There is strong scientific evidence that the gut microbiota exerts pivotal functions by regulating food digestion, maintaining the intestinal epithelial barrier, and contributing to immune system functions and development (Thursby and Juge 2017; Cani 2018). Moreover, commensal bacteria help to protect the host against pathogens through mechanisms that include secretion of antimicrobial factors, competition for binding sites, reduction of intraluminal pH and enhancing host immune responses (Llewellyn and Foey 2017). Alterations in normal gut microbiota biodiversity (dysbiosis) have been associated with a wide range of illnesses including inflammatory bowel disease (IBD), allergic and immune disorders, metabolic diseases (insulin resistance and obesity), and cancer (Vindigni et al. 2016; Baothman et al. 2016; Gopalakrishnan et al. 2018). The high plasticity of the human microbiome provides new opportunities for therapeutic strategies aimed at modulating the composition of the gut microbiota that is altered in certain pathologies (Shanahan 2011; Maguire and Maguire 2019). One such approach to modulate the host microbiota is the administration of probiotics.

The intestinal ecosystem is characterized by dynamic and reciprocal interactions between the microbiota, the epithelium, and the host immune system. The capacity of cells of the intestinal mucosa to discriminate between pathogens and commensal bacteria is a key issue. The host response to pathogens is characterized by rapid recognition of the pathogen combined with innate (inflammatory) and adaptive immune responses. This leads to pathogen eradication, at the cost of significant tissue damage. The response to symbiotic microbiota is known as tolerance: a process that encompasses a complex combination of microbe recognition and highly regulated innate and adaptive immune responses. This dichotomy in the host response is fundamental on the surface of the intestinal mucosa that is massively colonized by a diverse population of bacteria (Bron et al. 2011).

The gastrointestinal epithelial layer is the surface where the host interacts with microbiota. This epithelium creates a physical and biochemical barrier between gut microbiota and host. Several mechanisms are involved in the epithelial barrier function such as: (1) secretion of the mucin layer that covers the epithelial surface and avoids direct contact with gut microbes, (2) secretion of antimicrobial peptides, and (3) formation of tight junctions between intestinal epithelial cells that separate the host tissue from the luminal ecosystem.

Intestinal epithelial cells (IECs) play an important role in sensing microbial signals. Upon activation, these cells release signaling molecules that communicate the information to the intestinal immune cells, which trigger appropriate immune responses (reviewed by Turner 2009; Wells et al. 2011; Peterson and Artis 2014). Detection of gut microbes by IECs depends on specific immune receptors, known as pattern recognition receptors (PRRs) that specifically recognize conserved microbial-associated molecular patterns (MAMPs). Toll-like receptors (TLRs) are transmembrane proteins. TLRs located at the plasma membrane (TLR1, TLR2, TLR4, and TLR5) recognize extracellular components of bacteria, viruses, or fungi, whereas intracellular TLRs (TLR3, TLR7, TLR8, and TLR9), mainly found in endosomes and lysosomes, recognize internalized microbial DNA or RNA (Kawai and Akira 2010). There are also cytosolic receptors such as nucleotide-binding oligomerization domain protein 1 (NOD1) and NOD2 that are activated by peptidoglycan (PGN) fragments of Gram-negative and/or Gram-positive bacteria (Philpott et al. 2014). Other cytosolic receptors are part of the protein complex known as inflammasome, whose main function is the activation of caspase-1, which in turn activates the inflammatory cytokines IL-1β and IL-8 (reviewed in Thaiss et al.

2014). Interaction of PRRs with their specific ligand activates signaling cascades that lead not only to the release of chemokines and cytokines that communicate the information to the intestinal immune cells, but also to the secretion of antimicrobial peptides that help to control the gut microbial population. This feedback control is essential in limiting immune activation and maintaining mutualistic associations between bacteria and the host.

In addition to the intestinal epithelium, other cells of the innate immune system such as dendritic cells (DCs) of the lamina propria contribute to the sampling of gut microbes. DCs are antigen-presenting cells that can contact the luminal environment through the inner mucosal lining. These phagocytic cells sense intestinal microbes through their PRRs and act as messengers for the rest of the immune system through antigen presentation and release of immune mediators. DCs ensure intestinal homeostasis by tuning host immune responses in the gut and they are involved in immunological tolerance to gut microbiota (reviews of the topic in Belkaid and Hand 2014; Caballero and Pamer 2015). In response to beneficial gut microbes, DCs induce proliferation of anti-inflammatory regulatory FoxP3 T cells (Treg), which contribute to maintaining immune intestinal homeostasis (Geuking et al. 2011; Jia et al. 2018). Alterations in microbiota may have an impact on immune mucosal tolerance by negatively affecting the Treg response, which in turn leads to the development of intestinal inflammatory pathologies.

Given the complexity of the gut microbiota and its interaction with host intestinal cells, elaborated regulatory mechanisms are required to ensure symbiosis and avoid aberrant responses that lead to pathological states. Many studies have focused on the signaling pathways, regulatory proteins, and transcription factors activated by microbiota to modulate intestinal homeostasis (Caballero and Pamer 2015). An emerging topic is the role of host microRNAs (miRNAs) as key players in the host–microbiota interplay and in cell-to-cell communication. The miRNAs are small noncoding RNAs (20–25 nucleotides) that, after maturation, associate with the RNA-induced silencing complex and regulate the expression of target mRNAs through binding to sequences at the 3′-UTR region. This interaction triggers mRNA degradation or blocks translation. Therefore, miRNA are posttranscriptional regulators that allow signaling pathways to be tightly controlled. The miRNAs are involved in the control of multiple cellular processes, including the immune response. In this context, many studies indicate that microbiota and miRNAs regulate each other. Gut bacteria (either commensal or pathogens) have a great impact on miRNAs expression, and host miRNAs shape and regulate gut microbiota (Masotti 2012; Runtsch et al. 2014; Celluzzi and Masotti 2016; Feng et al. 2018; Aguilar et al. 2018).

## 9.2 Role of Microbiota-Secreted Membrane Vesicles in Interspecies Communication in the Gut

The gut microbiota is not in direct contact with the epithelium. Both cell types are physically separated by the mucus layer, which is structured in two sections. The inner dense mucus layer is closely linked to IECs and acts as a highly efficient barrier that prevents bacteria from reaching the intestinal epithelium (Johansson et al. 2011; Vaishnava et al. 2011). In addition to this protective role, the inner mucus layer also contributes to maintaining the outer mucus layer, which is highly dynamic and in close contact with the microbiota. The external mucus layer can be degraded by specific bacteria of the gut microbiota and thus needs to be constantly renewed. Goblet cells are the main source of secreted mucin (MUC2), whose production is upregulated by TLR signaling in response to its degradation by commensals or other mechanical sources (Faderl et al. 2015). In addition to the MUC2 structure, spatial separation between the microbiota and the intestinal epithelium is maintained by soluble factors with antimicrobial activity that are secreted by the epithelium, such as β-defensin which is active against Gram-negative bacteria, RegIIIγ lectin that is active against Gram-positive bacteria, and IgA that is secreted by immune cells. Since access of the microbiota to the intestinal epithelium is limited by the inner mucin layer, communication with the host mainly depends on microbiota-secreted factors (metabolites, proteins, and vesicles) that can go through the mucin layer and reach host cells at the intestinal mucosal surface.

All bacteria release extracellular membrane vesicles (MVs) as a mechanism of communication between species. MVs are nanoscale bilayer structures derived from the bacterial membranes (see Chaps. 2 and 3). They are part of a secretion mechanism that allows long-distance delivery of bacterial active compounds in a protected environment, avoiding direct intercellular contact. MVs comprise components of the bacterial membrane, cytosolic proteins, metabolites, DNA, and RNA. MVs from Gram-negative bacteria also include outer membrane and periplasmic biomolecules (Guerrero-Mandujano et al. 2017). The functions of MVs are versatile, including bacterial response to stress, quorum sensing, biofilm formation, and interspecies communication (bacteria–bacteria and bacteria–host dialog) (Schwechheimer and Kuehn 2015). In the last 20 years, numerous studies have focused on MVs from Gram-negative pathogens, showing that these structures act as vehicles for the delivery of cytotoxic/virulence factors and mediators that alter the host immune response (reviewed in Kaparakis-Liaskos and Ferrero 2015).

In this chapter, we will focus on gut microbiota-derived vesicles. This is an emerging topic, with the first reports of microbial-derived MVs dating from 2012 (Shen et al. 2012; López et al. 2012). Most studies deal with Gram-negative commensals of the genus Bacteroides (*B. fragilis* and *B. thetaiotaomicron*), *Akkermansia muciniphila*, and intestinal *Escherichia coli* isolates including the probiotic strain *E. coli* Nissle 1917 (EcN). Studies on MVs from Gram-positive beneficial gut microbes are limited and mostly centered on probiotics of the genus Lactobacillus and Bifidobacterium (reviewed in Liu et al. 2018). Due to the membrane structure of

Gram-positive bacteria, their MVs lack lipopolysaccharide (LPS) and periplasmic components, although they carry similar types of cargo molecules as Gram-negative MVs including PGN, lipids, proteins, and nucleic acids (Brown et al. 2015). The effects exerted by microbiota-derived MVs depend on their bacterial origin and cargo. Therefore, some of their effects are strain-specific. Proteomic studies have revealed that vesicles released by probiotic and commensal strains contain proteins that contribute to intestinal barrier and immune modulation and proteins that help to compete for colonization and bacterial survival in the harmful environment of the GI tract (Lee et al. 2007; Aguilera et al. 2014; Elhenawy et al. 2014; Domínguez Rubio et al. 2017; Li et al. 2017; Zakharzhevskaya et al. 2017). The first proteomic analysis of MVs released by beneficial gut bacteria was performed with the probiotic EcN (Aguilera et al. 2014). Some of the identified proteins are encoded by strain-specific genes, and most of them are fitness factors that contribute to adhesion to host tissues and to bacterial survival in the GI tract. In this context, EcN vesicles harbor the porin NanC, whose expression is induced by N-acetylneuraminic acid, one of the most abundant sialic acids of eukaryote cell membranes. Sialic acids can be used by enteric bacterial pathogens as carbon and nitrogen sources (Vimr et al. 2004; Severi et al. 2007). By this mechanism EcN MVs could compete with enteric pathogens for colonization of the intestinal tract. Furthermore, there is a set of identified proteins that are common to MVs secreted by Gram-negative pathogens and probiotic bacteria. The main group of these common probiotic/pathogen vesicular proteins is cytoplasmic proteins, and a high number of them are metabolic enzymes classified as moonlighting proteins that have different functions depending on the cell location (Aguilera et al. 2014). Examples of these moonlighting proteins are enolase, glyceraldehyde-3-phosphate dehydrogenase, fructose-bisphosphate aldolase, or succinyl-CoA synthase.

Regarding the presence of metabolic enzymes in bacterial MVs, a comparative study based on proteomics and metabolomics approaches has found great differences between vesicles isolated from genetically related pathogenic and commensal *B. fragilis* strains (Zakharzhevskaya et al. 2017). MVs from the nonpathogenic strain are enriched in enzymes required for polysaccharide utilization. In contrast, MVs from the pathogenic strain contain, in addition to virulence factors, a larger number of enzymes involved in energy-producing metabolic pathways such as glycolysis and the tricarboxylic acid cycle. The activity of the vesicular enzymes and transporters was validated by fluxomic experiments with isotope-labeled glucose ($^{13}$C-glucose), thus confirming that these pathways are fully operative in pathogen-derived MVs. The associated metabolic activity provides vesicles released by pathogens with energy for long persistence in the human GI tract (Zakharzhevskaya et al. 2017). The proteomic analysis of MVs isolated from the Gram-positive probiotic *Lactobacillus casei* BL23 corroborates the presence of numerous proteins that fit the profile of Gram-negative derived vesicles, including envelope associated proteins, metabolic enzymes, transporters, and ribosomal components. The presence of adhesins and proteins known to mediate the effects of this probiotic reinforces the role of MVs in the bacteria–gut interaction (Dominguez Rubio et al. 2017).

### 9.2.1 Contribution of Gut Microbiota-Derived MVs to the Intestinal Ecosystem

Several studies highlight the role of enzymes released through microbiota MVs in the colonization of the human gut (Donaldson et al. 2016). Once released into the intestinal lumen, MVs can affect the surroundings distantly from their parent cells. These MV triggered changes can be used by the parental bacterium to their own benefit or even to help other members of the microbiota community. As an example, MVs secreted by commensals of the Bacteroides genus deliver to the intestinal environment enzymes of the hydrolase class (glycosylases and proteases) that catalyze the breakdown of complex polysaccharides (Fig. 9.1a). This metabolic activity generates products that can be used as a source of nutrients by other members of the gut microbiota, which in turn produce short chain fatty acids that are beneficial to the host (Elhenawy et al. 2014). MVs from certain Bacteroides strains can also distribute sulfatases that help in the degradation of mucin glycans by other bacterial hydrolases (Hickey et al. 2015). As Bacteroides MVs are equipped with a wide range of hydrolytic enzymes, they can assist the gut microbial community in the acquisition of nutrients and favor symbiosis between the gut microbiota and the host (Rakoff-Nahoum et al. 2014; Elhenawy et al. 2014). Other vesicle associated-enzymes can be detrimental to the host. For instance, packaging of β-lactamases into Bacteroides MVs provides a mechanism for spreading antibiotic resistance to other microbiota members and to enteric pathogens (Stentz et al. 2015).

Studies performed in *B. thetaiotaomicron* showed that released MVs carry enzymes involved in the assimilation of dietary inositol polyphosphates (InsP6) present in vegetables (Stentz et al. 2014). In the luminal GI tract, InsP6 chelate divalent cations and inhibit polysaccharide digestion. The human gut lacks enzymes (phytases) that can dephosphorylate this kind of molecules. Therefore, assimilation of luminal InsP6 depends on enzymes provided by the gut microbiota (Fig. 9.1a). These bacterial enzymes are homologs of the mammalian InsP6 phosphatase and are widely distributed among resident gut bacteria (Stentz et al. 2014). In addition to this metabolic role in dietary InsP6, the bacterial enzyme also participates in inter-kingdom signaling (see Sect. 9.2.2).

Besides metabolic roles, microbiota-derived MVs can promote competitive interference among related bacterial species. It is known that gut microbiota releases a wide range of antimicrobial peptides as a mechanism to persist and compete with other members of the microbial community. A study performed with *B. fragilis* strains showed that secretion of the antimicrobial peptide known as BSAP-1 (*B*acteroidales *s*ecreted *a*ntimicrobial *p*rotein *1*), which displays inhibitory activity against other Bacteroidales of the human gut, is mediated by MVs (Chatzidaki-Livanis et al. 2014). This peptide contains the membrane attack complex/perforin (MACPF) domain present in host immune mediators that can kill pathogens and virus-infected cells. Proteins with the MACF domain generate pores in the membrane of sensitive target cells, causing increased membrane permeability (Fig. 9.1a). Furthermore, MVs secreted by pathogens also serve as a vehicle to secrete

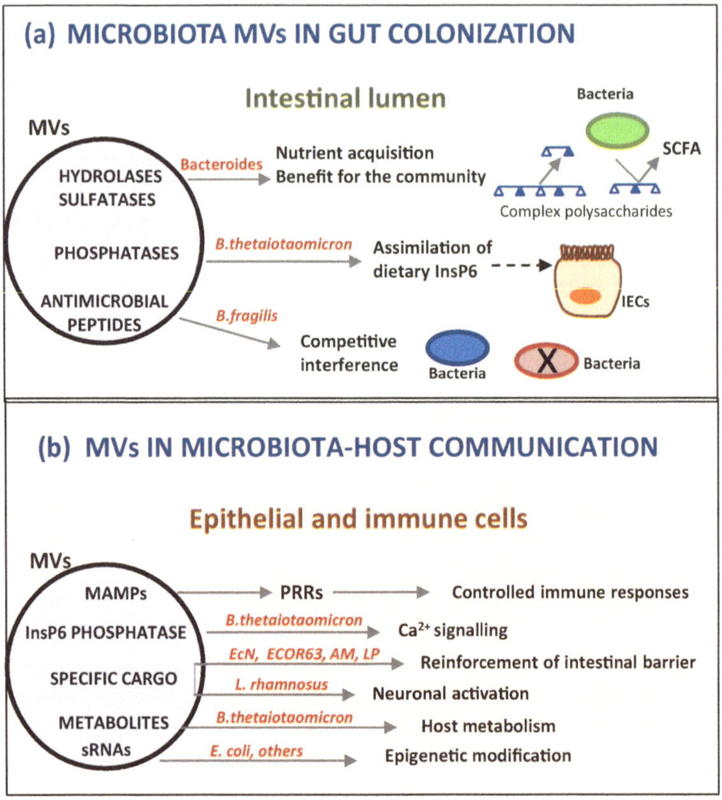


**Fig. 9.1** Overview of interspecies interaction in the gut mediated by microbiota-derived MVs. MVs are represented by circles, and the specific molecules responsible for the known effects on (**a**) gut colonization and (**b**) microbiota–host communication are indicated inside. *Arrows* link each vesicular mediator with the specific role and the producer bacterial strain (in red). *Bacteria* are represented by ellipses and *complex polysaccharides* by chain-connected triangles. A *cross* inside a bacterium indicates inhibitory activity on cell growth. *IECs* intestinal epithelial cells, *SCFA* short chain fatty acids, *InsP6* inositol phosphates, *MAMPs* microbial-associated molecular patterns, *PRRs* pattern-recognition receptors, *sRNAs* small RNAs, *EcN E. coli* Nissle 1917, AM *Akkermansia muciniphila*, *LP Lactobacillus plantarum*

hydrophobic quorum sensing molecules (Brameyer et al. 2018). However, no studies have been reported to date concerning this role for microbiota-derived MVs.

### *9.2.2 Contribution of Gut Microbiota-Derived MVs to Inter-Kingdom Signaling*

Microbiota MVs enclose biological components that exist in their parent bacteria. In particular, these vesicles contain a high number of MAMPs that are recognized by

PRRs expressed by epithelial and immune cells. As stated above, these receptors are key components of innate immunity as they sense gut microbes and trigger appropriate immune responses (Turner 2009). In the intestinal lumen, microbiota-released vesicles diffuse through the mucus layer and reach the intestinal epithelium. Surface-associated MAMPs (LPS, lipoproteins, and extracellular polysaccharides) can interact with extracellular PRRs to trigger the activation of signaling cascades that ultimately regulate defense and immune responses (Kawai and Akira 2010; Kaparakis-Liaskos and Ferrero 2015). In addition, bacterial MVs are internalized by IECs through endocytic pathways (O'Donoghue and Krachler 2016). By this mechanism MVs allow the intracellular delivery of MAMPs (DNA, RNA, and PGN) that bind intracellular receptors activating signaling pathways that control host responses (Kaparakis-Liaskos and Ferrero 2015). Microbiota MVs can also interact with cells of the innate immune system, especially DCs, which in turn coordinate appropriate immune responses transmitting the information to cells of the adaptive immune system (Belkaid and Hand 2014; Caballero and Pamer 2015) (Fig. 9.1b). A detailed explanation of the mechanisms whereby bacterial membrane vesicles interact with the innate immune system can be found in Chap. 8.

Modulation of the innate immune system by gut microbiota plays an essential role in maintaining gut homeostasis by favoring quick responses against pathogens in addition to preserving tolerance to commensal bacteria. MVs released by resident gut microbiota share a number of MAMPs with vesicles secreted by enteric pathogens, and hence can activate the same downstream signaling pathways. Therefore, inflammatory responses should be tightly controlled to avoid aberrant responses against commensal microbiota. Some specific microbiota vesicular components such as *B. fragilis* polysaccharide A (PSA) trigger TLR-mediated signaling events that restrain host immune responses and allow commensal gut colonization (Round et al. 2011). Concerning cytosolic immune receptors, NOD receptors were first discovered as a defense mechanism against bacterial pathogens and the subsequent host inflammatory responses. However, it is suggested that NODs may also play a role in maintaining intestinal homeostasis as mutations that impair NOD2 activity or expression have been associated with chronic inflammatory and autoimmune diseases (Philpott et al. 2014; Feerick and McKernan 2016; Chu 2017).

The integrity of the epithelial barrier is also critical in maintaining homeostasis in the body. Several diseases are associated with the increased intestinal permeability that follows the disruption of gut epithelial tight junctions (Suzuki 2013). This mechanism is also used by certain enteric pathogens, either bacteria or viruses, to help their dissemination into host tissues (Lu et al. 2013). It is well known that the gut microbiota plays a relevant role in maintaining the intestinal barrier, by either modulating epithelial tight junctions or by enhancing host defense mechanisms (Jandhyala et al. 2015). The ability of microbiota MVs to reinforce epithelial tight junctions has been reported for certain commensal *E. coli* and *A. muciniphila* strains (Fig. 9.1b) (Alvarez et al. 2016; Chelakkot et al. 2018). Moreover, MVs released by commensal *E. coli* and *L. plantarum* strains trigger upregulation of host defense genes that encode secreted peptides with antimicrobial activity, such as β-defensin (Fábrega et al. 2016) or C-type lectins (Li et al. 2017). In the intestinal tract, microbiota-derived MVs allow the sensing and delivery of microbial products that steadily prime the host innate immune system (Shen et al. 2012; Fábrega et al. 2016; Cañas et al. 2018). In this context, constant stimulation of immune receptors by MVs released by beneficial gut microbes may result in controlled basal inflammation that contributes to appropriate defense and immune responses against pathogens and, ultimately, to intestinal homeostasis (Cañas et al. 2018; Molina-Tijeras et al. 2019).

Components of microbiota MVs other than MAMPs can also modulate signal transduction pathways. As stated above, MVs from the commensal *B. thetaiotaomicron* carry InsP6 phosphatase. In addition to a role in dietary InsP6 metabolism, the bacterial enzyme can also modulate cellular functions and gastrointestinal physiology (Fig. 9.1b). Upon internalization in IECs, MVs intracellularly deliver InsP6 phosphatase that interacts with the inositol signaling pathway of the host cell, leading to an increase in inositol-3-phosphate levels and the subsequent release of $Ca^{2+}$ from intracellular stores. Thus, OMVs contribute to inter-kingdom cell-to-cell cross talk triggering intracellular $Ca^{2+}$-signaling pathways in intestinal epithelial cells (Stentz et al. 2014).

The connection between gut commensal MVs and the host enteric nervous system has been reported for *L. rhamnosus* JB-1. In addition to immunomodulatory effects, released *L. rhamnosus* MVs mediate the functional effects of this commensal on peristalsis through nerve-dependent regulation of colon migrating motor complexes (Fig. 9.1b). The modulation of this neuronal response by *L. rhamnosus* JB-1 MVs was observed in ex vivo experiments performed with colonic explants but not by direct neuronal stimulation. This finding highlights the role of IECs in inter-kingdom signaling between bacterial MVs and the enteric neuronal system (Al-Nedawi et al. 2015). The study reveals the ability of certain gut commensal MVs to communicate with local neurons indirectly through signals released by the intestinal epithelium. The nature of both the vesicular bacterial effector and the epithelium-derived signal remains unknown.

Other studies also point to the connection of microbiota MVs with the host metabolism. As stated above, metabolomic approaches have revealed that bacterial MVs contain a set of metabolites (Zakharzhevskaya et al. 2017). The metabolite content of *B. thetaiotaomicron* MVs was analyzed, and the putative role of the packaged metabolites was predicted by in silico approaches. This study revealed that *B. thetaiotaomicron* MVs are enriched with metabolites known to facilitate intestinal colonization, and interestingly with metabolites that can be incorporated into mouse metabolic pathways (Bryant et al. 2017). This is the first study showing that vesicles from a prominent gut commensal selectively contain metabolites that are useful for the mammalian host, although the specific effects on the host metabolism have yet to be confirmed (Fig. 9.1b).

The RNA content of microbiota-derived MVs has been linked with regulatory functions affecting host epigenome and gene expression (Celluzzi and Masotti 2016). The first data on RNA sequences found in *E. coli* MVs revealed that the associated RNA is enriched in noncoding small RNA molecules (ncRNAs), which differ from bacterial intracellular RNAs (Ghosal et al. 2015). Interestingly, many extracellular ncRNA sequences align to the human genome, mostly in regions related to epigenetic mechanisms such as histone modification and chromatin remodeling or with cell-specific transcriptional control (Celluzzi and Masotti 2016). Changes in the epigenetic profile induced by the environment (diet, physical activity, drugs, etc.) have a great impact on gene expression and disease development. In this context, delivery of bacterial ncRNA through MVs might be used as a mechanism to exert multiple effects on host gene expression, contributing to host health in the case of gut beneficial microbes, or the onset of diseases in the case of pathogens or imbalanced microbiota (Fig. 9.1b). The role of bacterial small RNAs secreted through MVs in the dysregulation of host immune responses has been reported in pathogenic bacteria (Koeppen et al. 2016; Choi et al. 2017), suggesting that this may also be possible for microbiota-derived MVs.

## 9.3 MVs in Microbiota–Host Interaction at the Intestinal Mucosa

### *9.3.1 Interaction with Intestinal Epithelial Cells*

**Internalization Pathways** Direct and indirect evidence proves that bacterial MVs are taken up by host epithelial cells. Most studies on this topic have been performed with MVs from Gram-negative pathogens. The uptake of pathogen-derived MVs into epithelial cells is mainly driven by endocytosis. This process involves invagination of the cell membrane and occurs through different pathways depending on the surface and cargo of the vesicles. Endocytic pathways can be classified into two main groups: clathrin-mediated endocytosis (CME) and clathrin-independent pathways that include lipid raft-mediated processes (O'Donoghue and Krachler 2016). These pathways involve endosomal compartments with different surfaces that allow sorting of internalized vesicle cargo to various subcellular locations. CME depends on a complex protein network including clathrin and dynamin as key components (Vercauteren et al. 2010). Lipid rafts are dynamic membrane microdomains rich in cholesterol, sphingolipids, and proteins such as caveolin or flotillin and are associated with distinct internalization pathways that are cholesterol sensitive (O'Donoghue and Krachler 2016). The endocytic pathway of MVs depends on the presence of vesicular components (proteins, toxins, or surface structures) that target the MVs to specific receptors in the host cell membrane (Kesty et al. 2004; O'Donoghue et al. 2017). More than one pathway mediates the internalization of some MVs, like those from *Helicobacter pylori* (Olofsson et al. 2014; Turner et al. 2018). Membrane fusion and micropinocytosis have also been implicated in the uptake of certain bacterial MVs by host cells (O'Donoghue and Krachler 2016). Micropinocytosis is an actin-driven process shown to be involved in the uptake of large MVs from *H. pylori* (Turner et al. 2018). Despite the structural differences between the membrane of bacterial vesicles and that of host eukaryotic cells, membrane fusion has been shown to direct entry of *Listeria monocytogenes* MVs into host cells (Jager et al. 2014). This mechanism has also been described for *Pseudomonas aeruginosa* MVs (Bomberger et al. 2009). An assay based on fluorescence resonance energy transfer has been used to study bacterial and host factors that determine the vesicle internalization pathway, kinetics, and efficiency (O'Donoghue et al. 2017). One factor that has a great impact on the selection and kinetics of the entry route is the lipopolysaccharide O antigen present in the MV surface. Gram-negative MVs lacking the O antigen need surface protein receptors in the host cell membrane for their entry, whereas the presence of the O antigen allows MVs entry through receptor-independent uptake pathways (O'Donoghue et al. 2017).

In the first decade of the twenty-first century, there have been numerous studies on the internalization pathway of pathogen-derived MVs in epithelial cells. The study of microbiota-derived MVs uptake started later and was principally focused on *Escherichia coli*, which is found as part of normal human gut microbiota. The first study on this topic was performed with MVs from the probiotic EcN and the commensal ECOR12 strains (Cañas et al. 2016). EcN is a good colonizer of the human gut with beneficial effects on intestinal homeostasis and microbiota balance. This probiotic strain belongs to the phylogenetic group B2, associated with virulent strains that cause extraintestinal infections. In fact, EcN shares a common ancestor with *E. coli* uropathogenic strains. During evolution, EcN lost virulence factors but preserved fitness factors that confer competence to survive in the human gut (Vejborg et al. 2010; Toloza et al. 2015). ECOR12 is an intestinal isolate that belongs to phylogenetic group A, which is mostly associated with nonpathogenic *E. coli* strains (Ochman and Selander 1984). Uptake analysis of fluorescent-labeled MVs in human intestinal epithelial cell lines in the presence of endocytosis inhibitors showed that EcN and ECOR12 MVs enter epithelial cells via CME (Fig. 9.2). Disruption of lipid rafts and caveolae domains by cholesterol-sequestering agents have no effect on the vesicle uptake by HT-29 or Caco-2 cells, whereas vesicle internalization is severely impaired by CME inhibitors (Cañas et al. 2016). Following the intracellular trafficking through CME pathway, internalized MVs reach lysosomal compartments. Consistently intracellular MVs colocalize with clathrin and specific markers of early endosomes and lysosomes (Cañas et al. 2016). A study performed with an EcN *tolR* mutant that displays a hypervesiculating phenotype evidenced that typical MVs are internalized by epithelial cells, whereas aberrant membranous structures are not. Cryo-transmission electron microscopy analysis of isolated MVs from wild-type EcN and the *tolR* isogenic mutant showed substantial structural heterogeneity in EcN *tolR* samples. In addition to common MVs (outer membrane vesicles and outer-inner membrane vesicles), aberrant vesicular structures were observed. Analysis of MV uptake in Caco-2 cells evidenced that mutant-derived MVs exhibit lower internalization levels than wild-type MVs due to the reduced capacity of EcN *tolR*-derived MVs to bind host cell membrane (Pérez-Cruz et al. 2016). These findings prove that EcN MVs interact with their target(s) in the host cell membrane, a key step before being taken up by epithelial cells through CME. In contrast, circularized broken membranes or artifacts generated during

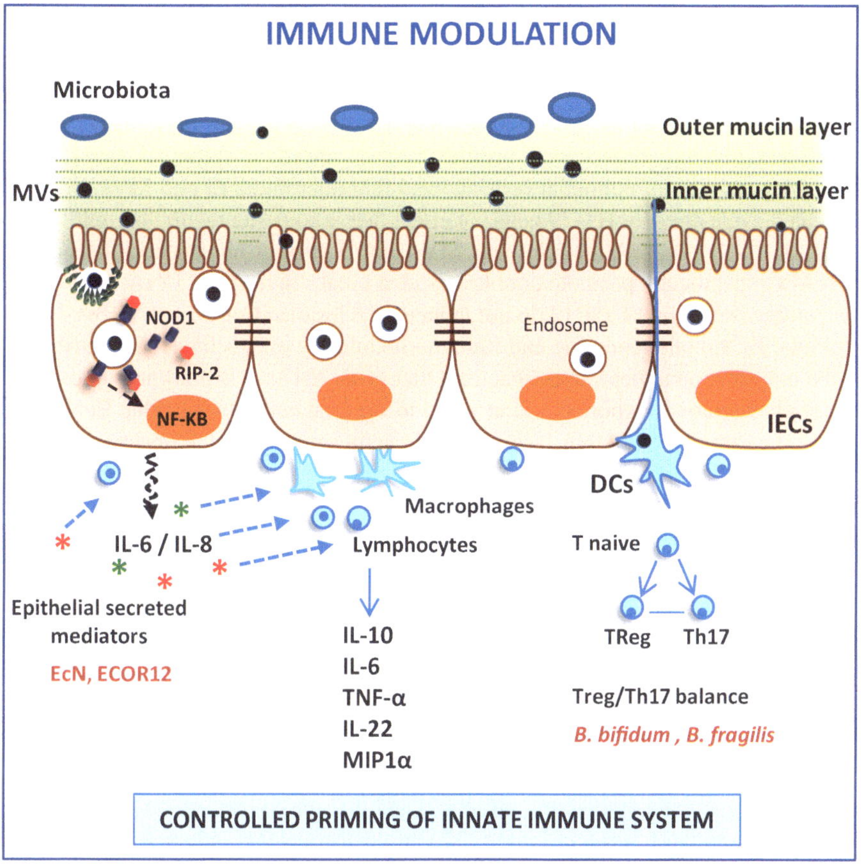


**Fig. 9.2** Schematic representation of immunomodulatory effects elicited by microbiota-derived MVs in the gut. The drawing shows the structure of the intestinal barrier in which the mucin layer maintains segregation between luminal microbes and the intestinal epithelium. *Gut microbiota* (represented by ellipses) reside in the outer mucus layer. *MVs* (small black circles) can diffuse through the *inner mucus layer* (dotted lines) and reach the epithelium. Immune cells (*lymphocytes, macrophages, and dendritic cells*) in the lamina propria are shown below the epithelial monolayer. Microbiota MVs exert immune modulation through two main mechanisms. Indirect activation of immune cells through the intestinal epithelium. MVs from EcN and ECOR12 are taken up by intestinal epithelial cells (*IECs*) and trigger secretion of *epithelial mediators* (asterisks), which in turn activate underlying immune cells that secrete a wide range of cytokines. Activation of the NOD1 signaling pathway by internalized EcN and ECOR12 MVs is presented in more detail. MVs enter IECs by clathrin-mediated endocytosis and recruit *NOD1* (dark blue cylinders) to early endosomes. Activated NOD1 interacts with the specific kinase *RIP2* (red hexagons), leading to NF-κB activation and its translocation into the nucleus. This transcription factor upregulates host genes involved in the inflammatory response (IL-6, IL-8). Direct activation of dendritic cells (*DCs*) by MVs from *B. bifidum* and *B. fragilis* promotes a Treg response that contributes to immune tolerance. Activation of IECs and immune cells by microbiota MVs provides a controlled physiological mechanism for priming of the innate immune system

bacterial cell lysis are not properly targeted to these sites, and therefore cannot mediate the functional effects attributed to conventional MVs.

**Effects on Cell Viability** In contrast to MVs from pathogenic *E. coli* strains MVs derived from EcN and ECOR12 do not affect cell viability nor promote oxidative stress, but they do reduce cell proliferation of intestinal epithelial cells. Flow cytometry analysis of epithelial cells showed that microbiota *E. coli* MVs promote S/G2 cell cycle arrest in HT-29 cells, an effect that is consistent with their inhibitory effect on cell growth. Although these MVs have not been observed in the nuclei, EcN MVs specifically promote double-stranded breaks in host cell DNA. MVs from the commensal strain ECOR12 do not induce such lesions (Cañas et al. 2016). In the probiotic EcN, both genotoxic and immunomodulatory effects have been attributed to the non-ribosomal peptide colibactin (Olier et al. 2012). It is not known how this bacterial mediator is exported and targeted to the host cell. The fact that EcN MVs induce the same type of DNA lesions as colibactin suggests that colibactin could be delivered to mammalian cells by MVs. This emphasizes the role of MVs derived from beneficial gut microbes in communication with the host. Secretion of colibactin through MVs is an open question that requires further study.

**Activation of Cytosolic NOD Receptors** NOD1 and NOD2 cytosolic receptors sense PGN, a component of the bacterial cell wall. NOD1 detects D-glutamyl-meso-diaminopimelic acid, which principally exists in PGN of Gram-negative bacteria while NOD2 detects muramyl dipeptide, which is common to Gram-negative and Gram-positive bacteria (Girardin et al. 2003a; Girardin et al. 2003b; Chamaillard et al. 2003). PGN interaction with NOD triggers receptor oligomerization, the initial step in the downstream signaling cascade that leads to recruitment of the specific kinase RIP2 (receptor-interacting protein 2) and the subsequent activation of NF-κB and mitogen-activated protein kinase (MAPK) pathways that induce the expression of inflammatory genes (Inohara et al. 2000; Hasegawa et al. 2008; Allison et al. 2009). As indicated above, NOD receptors are essential in maintaining intestinal homeostasis. They are involved in defense responses against bacterial infection and in regulation of the intestinal inflammatory response to microbiota (reviewed Kaparakis-Liaskos 2015). Since gut microbiota is composed of noninvasive bacteria, a key matter is to decipher the mechanisms for PGN delivery into host cells. One mechanism involves PGN fragments released into the gut lumen during bacterial cell division, which can be internalized in epithelial cells by endocytosis or through oligopeptide transporters (Swaan et al. 2008; Philpott et al. 2014). Another pathway for intracellular PGN delivery is through MVs. This is a well-studied mechanism in Gram-negative pathogens. Studies performed with *H. pylori* revealed that OMVs internalized through CME reach endosomal compartments, and that interaction of vesicle-associated PGN with NOD1 occurs at early endosomes. (Irving et al. 2014). Studies performed with EcN and ECOR12 MVs showed that this pathway is also effective for microbiota-derived MVs, which activate NOD1 signaling pathways in IECs (Cañas et al. 2018). In Caco-2 cells, both RIP2 inhibition and NOD1-specific siRNA knockdown (but not that of NOD2) decrease vesicle-mediated activation of NF-κB and subsequent expression of the pro-inflammatory cytokines IL-6 and IL-8

(Fig. 9.2). Results concerning IL-8 secretion revealed that in addition to PGN, other MAMPs included in MVs can activate signaling pathways that lead to the expression of this cytokine (Cañas et al. 2018). EcN and ECOR12 internalized MVs colocalize with NOD1, activate NOD1 aggregation, and promote association of NOD1 with early endosomes. Although MVs from both strains activate the NOD1 pathway, kinetics of NOD1 aggregation and NF-κB activation reveal significant differences between them. The cell response to internalized MVs is faster for the probiotic EcN than for the commensal ECOR12 strain (Cañas et al. 2018). This study revealed that MVs released by beneficial gut bacteria modulate NOD-mediated host immune responses and contribute to priming of the immune system.

**Immune Modulation Through the Intestinal Epithelial Barrier** Studies performed with EcN and ECOR12 MVs also corroborate the role of MVs as an active mechanism used by beneficial gut bacteria to activate signaling pathways through the intestinal epithelial barrier, which result in modulation of immune responses at the intestinal mucosa (Fábrega et al. 2016). In this study several models were used to assess the cross talk between bacterial MVs, intestinal epithelial cells, and immune cells. This involved: (1) stimulation of human peripheral blood mononuclear cells (PBMCs) as a model of intestinal barrier disruption, where microbiota MVs interact directly with immune cells, (2) apical stimulation of polarized Caco-2/PMBCs cocultures as a model of healthy, undamaged intestinal mucosa, and (3) normal human colon tissue as an ex vivo model that is more similar to in vivo gut conditions. To prove that MVs exert specific effects, stimulations with bacterial lysates were also performed in all these models. Analysis of cell responses revealed that bacterial MVs and lysates trigger the expression of immune mediators in PBMCs. In contrast, only MVs induce expression and secretion of cytokines to the basolateral compartment in Caco-2/PBMCs cocultures. In this model, the epithelial barrier made by differentiated Caco-2 cells prevents direct access of bacterial effectors to PBMCs. Under these conditions, MVs taken up by epithelial cells interact with immune receptors and activated epithelial cells in turn release soluble mediators that elicit a response in immune cells (Fig. 9.2). This intercellular cross talk was corroborated in human colonic explants. In this ex vivo model MVs trigger upregulation of MIP1α, IL-10, IL-8, IL-6, and TNF-α and downregulation of the pro-inflammatory cytokine IL-12 (Fábrega et al. 2016). Remarkably, values of the IL-10/TNF-α and IL-10/IL-12 expression ratio indicate that MVs from the probiotic EcN elicit better anti-inflammatory balance than ECOR12 MVs.

Additional analysis in human colonic explants revealed that EcN and ECOR12 MVs also promote upregulation of IL-22 and the antimicrobial peptide β-defensin-2 (Fábrega et al. 2016). These two mediators are interconnected. Epithelial cells are targets of IL-22, a cytokine mainly expressed by immune cells. In the gut, the local innate lymphoid cells in the lamina propria integrate microbiota-derived signals and regulate adaptive immune responses. In particular, IL-22 released by intestinal lymphoid cells helps to maintain the integrity of the epithelial barrier through several mechanisms, one of which is the induction of β-defensins (Nikoopour et al. 2015). In addition, these bacterial MVs elicit downregulation of genes encoding TGF-β and the membrane-anchored mucin 1 (MUC-1), both linked to IL-17 responses. TGF-β is a pleiotropic cytokine, whose inflammatory and regulatory activities depend on other cellular factors. It is known that TGF-β triggers differentiation of Treg cells (anti-inflammatory), but in the presence of IL-6, this factor can promote differentiation of T helper 17 (Th17) cells (pro-inflammatory) (Sanjabi et al. 2009). Cooperation between Th17 and Treg cells is essential to preserve intestinal homeostasis. Imbalances in these cell populations toward high production of Th17 cells result in IBD. Since the pro-inflammatory effects of TGF-β are linked to Th17 cell differentiation, downregulation of TGF-β promoted by the probiotic EcN is consistent with its effectiveness in the remission of ulcerative colitis by restoring the Th17/Treg balance (Fábrega et al. 2016). MUC-1 and TGF-β are overexpressed in several cancer types (Apostolopoulos et al. 2015). In this context, the ability of EcN and ECOR12 MVs to downregulate these markers links the beneficial effects of certain gut bacteria to cancer progression or treatment effectiveness, especially in immunotherapy-based strategies in which the individual response depends on gut microbiota (Vétizou et al. 2015; Gopalakrishnan et al. 2018).

**Modulation of Tight Junction Proteins** The integrity of the intestinal epithelial barrier is critical in maintaining homeostasis. Adjacent IECs are connected by a set of proteins that establish tight junctions. The tight junction complexes include integral membrane proteins (occludin, several claudins, tricellulin, and junctional adhesion molecules) and peripheral membrane proteins of the zonula occludens (ZO), which bind to claudins and act as scaffolds anchoring the transmembrane proteins to the actin cytoskeleton (Turner 2009). Claudins are a large family of tight junction proteins that regulate barrier integrity and paracellular permeability. Besides claudins that have a sealing function (like claudin-1 or claudin-14), others act as selective channels for ions or water (like claudin-2). Certain beneficial gut microbes positively influence barrier integrity by strengthening tight junctions and reducing gut permeability. In some cases, these effects are mediated, at least partially, by secreted bacterial metabolites (Ewaschuk et al. 2008), proteins (Hering et al. 2014) or MVs (Alvarez et al. 2016; Chelakkot et al. 2018).

The probiotic EcN reinforces the intestinal epithelial barrier through upregulation and redistribution of the tight junction proteins ZO-1, ZO-2, and claudin-14 (Ukena et al. 2007; Zyrek et al. 2007; Hering et al. 2014). Induction of claudin-14 is mediated by the secreted protein TcpC, an immunomodulatory protein that inhibits the TLR4 signaling cascade (Hering et al. 2014). The contribution of EcN MVs to the regulation of tight junction proteins was studied in the intestinal epithelial cell lines T-84 and Caco-2 (Alvarez et al. 2016). This study included other *E. coli* strains of human intestinal origin such as ECOR12 (*tcpC* negative) and ECOR63 (*tcpC* positive). Secreted MVs and soluble factors were separated from culture supernatants of the wild-type strains and the isogenic EcN and ECOR63 *tcpC* mutants, and their effect on the epithelial barrier was assessed by measuring transepithelial resistance, and gene and protein expression analyses of several tight junction proteins in polarized cell monolayers. This analysis revealed that MVs from the commensal ECOR12 do not have any positive effect on these parameters. In

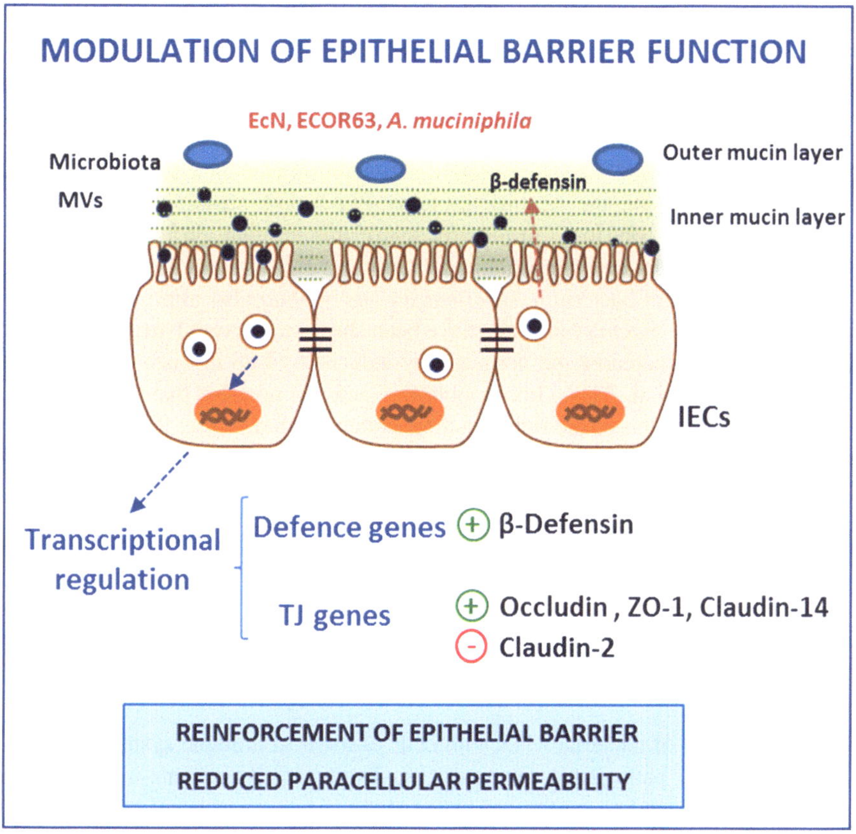


**Fig. 9.3** Modulation of the gut epithelial barrier by microbiota MVs. The drawing shows the structure of the intestinal barrier in which the mucin layer maintains segregation between luminal microbes and the intestinal epithelium. *Gut microbiota* (represented by ellipses) reside in the outer mucus layer. *MVs* (small black circles) can diffuse through the *inner mucus layer* (dotted lines) and reach the epithelium. Interaction of MVs from EcN, ECOR63, and *A. muciniphila* with intestinal epithelial cells (*IECs*) elicits transcriptional regulation of genes encoding tight junction (*TJ*) proteins. Upregulation is indicated by (+) and downregulation by (−). MVs from EcN also induce expression of the antimicrobial peptide *β-defensin*. The regulatory effects mediated by microbiota MVs on TJ proteins lead to the reinforcement of the gut epithelial barrier and, consequently, to a reduction in paracellular permeability

contrast, both extracellular fractions (MVs and soluble factors) from EcN and ECOR63 promote upregulation of ZO-1 and claudin-14 without affecting the expression of ZO-2 (Fig. 9.3). The strengthening barrier effects mediated by EcN and ECOR63 MVs are independent of TcpC, which is secreted as a soluble protein. In addition to tight junction proteins known to be regulated by EcN, this study also examined the leaky protein claudin-2. It is known that regulation of claudin-2 results in increased intestinal permeability (Luettig et al. 2015). Pathogens like Salmonella or microbial toxins increase claudin-2 expression to disrupt the intestinal epithelial barrier and facilitate bacterial invasion (Zhang et al. 2013; Liu et al. 2013), whereas some probiotic strains enhance the intestinal barrier by downregulating claudin-2 (Ewaschuk et al. 2008). The probiotic EcN and the intestinal isolate ECOR63 also exploit this mechanism to improve barrier function (Fig. 9.3). In these strains, downregulation of claudin-2 is mediated by released MVs and soluble factors that have not been identified yet. The negative regulation of claudin-2 together with the positive regulation of the sealing proteins ZO-1 and claudin-14 exerted by EcN-secreted MVs contribute to the efficacy of this probiotic in the treatment of inflammatory diseases and intestinal infections.

The beneficial gut bacterium *Akkermansia muciniphila* also improves intestinal barrier function. *A. muciniphila* MVs have been shown to prevent barrier disruption and neutralize the increased gut permeability induced by LPS in Caco-2 cell monolayers (Chelakkot et al. 2018). Immunoblotting analysis revealed that these vesicles increase the expression of the tight junction protein occludin (Fig. 9.3). The mechanism involved in these vesicle-mediated effects is the phosphorylation of AMPK, a kinase involved in the regulation of tight junction assembly (Chelakkot et al. 2018). The ability of *A. muciniphila* MVs to improve gut permeability in vivo has also been evaluated in mouse models (see Sect. 9.3.3).

### *9.3.2 Interaction with Immune Cells*

Direct interaction of bacterial MVs with cells of the host immune system has been widely studied for pathogen-derived vesicles. These vesicles activate immune cells via PRRs and, depending on the producing strain, act as pro- or anti-inflammatory mediators and, in some cases, subvert the immune system to promote pathogen survival in the host. Although there is extensive information on the interactions and effects of beneficial gut microbes in immune cells, few studies are focused on their MVs. Strains of the genus *Bifidobacterium* elicit strain-specific effects on the maturation of DCs and their ability to polarize naïve $CD4^+$ T cells to an effector response (López et al. 2010). Specifically, the probiotic *Bifidobacterium bifidum* LMG13195 can elicit a T regulatory (Treg) response. Analysis of several probiotic subcellular fractions on human DCs showed that isolated MVs activate DCs to promote differentiation of Treg functional cells ($FoxP3^+$) with a suppressor balance, supported by the highest IL-10/pro-inflammatory cytokines ratio (López et al. 2012).

The gut symbiont *Bacteroides fragilis* can also induce Treg cells and mucosal tolerance. In this case the immunomodulatory molecule is PSA, which is secreted through MVs. Extracellular vesicles from this commensal are taken up by DCs in an actin-dependent manner and induce anti-inflammatory immune responses in cell cocultures of mouse bone marrow-derived DCs (BMDCs) and naïve $CD4^+$ T cells (Shen et al. 2012). PSA delivered through MVs is sensed by DCs via TLR2. Activation of this receptor triggers signal transduction events that induce tolerogenic DCs, which promote the development of IL-10 secreting Treg cells (Fig. 9.2). MVs isolated from a defective PSA mutant failed to mediate these effects. A microarray-based transcriptomic analysis in BMDCs challenged with MVs from the wild-type or the ΔPSA mutant strain revealed that transcripts specifically regulated by PSA essentially correspond to genes known to be regulated in a TLR2 dependent manner. In addition, *B. fragilis* MVs elicit gene expression changes that are PSA independent (Shen et al. 2012).

### 9.3.3 Immunomodulatory and Barrier Protective Effects in Animal Models of Human Diseases

Several in vivo studies in mouse models of IBD and food allergy have confirmed the immunomodulatory and barrier protective effects of gut microbiota-derived MVs (Shen et al. 2012; Kang et al. 2013; Kim et al. 2015; Fábrega et al. 2017). The term IBD encompasses chronic inflammatory disorders of the intestinal tract such as ulcerative colitis or Crohn's disease. These are multifactorial diseases that involve an imbalanced immune response to commensal gut microbes in genetically susceptible individuals, leading to inflammation and reduced intestinal epithelial integrity (Zhang and Li 2014). Dysbiosis is a common feature in IBD. For this reason, therapeutic approaches targeting modulation of the gut microbiota have been explored, and the therapeutic potential of certain beneficial gut bacteria has been confirmed in clinical trials or animal models of experimental colitis. Experimental colitis in mice can be chemically induced by rectal administration of trinitrobenzene sulphonic acid (TNBS) or oral treatment with dextran sulfate sodium (DSS).

Concerning *B. fragilis*, the immunomodulatory potential of PSA has been proved in the TNBS-induced colitis model. Oral feeding of purified PSA (Mazmanian et al. 2008) or *B. fragilis* MVs (Shen et al. 2012) ameliorates colitis progression by reducing animal weight loss, histological damage, and inflammation. Consistently, mice treated with PSA-containing MVs display reduced leukocytic infiltration at the intestinal mucosa, downregulation of pro-inflammatory cytokines, and increased IL-10 expression. In contrast, MVs isolated from a PSA-defective mutant fail to protect mice against TNBS-induced colitis (Shen et al. 2012). Since the effect of PSA on DCs is mediated through TLR2, the function of Gadd45α (Growth arrest and DNA-damage-inducible protein), a downstream factor of the TLR2 signaling pathway that stimulates T cell responses, was analyzed in *Gadd45α*$^{-/-}$ knockout mice. In contrast to wild-type mice, *B. fragilis* MVs do not protect *Gadd45α*$^{-/-}$ knockout mice from TNBS-induced colitis. Therefore, this study showed that MVs from this resident gut microbe prevent colitis development by essentially activating tolerogenic DCs (Shen et al. 2012).

The therapeutic efficacy of EcN on the remission of ulcerative colitis has been supported in several clinical assays (Losurdo et al. 2015) and experimental colitis models. Like viable probiotic suspensions, daily oral administration of EcN MVs significantly reduces DSS-induced weight loss and ameliorates clinical symptoms, mucosal injury, and inflammation in the gut (Fábrega et al. 2017). Treatment with

EcN MVs counteracts altered expression of cytokines and markers of intestinal barrier function. Several mechanisms are exploited by EcN vesicles to ameliorate disease progression. Regarding inflammatory markers of colitis, EcN MVs decrease the expression of several pro-inflammatory cytokines (IL-6, IL-8, IL-1β, TNF-α, IL-12 and IL-17) and counteract the lower expression of IL-10 associated with DSS treatment. In addition, EcN MVs elicit compensatory effects on expression of the inflammatory enzymes cyclooxygenase-2 and inducible nitric oxide synthase (iNOS) (Fábrega et al. 2017). Increased iNOS expression in infiltrating macrophages in the intestinal mucosa is a common feature in clinical IBD. The consequent excessive nitric oxide production results in tissue injury in IBD patients and colitic mice (Palatka et al. 2005). Consequently, reduction of iNOS expression by EcN MVs helps to attenuate colitis in MVs-treated mice (Fábrega et al. 2017). Concerning markers of intestinal barrier function, EcN MVs cannot counteract DSS-induced downregulation of ZO-1. These findings do not match the effects observed in in vitro models of epithelial barrier integrity (Alvarez et al. 2016), indicating that different regulatory mechanisms could be activated by EcN MVs in the presence of highly expressed inflammatory mediators. Reinforcement of the epithelial barrier can also be triggered by posttranslational modification mechanisms that direct ZO-1 to the cell boundaries and allow its association with tight junction structures. The intestinal trefoil factor 3 (TFF-3), a bioactive peptide involved in epithelial protection and repair, is one of the mediators that promote ZO-1 redistribution from the cytosol to intercellular tight junctions in intestinal epithelial monolayers without changes in ZO-1 protein levels. Expression of TFF-3 is downregulated in DSS-inflamed colonic tissue. Oral administration of EcN MVs restores the mRNA levels of TFF-3 to values similar to those of healthy mice, thus preserving the colonic mucosa against DSS-induced damage (Fábrega et al. 2017). In the context of tissue remodeling, the matrix metalloproteinases (MMPs) MMP-9 and MMP-2 are also relevant. Upregulation of MMP-9 in DSS-treated mice promotes tissue injury by disrupting intestinal tight junctions, which in turn results in increased intestinal permeability and subsequent inflammation (Nighot et al. 2015). In contrast, MMP-2 plays a protective role in maintenance of gut barrier function. Treatment with EcN MVs downregulates MMP-9 and tends to preserve MMP-2 expression, thus pointing to another mechanism used by EcN vesicles to protect intestinal barrier function in DSS-experimental colitis (Fábrega et al. 2017).

MVs released by the beneficial gut bacterium *A. muciniphila* also protect DSS-induced experimental colitis in mice. The beneficial effects of these vesicles have been evidenced by the preservation of body weight and colon length, and reduced infiltration of immune cells in the colonic tissue (Kang et al. 2013). The role of *A. muciniphila* MVs in the modulation of gut permeability has been examined in a mouse model of diabetes induced by a high-fat diet (HFD) (Chelakkot et al. 2018). There is mounting evidence of a close association between intestinal permeability and metabolic diseases. Metagenomics studies have found diminished abundance of *A. muciniphyla* in patients with obesity and type 2 diabetes, and their fecal samples also contain less *A. muciniphila*-derived MVs than healthy controls (Chelakkot et al. 2018). The impact of orally administered MVs isolated from this gut symbiont in

HFD-fed mice was compared with control mice fed a normal diet. Treatment with *A. muciniphila* MVs improved all the alterations caused by the HFD. Specifically, MVs reduced body weight gain and HFD-increased intestinal permeability, restored the intestinal barrier from HFD-induced damage, reduced the subsequent recruitment of immune cells, and increased the expression of the tight junction proteins downregulated by HFD (Chelakkot et al. 2018). Consistently, improvement of the intestinal barrier function by *A. muciniphila* MVs resulted in less endotoxemia and improved glucose tolerance in HFD-fed mice.

Food allergy, a disease with increasing incidence over the last decade, results from abnormal immune responses to food components. Apart from avoiding allergenic foods, administration of probiotics has been proposed as an alternative treatment for patients with food allergy. In this context, the therapeutic potential of *Bifidobacterium longum* has been evaluated in a mouse model of allergen-induced food allergy that causes acute diarrhea. Allergy was induced by intraperitoneal injection of ovalbumin in aluminum potassium sulfate adjuvant followed by oral administration of the allergen for several days (Kim et al. 2015). During the induction of food allergy, animals were orally treated with two resident gut bacteria, either *B. longum* or *Enterococcus faecalis*. Only *B. longum* alleviated diarrhea without counteracting the allergen-induced Th2 response. The specific mechanism used by this probiotic to limit inflammation is the reduction of mast cell number in the small intestine. *B. longum*-derived MVs selectively induce apoptosis of bone marrow-derived mast cells. The effector molecule that triggers cell-death is the main vesicular protein ESBP (family 5 extracellular solute-binding protein) (Kim et al. 2015).

## 9.4 Spreading of Microbiota MVs through the Body

As stated above, gut dysbiosis has been associated with a wide range of diseases that affect distal body tissues (Maguire and Maguire 2019). Pathohistological effects have traditionally been attributed to the increased gut permeability that follows an imbalance in the microbiota. This condition, known as leaky gut, impairs epithelial barrier function, leading to the translocation of microbiota-derived products or luminal bacteria to the bloodstream. Once in the general circulation, these bacterial effectors can be distributed throughout the body and reach any tissue. First studies in this field connected LPS in human blood with endotoxemia, a condition that has been associated with obesity and insulin resistance (Cani et al. 2007). However, recent reports prove that gut microbiota-derived compounds are also found in blood, urine, and distal tissues in healthy subjects (Clarke et al. 2010; Païssé et al. 2016; Lee et al. 2017). One example is bacterial PGN derived from gut microbiota, which crosses the blood-brain barrier and regulates brain development and behavior through specific sensing molecules and activation of NOD signaling pathways in normal mice (Arentsen et al. 2017). Moreover, perturbation of gut microbiota leads to dysregulation of proteins involved in PGN detection in the developing brain

(Arentsen et al. 2017). Bacterial genomic DNA has also been found in blood samples of healthy donors (Nikkari et al. 2001; Païssé et al. 2016). The mechanism by which this bacterial DNA reaches the bloodstream has been recently revealed in mice (Park et al. 2017). This study showed that blood-associated bacterial DNA is transported by MVs, predominantly from gut microbiota. To reach the bloodstream, gut microbiota MVs have to cross intestinal epithelium and the vascular endothelium. This passage has been proposed to be mediated by paracellular or transcellular pathways (Stentz et al. 2018).

The blood-brain barrier efficiently regulates cellular and molecular trafficking between the blood and the neural tissue. The permeability of this barrier to circulating compounds increases under inflammatory conditions. In this context, studies based on sequencing approaches provide evidence on the presence of ribosomal DNA from commensal bacteria in the brain (Zhan et al. 2016; Emery et al. 2017). How this bacterial DNA comes from translocated bacteria or circulating MVs is an open question (Stentz et al. 2018).

The role of gut microbiota-derived MVs as vehicles for the distribution and delivery of many bacterial effectors to distal tissues is an emerging topic. Due to the existence of the gut–brain axis and the great impact of gut microbiota on neurological diseases (Marin et al. 2017; Van Den Elsen et al. 2017) such as depression, stress, anxiety, autism, or Alzheimer's, studies on circulating microbiota-derived vesicles are principally focused on human patients or animal models of such diseases (Park et al. 2017; Lee et al. 2017). In this context, metagenomic analysis of MVs isolated from human urine samples has been proved an efficient method to assess changes in gut microbiota composition in autism. Sequencing of 16S rRNA gene variable regions in DNA samples extracted from bacterial MVs isolated from urine have revealed markedly altered microbiota profiles in people with autism disorder relative to healthy control individuals, even at the genus level (Lee et al. 2017). These differences correlate with the changes in gut microbiome previously described in autism, thus reinforcing the utility of urine MVs for diagnostic purposes. Likewise, in a mouse model of Alzheimer's disease, great differences in the genomic profile of bacterial MVs isolated from blood have also been found compared to wild-type mice (Park et al. 2017).

There is growing evidence of the gut microbiome's influence on cancer onset, development, and treatment (reviewed in Gopalakrishnan et al. 2018). However, very few studies have focused on the impact of microbiota-derived MVs. As stated above, MVs released in the GI tract by resident gut bacteria can be disseminated throughout the body, reach any tissue, and induce either beneficial or harmful effects depending on the producer bacterial strain (Liu et al. 2018). The beneficial effects of MVs secreted by *Lactobacillus rhamnosus* on hepatic cancer cells have been assessed in HepG2 cell cultures. Specifically, MVs from the probiotic *L. rhamnosus* induce apoptosis in this cancer cell line (Behzadi et al. 2017).

## 9.5 Conclusions

We are currently living in the microbiome age. Clinical studies have provided insight into the influence of the microbiome on immunity and a wide range of diseases. However, there is still much to learn about the intrinsic mechanisms of this action. Further research is also required to develop optimal strategies to modulate gut microbiota for therapeutic purposes. Microbiota-secreted MVs can deliver functional molecules to distant cells and, nowadays, are being revealed as key players in microbiota–host communication. Due to their nano-size structure and the great variety of cargo molecules, microbiota-derived vesicles can diffuse through the intestinal mucus layer and modulate host metabolism, immune and defense responses. Consequently, microbiota-derived MVs may have great influence on health and disease. Gut microbiome profiling is also seen as a potentially useful tool for diagnosis or improvement of the immunotherapy response. However, the method selected for sampling (stools or gut luminal content) and sequencing (metagenomics or 16SRNA sequencing) greatly affect the final output and reproducibility of the results. The recent discovery of gut-derived MVs in blood and urine samples from either patients or healthy individuals with a profile that strongly correlates with the real gut microbiome opens novel strategies for diagnostic purposes.

## References


Aguilar C, Mano M, Eulalio A (2018) MicroRNAs at the host-bacteria interface: host defense or bacterial offense. Trends Microbiol 27:206–218

Aguilera L, Toloza L, Giménez R et al (2014) Proteomic analysis of outer membrane vesicles from the probiotic strain *Escherichia coli* Nissle 1917. Proteomics 14(2–3):222–229

Allison CC, Kufer TA, Kremmer et al (2009) *Helicobacter pylori* induces MAPK phosphorylation and AP-1 activation via a NOD1-dependent mechanism. J Immunol 183(12):8099–8109

Al-Nedawi K, Main MF, Hossain N et al (2015) Gut commensal microvesicles reproduce parent bacterial signals to host immune and enteric nervous systems. FASEB J 29(2):684–695

Alvarez CS, Badia J, Bosch M et al (2016) Outer membrane vesicles and soluble factors released by probiotic *Escherichia coli* Nissle 1917 and commensal EcoR63 enhance barrier function by regulating expression of tight junction proteins in intestinal epithelial cells. Front Microbiol 7:1981

Apostolopoulos V, Stojanovska L, Gargosky SE (2015) MUC1 (CD227): a multi-tasked molecule. Cell Mol Life Sci 72:4475–4500

Arentsen T, Qian Y, Gkotzis S et al (2017) The bacterial peptidoglycan-sensing molecule Pglyrp2 modulates brain development and behavior. Mol Psychiatry 22:257–266

Baothman OA, Zamzami MA, Taher I et al (2016) The role of gut microbiota in the development of obesity and diabetes. Lipids Health Dis 15:108

Belkaid Y, Hand TW (2014) Role of the microbiota in immunity and inflammation. Cell 157 (1):121–141

Behzadi E, Mahmoodzadeh Hosseini H, Imani Fooladi AA (2017) The inhibitory impacts of *Lactobacillus rhamnosus* GG-derived extracellular vesicles on the growth of hepatic cancer cells. Microb Pathog 110:1–6

Bomberger JM, Maceachran DP, Coutermarsh BA et al (2009) Long-distance delivery of bacterial virulence factors by *Pseudomonas aeruginosa* outer membrane vesicles. PLoS Pathog 5: e1000382. https://doi.org/10.1371/journal.ppat.1000382

Brameyer S, Plener L, Muller A et al (2018) Outer membrane vesicles facilitate trafficking of the hydrophobic signaling molecules CAI-1 between *Vibrio harveyi* cells. J Bacteriol 200:e00740–e00717

Bron PA, van Baarlen P, Kleerebezem M (2011) Emerging molecular insights into the interaction between probiotics and the host intestinal mucosa. Nat Rev Microbiol 10(1):66–78

Brown L, Wolf JM, Prados-Rosales R et al (2015) Through the wall: extracellular vesicles in Gram-positive bacteria, mycobacteria and fungi. Nat Rev Microbiol 13(10):620–630. https://doi.org/10.1038/nrmicro3480

Bryant WA, Stentz R, Le Gall G et al (2017) In silico analysis of the small molecule content of outer membrane vesicles produced by *Bacteroides thetaiotaomicron* indicates an extensive metabolic link between microbe and host. Front Microbiol 8:2440. https://doi.org/10.3389/fmicb.2017.02440

Caballero S, Pamer EG (2015) Microbiota-mediated inflammation and antimicrobial defense in the intestine. Annu Rev Immunol 33:227–256

Cani PD, Amar J, Iglesias MA et al (2007) Metabolic endotoxemia initiates obesity and insulin resistance. Diabetes 56:1761–1772

Cani PD (2018) Human gut microbiome: hopes, threats and promises. Gut 67(9):1716–1725

Cañas MA, Giménez R, Fábrega MJ et al (2016) Outer membrane vesicles from the probiotic *Escherichia coli* Nissle 1917 and the commensal ECOR12 enter intestinal epithelial cells via clathrin-dependent endocytosis and elicit differential effects on DNA damage. PLoS One 11(8): e0160374

Cañas MA, Fábrega MJ, Giménez R et al (2018) Outer membrane vesicles from probiotic and commensal *Escherichia coli* activate Nod1-mediated immune responses in intestinal epithelial cells. Front Microbiol 9:98

Celluzzi A, Masotti A (2016) How our other genome controls our epi-genome. Trends Microbiol 24 (10):777–787

Chamaillard M, Hashimoto M, Horie Y et al (2003) An essential role for NOD1 in host recognition of bacterial peptidoglycan containing diaminopimelic acid. Nat Immunol 4:702–707

Chatzidaki-Livanis M, Coyne MJ, Comstock LE (2014) An antimicrobial protein of the gut symbiont *Bacteroides fragilis* with a MACPF domain of host immune proteins. Mol Microbiol 94(6):1361–1374

Chelakkot C, Choi Y, Kim DK et al (2018) *Akkermansia muciniphila*-derived extracellular vesicles influence gut permeability through the regulation of tight junctions. Exp Mol Med 50(2):e450

Choi JW, Kim SC, Hong SH et al (2017) Secretable small RNAs via outer membrane vesicles in periodontal pathogens. J Dent Res 96(4):458–466

Chu H (2017) Host gene–microbiome interactions: molecular mechanisms in inflammatory bowel disease. Genome Med 9:69

Clarke TB, Davis KM, Lysenko ES et al (2010) Recognition of peptidoglycan from the microbiota by Nod1 enhances systemic innate immunity. Nat Med 16(2):228–231

Domínguez Rubio AP, Martínez JH, Martinez Casillas DC et al (2017) *Lactobacillus casei* BL23 produces microvesicles carrying proteins that have been associated with its probiotic effect. Front Microbiol 8:1783

Donaldson GP, Lee SM, Mazmanian SK (2016) Gut biogeography of the bacterial microbiota. Nat Rev Microbiol 14(1):20–32

Emery DC, Shoemark DK, Batstone TE et al (2017) 16S rRNA next generation sequencing analysis shows bacteria in Alzheimer's post-mortem brain. Front Aging Neurosci 9:195

Elhenawy W, Debelyy MO, Feldman MF (2014) Preferential packing of acidic glycosidases and proteases into Bacteroides outer membrane vesicles. MBio 5(2):e00909–e00914

Ewaschuk JB, Diaz H, Meddings L et al (2008) Secreted bioactive factors from *Bifidobacterium infantis* enhance epithelial cell barrier function. Am J Physiol Gastrointest Liver Physiol 295: G1025–G1034

Fábrega MJ, Aguilera L, Giménez R et al (2016) Activation of immune and defense responses in the intestinal mucosa by outer membrane vesicles of commensal and probiotic *Escherichia coli* strains. Front Microbiol 7:705

Fábrega MJ, Rodríguez-Nogales A, Garrido-Mesa J et al (2017) Intestinal anti-inflammatory effects of outer membrane vesicles from *Escherichia coli* Nissle 1917 in DSS-experimental colitis in mice. Front Microbiol 8:1274

Faderl M, Noti M, Corazza N et al (2015) Keeping bugs in check: the mucus layer as a critical component in maintaining intestinal homeostasis. IUBMB Life 67(4):275–285

Feerick CL, McKernan DP (2016) Understanding the regulation of pattern recognition receptors in inflammatory diseases - a 'Nod' in the right direction. Immunology 150(3):237–247

Feng Q, Chen WD, Wang YD (2018) Gut microbiota: an integral moderator in health and disease. Front Microbiol 9:151

Geuking MB, Cahenzli J, Lawson MA et al (2011) Intestinal bacterial colonization induces mutualistic regulatory T cell responses. Immunity 34(5):794–806

Girardin SE, Boneca IG, Carneiro LA et al (2003a) Nod1 detects a unique muropeptide from gram-negative bacterial peptidoglycan. Science 300:1584–1587

Girardin SE, Boneca IG, Viala J et al (2003b) Nod2 is a general sensor of peptidoglycan through muramyl dipeptide (MDP) detection. J Biol Chem 278:8869–8872

Gopalakrishnan V, Helmink BA, Spencer CN et al (2018) The influence of the gut microbiome on cancer, immunity and cancer immunotherapy. Cancer Cell 33:570–580

Ghosal A, Upadhyaya BB, Fritz JV et al (2015) The extracellular RNA complement of *Escherichia coli*. Microbiology 4(2):252–266

Guerrero-Mandujano A, Hernández-Cortez C, Ibarra JA (2017) The outer membrane vesicles: secretion system type zero. Traffic 18:425–432

Hasegawa M, Fujimoto Y, Lucas PC et al (2008) A critical role of RICK/RIP2 polyubiquitination in nod-induced NF-kappa B activation. EMBO J 27:373–383

Hering NA, Richter JF, Fromm A et al (2014) TcpC protein from *E. coli* Nissle improves epithelial barrier function involving PKCʒ and ERK1/2 signaling in HT-29/B6 cells. Mucosal Immunol 7:369–378

Hickey CA, Kuhn KA, Donermeyer DL et al (2015) Colitogenic *Bacteroides thetaiotaomicron* antigens access host immune cells in a sulfatase-dependent manner via outer membrane vesicles. Cell Host Microbe 17(5):672–680

Inohara N, Koseki T, Lin J et al (2000) An induced proximity model for NF-kappa B activation in the Nod1/RICK and RIP signaling pathways. J Biol Chem 275(36):27823–27831

Irving AT, Mimuro H, Kufer TA et al (2014) The immune receptor NOD1 and kinase RIP2 interact with bacterial peptidoglycan on early endosomes to promote autophagy and inflammatory signalling. Cell Host Microbe 15(5):623–635

Jager J, Keese S, Roessle M et al (2014) Fusion of *Legionella pneumophila* outer membrane vesicles with eukaryotic membrane systems is a mechanism to deliver pathogen factors to host cell membranes. Cell Microbiol 17:607–620. https://doi.org/10.1111/cmi.12392

Jandhyala SM, Talukdar R, Subramanyam C et al (2015) Role of the normal gut microbiota. World J Gastroenterol 21(29):8787–8803

Jia L, Lu J, Zhou Y et al (2018) Tolerogenic dendritic cells induced the enrichment of CD4+Foxp3+ regulatory T cells via TGF-β in mesenteric lymph nodes of murine LPS-induced tolerance model. Clin Immunol 197:118–129

Johansson MEV, Holmén Larsson JM, Hansson GC (2011) The two mucus layers of colon are organized by the MUC2 mucin, whereas the outer layer is a legislator of host-microbial interactions. Proc Natl Acad Sci U S A 108(3Suppl 1):4659–4665

Kang CS, Ban M, Choi EJ et al (2013) Extracellular vesicles derived from gut microbiota, especially *Akkermansia muciniphila*, protect the progression of dextran sulfate sodium-induced colitis. PLoS One 8(10):e76520. https://doi.org/10.1371/journal.pone.0076520
Kaparakis-Liaskos M (2015) The intracellular location, mechanisms and outcomes of NOD1 signalling. Cytokine 74:207–212
Kaparakis-Liaskos M, Ferrero RL (2015) Immune modulation by bacterial outer membrane vesicles. Nat Rev Immunol 15(6):375–387
Kawai T, Akira S (2010) The role of pattern-recognition receptors in innate immunity: update on toll-like receptors. Nat Immunol 11(5):373–384
Kesty NC, Mason KM, Reedy M et al (2004) Enterotoxigenic *Escherichia coli* vesicles target toxin delivery into mammalian cells. EMBO J 23:4538–4549
Kim JH, Jeun EJ, Hong CP et al (2015) Extracellular vesicle-derived protein from *Bifidobacterium longum* alleviates food allergy through mast cell suppression. J Allergy Clin Immunol 137 (2):507–516.e8
Koeppen K, Hampton TH, Jarek M et al (2016) Novel mechanism of host-pathogen interaction through sRNA in bacterial outer membrane vesicles. PLoS Pathog 12(6):e1005672
Lee E-Y, Bang JY, Park GW et al (2007) Global proteomic profiling of native outer membrane vesicles derived from *Escherichia coli*. Proteomics 7:3143–3153
Lee Y, Park JY, Lee EH et al (2017) Rapid assessment of microbiota changes in individuals with autism spectrum disorder using bacteria-derived membrane vesicles in urine. Exp Neurobiol 26 (5):3017–3317
Li M, Lee K, Hsu M et al (2017) *Lactobacillus*-derived extracellular vesicles enhance host immune responses against vancomycin-resistant enterococci. BMC Microbiol 17(1):66. https://doi.org/10.1186/s12866-017-0977-7
Liu X, Yang G, Geng XR et al (2013) Microbial products induce claudin-2 to compromise gut epithelial barrier function. PloS ONE 8:e68547. https://doi.org/10.1371/journal.pone.0068547
Liu Y, Defourny KAY, Smid EJ et al (2018) Gram-positive bacterial extracellular vesicles and their impact on health and disease. Front Microbiol 9:1502. https://doi.org/10.3389/fmicb.2018.01502
Llewellyn A, Foey A (2017) Probiotic modulation of innate cell pathogen sensing and signaling events. Nutrients 9(10):E1156
López P, Gueimonde M, Margolles A et al (2010) Distinct Bifidobacterium strains drive different immune responses in vitro. Int J Food Microbiol 138(1–2):157–165
López P, González-Rodríguez I, Sánchez B et al (2012) Treg-inducing membrane vesicles from *Bifidobacterium bifidum* LMG13195 as potential adjuvants in immunotherapy. Vaccine 30 (5):825–829
Losurdo G, Iannone A, Contaldo A et al (2015) *Escherichia coli* Nissle 1917 in ulcerative colitis treatment: systematic review and meta-analysis. J Gastrointestin Liver Dis 24(4):499–505
Lu Z, Ding L, Lu Q et al (2013) Claudins in intestines: distribution and functional significance in health and diseases. Tissue Barriers 1(3):e24978
Luettig J, Rosenthal R, Barmeyer DC et al (2015) Claudin-2 as a mediator of leaky gut barrier during intestinal inflammation. Tissue Barriers 3(1–2):e977176
Maguire M, Maguire G (2019) Gut dysbiosis, leaky gut, and intestinal epithelial proliferation in neurological disorders: towards the development of a new therapeutic using amino acids, prebiotics, probiotics, and postbiotics. Rev Neurosci 30(2):179–201
Marin IA, Goertz JE, Ren T et al (2017) Microbiota alteration is associated with the development of stress-induced despair behavior. Sci Rep 7:43859
Masotti A (2012) Interplays between gut microbiota and gene expression regulation by miRNAs. Front Cell Infect Microbiol 2:137
Mazmanian SK, Round JL, Kasper DL (2008) A microbial symbiosis factor prevents intestinal inflammatory disease. Nature 453(7195):620–625

Molina-Tijeras JA, Gálvez J, Rodríguez-Cabezas ME (2019) The immunomodulatory properties of extracellular vesicles derived from probiotics: a novel approach for the management of gastrointestinal diseases. Nutrients 11(5):E1038
Nighot PK, Al-Sadi R, Rawat M et al (2015) Matrix metalloproteinase 9-induced increase in intestinal epithelial tight junction permeability contributes to the severity of experimental DSS colitis. Am J Physiol Gastrointest Liver Physiol 309(12):G988–G997
Nikkari S, McLaughlin IJ, Bi W et al (2001) Does blood of healthy subjects contain bacterial ribosomal DNA? J Clin Microbiol 39:1956–1959
Nikoopour E, Bellemore SM, Singh B (2015) IL-22, cell regeneration and autoimmunity. Cytokine 74:35–42
Ochman H, Selander RK (1984) Standard reference strains of *Escherichia coli* from natural populations. J Bacteriol 157:690–693
O'Donoghue EJ, Krachler AM (2016) Mechanisms of outer membrane vesicle entry into host cells. Cell Microbiol 18(11):1508–1517
O'Donoghue EJ, Sirisaengtaksin N, Browning DF et al (2017) Lipopolysaccharide structure impacts the entry kinetics of bacterial outer membrane vesicles into host cells. PLoS Pathog 13(11):e1006760. https://doi.org/10.1371/journal.ppat.1006760
Olier M, Marcq I, Salvador-Cartier C et al (2012) Genotoxicity of *Escherichia coli* Nissle 1917 strain cannot be dissociated from its probiòtic activity. Gut Microbes 3:501–509
Olofsson A, Nygard Skalman L, Obi I et al (2014) Uptake of *Helicobacter pylori* vesicles is facilitated by clathrin-dependent and clatrin-independent endocytic pathways. mBio 5(3): e00979-14. https://doi.org/10.1128/mBio.00979-14
Païssé S, Valle C, Servant F et al (2016) Comprehensive description of blood microbiome from healthy donors assessed by 16S targeted metagenomic sequencing. Transfusion 56:1138–1147
Palatka K, Serfozo Z, Veréb Z et al (2005) Changes in the expression and distribution of the inducible and endothelial nitric oxide synthase in mucosal biopsy specimens of inflammatory bowel disease. Scand J Gastroenterol 40(6):670–680
Park JY, Choi J, Lee Y et al (2017) Metagenome analysis of bodily microbiota in a mouse model of Alzheimer disease using bacteria-derived membrane vesicles in blood. Exp Neurobiol 26:369–379
Pérez-Cruz C, Cañas MA, Giménez R et al (2016) Membrane vesicles released by a hypervesiculating *Escherichia coli* Nissle 1917 *tolR* mutant are highly heterogeneous and show reduced capacity for epithelial cell interaction and Entry. PLoS One 11(12):e0169186. https://doi.org/10.1371/journal.pone.0169186
Peterson LW, Artis D (2014) Intestinal epithelial cells: regulators of barrier function and immune homeostasis. Nat Rev Immunol 14(3):141–153
Philpott DJ, Sorbara MT, Robertson SJ et al (2014) NOD proteins: regulators of inflammation in health and disease. Nat Rev Immunol 14(1):9–23
Qin J, Li R, Raes J et al (2010) A human gut microbial gene catalogue established by metagenomic sequencing. Nature 464(7285):59–65
Rakoff-Nahoum S, Coyne MJ, Comstock LE (2014) An ecological network of polysaccharide utilization among human intestinal symbionts. Curr Biol 24(1):40–49
Round JL, Lee SM, Li J et al (2011) The toll-like receptor 2 pathway establishes colonization by a commensal of the human microbiota. Science 332(6032):974–977
Runtsch MC, Round JL, O'Connell RM (2014) MicroRNAs and the regulation of intestinal homeostasis. Front Genet 5:347
Sanjabi S, Zenewicz LA, Kamanaka M et al (2009) Anti-inflammatory and pro-inflammatory roles of TGF-beta, IL-10 and IL-22 in immunity and autoimmunity. Curr Opin Pharmacol 9:447–453
Schwechheimer C, Kuehn MJ (2015) Outer-membrane vesicles from Gram-negative bacteria: biogenesis and functions. Nat Rev Microbiol 13(10):605–619
Severi E, Hood DW, Thomas GH (2007) Sialic acid utilization by bacterial pathogens. Microbiology 153:2817–2822

Shanahan F (2011) The gut microbiota in 2011: translating the microbiota to medicine. Nat Rev Gastroenterol Hepatol 9(2):72–74

Shen Y, Giordano Torchia ML, Lawson GW et al (2012) Outer membrane vesicles of a human commensal mediate immune regulation and disease protection. Cell Host Microbe 12 (4):509–520. https://doi.org/10.1016/j.chom.2012.08.004

Stentz R, Osborne S, Horn N et al (2014) A bacterial homolog of a eukaryotic inositol phosphate signaling enzyme mediates cross-kingdom dialog in the mammalian gut. Cell Rep 6 (4):646–656. https://doi.org/10.1016/j.celrep.2014.01.021

Stentz R, Horn N, Cross K et al (2015) Cephalosporinases associated with outer membrane vesicles released by Bacteroides spp. protect gut pathogens and commensals against β-lactam antibiotics. J Antimicrob Chemother 70(3):701–709. https://doi.org/10.1093/jac/dku466

Stentz R, Carvalho AL, Jones EJ et al (2018) Fantastic voyage: the journey of intestinal microbiota-derived microvesicles through the body. Biochem Soc Trans 46(5):1021–1027. https://doi.org/10.1042/BST20180114

Swaan PW, Bensman T, Bahadduri PM et al (2008) Bacterial peptide recognition and immune activation facilitated by human peptide transporter PEPT2. Am J Respir Cell Mol Biol 39 (5):536–542

Suzuki T (2013) Regulation of intestinal epithelial permeability by tight junctions. Cell Mol Life Sci 70(4):631–659

Thaiss CA, Levy M, Suez J et al (2014) The interplay between the innate immune system and the microbiota. Curr Opin Immunol 26:41–48

Thursby E, Juge N (2017) Introduction to the human gut microbiota. Biochem J 474 (11):1823–1836

Toloza L, Gimenez R, Fabrega MJ et al (2015) The secreted autotransporter toxin (Sat) does not act as a virulence factor in the probiotic *Escherichia coli* strain Nissle 1917. BMC Microbiol 15:250

Turner JR (2009) Intestinal mucosal barrier function in health and disease. Nat Rev Immunol 9 (11):799–809

Turner L, Bitto NJ, Steer DL et al (2018) *Helicobacter pylori* outer membrane vesicle size determines their mechanisms of host cell entry and protein content. Front Immunol 9:1466. https://doi.org/10.3389/fimmu.2018.01466

Ukena SN, Singh A, Dringenberg U et al (2007) Probiotic *Escherichia coli* Nissle 1917 inhibits leaky gut by enhancing mucosal integrity. PLoS One 2:e1308. https://doi.org/10.1371/journal.pone.0001308

Vaishnava S, Yamamoto M, Severson KM et al (2011) The antibacterial lectin RegIII-gamma promotes the spatial segregation of microbiota and host in the intestine. Science 334 (6053):255–258

Van Den Elsen LW, Poyntz HC, Weyrich LS et al (2017) Embrancing the gut microbiota: the new frontier for inflammatory and infectious diseases. Clin Transl Immunol 10:18–26

Vejborg RM, Friis C, Hancock V et al (2010) A virulent parent with probiotic progeny: comparative genomics of *Escherichia coli* strains CFT073, Nissle 1917 and ABU 83972. Mol Gen Genomics 283(5):469–484

Vercauteren D, Vandenbroucke RE, Jones AT et al (2010) The use of inhibitors to study endocytic pathways of gene carriers: optimization and pitfalls. Mol Ther 18:561–569

Vétizou M, Pitt JM, Daillère R et al (2015) Anticancer immunotherapy by CTLA-4 blockade relies on the gut microbiota. Science 350:1079–1084

Vimr ER, Kalivoda KA, Deszo EL et al (2004) Diversity of microbial sialic acid metabolism. Microbiol Mol Biol Rev 68(1):132–153

Vindigni SM, Zisman TL, Suskind DL et al (2016) The intestinal microbiome, barrier function, and immune system in inflammatory bowel disease: a tripartite pathophysiological circuit with implications for new therapeutic directions. Ther Adv Gastroenterol 9(4):606–625

Wells JM, Rossi O, Meijerink M et al (2011) Epithelial crosstalk at the microbiota-mucosal interface. Proc Natl Acad Sci U S A 108(Suppl 1):4607–4614

Zakharzhevskaya NB, Vanyushkina AA, Altukhov IA et al (2017) Outer membrane vesicles secreted by pathogenic and nonpathogenic *Bacteroides fragilis* represent different metabolic activities. Sci Rep 7(1):5008

Zhan X, Stamova B, Jin LW et al (2016) Gram-negative bacterial molecules associate with Alzheimer disease pathology. Neurology 87:2324–2332

Zhang YG, Wu S, Xia Y et al (2013) Salmonella infection upregulates the leaky protein claudin-2 in intestinal epithelial cells. PLoS One 8:e58606. https://doi.org/10.1371/journal.pone.0058606

Zhang YZ, Li YY (2014) Inflammatory bowel disease: pathogenesis. World J Gastroenterol 20:91–99. https://doi.org/10.3748/wjg.v20.i1.91

Zyrek AA, Cichon C, Helms S et al (2007) Molecular mechanisms underlying the probiotic effects of *Escherichia coli* Nissle 1917 involve ZO-2 and PKCzeta redistribution resulting in tight junction and epithelial barrier repair. Cell Microbiol 9:804–816

# Chapter 10
# Bacterial Membrane Vesicles and Their Applications as Vaccines and in Biotechnology

**Julie C. Caruana and Scott A. Walper**

**Abstract** Human society has coevolved with our invisible, microbial neighbors. From their use in the processing of food and drink to the manufacture of commercial products and therapeutics, bacteria, yeast, and fungi are invaluable tools to many aspects of our existence. Despite our long history together, researchers are only recently beginning to understand the complexities of these relationships and how the microbial world can fully be exploited. In the last 50 years alone, researchers have shown that bacteria can be used to manufacture drugs to aid in the treatment of medical disorders such as diabetes, that environmental microbes can be used to clean up chemical spills and disasters, and that the microbial communities that reside within our bodies are capable of influencing our mental and physical health. With this greater understanding comes new avenues of research to utilize these microbes to advance human society. Here we discuss efforts to utilize both native and engineered membrane vesicles shed by bacteria and their potential applications to several areas of biotechnology. We highlight the use of bacterial membrane vesicles in vaccine research and as emerging therapeutics as well as exploring their potential commercial applications and benefits.

## 10.1 Introduction

Society has a long and storied history with the microbial world. While most often thought of in the context of historical events of plague, disease, and infection, microbes such as bacteria, yeast, and fungi have benefited and advanced human society in countless ways. Microbes have become integral components in food and beverage production, in the manufacture of therapeutics, and as platforms for

J. C. Caruana
American Society for Engineering Education Postdoctoral Research Fellow, Washington, DC, USA

S. A. Walper (✉)
U.S. Naval Research Laboratory, Center for Bio/Molecular Science and Engineering, Washington, DC, USA
e-mail: Scott.walper@nrl.navy.mil



M. Kaparakis-Liaskos, T. A. Kufer (eds.), *Bacterial Membrane Vesicles*,
https://doi.org/10.1007/978-3-030-36331-4_10

synthesis of commercial enzymes and other products. In recent years, the ability to engineer biology has made exponential leaps forward allowing researchers to explore alternatives to using natural, "wild-type" bacteria. Advances in DNA sequencing, the accessibility of molecular engineering technologies, improvements in omics' instrumentation and analysis tools, and many other factors have opened new avenues of biological engineering and synthetic biology.

Since they were first observed under scanning electron microscopy, the vesicles shed by microbes have gone by many names: membrane vesicles, exosomes, outer membrane vesicles, microvesicles, and others (Knox et al. 1967). Often the nomenclature follows the characteristics of the parental organisms such as membrane vesicle (MV) for those particles originating from Gram-positive bacteria and outer membrane vesicle (OMV) for the vesicles produced by Gram-negative bacteria, though this is not always the case. Regardless of the nomenclature and the parental organism, these vesicles are shed from the outermost membrane of their parental microorganism (Fig. 10.1). The biomolecules' composition is highly variable, as are many of their morphological properties. Most Gram-negative OMVs typically range in size from 50 to 300 nm, though particles of 85–125 nm are by far the most prevalent (Deatherage et al. 2009; Schwechheimer et al. 2013). In comparison, the MVs of Gram-positive bacteria often exhibit a bimodal distribution of particle sizes with a significant number of vesicles in the 10–30 nm range (Dean et al. 2019; Grande et al. 2017). The exosomes of eukaryotic cells show a distribution similar to the OMVs, while microvesicles of mammalian cells have been reported from 100 nm to 1 μm in range (Raposo and Stoorvogel 2013; Tricarico et al. 2017; Zomer et al. 2010). Typically, vesicles are released from the parental bacteria into the surrounding environment, however, researchers have observed some elongated tubules extending from bacterial surfaces that appear as chains of vesicles as seen with some *Myxococcus* species (Berleman et al. 2014).

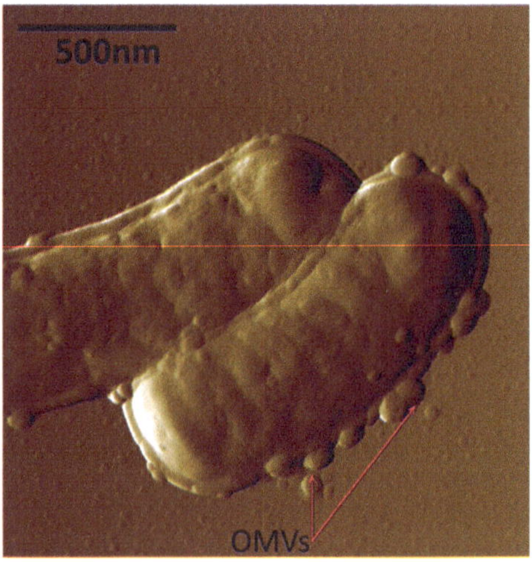


**Fig. 10.1** Nascent MVs of *Lactobacillus reuteri*. Atomic force microscopy was used to image the formation of MVs at the surface of *L. reuteri*. Cells were fixed at mid-log stage and immobilized to mica surfaces for imaging. Scale bar represents 500 nm

Beyond simple morphological differences, the composition of biological vesicles is highly variable, as would be expected. Proteomic analysis of several bacterial species of both Gram-positive and Gram-negative classification has shown that membrane proteins are often the most prevalent components of the structures (Kroniger et al. 2018; Lee et al. 2007). However, in these studies, researchers have also shown that in many instances the protein composition of the vesicles does not always mimic that of the parental microbe. Early studies of *Escherichia coli* OMVs by Lee et al. highlighted this phenomenon, reporting that highly abundant membrane and periplasmic proteins were often not found at detectable levels in OMV preparations (Lee et al. 2007, 2008). Similar observations have been made with regard to toxins, small molecules, and peptides that appear to be enriched in vesicles by some microbes, suggesting a cellular mechanism or signal that facilitates loading (Dean et al. 2019; Jan 2017). Cell membranes are very heterogeneous by nature with specific proteins and lipids distributed across the exterior of the cell (Barak and Muchova 2013). For most microbes, the outermost membrane is a dynamic surface that not only changes through the various stages of the cells' life cycle but also responds to environmental cues and conditions and changes in response to these stimuli. While there are many mechanisms proposed for vesicle biogenesis, evidence is mounting that formation and release may be a well-controlled cellular event in some bacteria (Elhenawy et al. 2016; Roier et al. 2016). The bacterial membrane vesicles therefore take on a variety of functions, serving as a defense mechanism from phages and antimicrobial peptides, facilitating toxin delivery in pathogenesis, and shuttling DNA and RNA between species allowing for horizontal gene transfer, among others.

In the subsequent sections, we will explore the potential biotechnology applications of both natural and engineered bacterial membrane vesicles. At present, membrane vesicles have seen their greatest acceptance in the area of vaccine development employing the OMVs of Gram-negative bacteria. Unlike many traditional strategies that require inactivation of the virus or toxin, OMV-based vaccines allow antigens to be maintained in their native state and delivered simultaneously with immunostimulatory epitopes that serve as the adjuvant. OMV-based vaccines offer an all-in-one vehicle that is relatively low cost and easy to manufacture. Beyond vaccines, the complexity of protein and other biomolecule composition opens numerous avenues of engineering bacterial membrane vesicles for specific applications that may have commercial value. Researchers have demonstrated that through careful design of molecular systems, specific proteins can be targeted to both the interior and exterior of these biological nanoparticles. These efforts have allowed for the development of OMV-based assays for antigen detection, enzyme-based systems for bioremediation, and tools for imaging cell–cell interactions. These early success in both health-related and commercial applications are foundational to ongoing efforts that may lead to new therapeutics, engineered probiotics, or tools for regulating complex microbial communities.

## 10.2 OMV Use in Vaccines

The design of vaccines is based on the concept that protective immunity against a given pathogen results from inducing the natural immune response against that pathogen without actually causing the associated disease. Therefore, the vaccine must resemble the pathogen enough to trigger the correct immune response, yet must not itself be infectious. Traditionally, this has been achieved through several strategies, including the use of attenuated pathogens (rendered nonpathogenic by various methods) or killed pathogens (inactivated by heat or chemical denaturation) (Zepp 2010). While numerous vaccines in use today are still produced by these strategies, there are drawbacks to both. Attenuated pathogens may potentially revert back to virulent forms or may pose a higher threat to individuals with compromised immune systems. On the other hand, killed pathogens may not stimulate the immune system enough to provide long-lasting protection, necessitating the use of adjuvants such as aluminum compounds that further stimulate the innate immune response (Coffman et al. 2010; Zepp 2010). Thus, modern vaccines are based on the use of carefully chosen antigens in combination with adjuvants to enhance the body's response to the antigen (Fig. 10.2).

OMVs possess a number of characteristics that make them ideal candidates as a vaccine platform. First, they are naturally produced by bacteria and have intrinsic immunostimulatory properties, as they contain species-specific antigens as well as pathogen-associated molecular patterns (PAMPs). Thus, OMVs induce both adaptive and innate immune responses, allowing them to function as antigen and adjuvant in one package (Ellis and Kuehn 2010; Kaparakis-Liaskos and Ferrero 2015). Second, OMVs are non-replicative and so do not need to be treated with inactivating agents, which preserves antigens and PAMPs in their native states. Third, while a number of nanoparticle-based vaccine delivery vehicles are in development today including OMVs, virus-like particles (VLPs), immunostimulatory complexes (ISCOMs), and inorganic nanoparticles (reviewed in Xiang et al. 2006), OMVs are uniquely suited to this purpose as their natural size range (20–250 nm) enables them to both drain freely into the lymph nodes to target immune cells residing there, as well as to be taken up by antigen-presenting cells such as dendritic cells (Fig. 10.3) (Gerritzen et al. 2017; Kulp and Kuehn 2010; Manolova et al. 2008). Finally, genetic engineering can be used in a number of ways to improve the production, immunostimulatory properties, and safety of OMV-based vaccines (discussed in detail below). As a result of their potential as a vaccine platform, a significant amount of research has been dedicated in recent decades to the engineering of OMV-based vaccines, including an FDA-approved vaccine against meningitis serogroup B.

There are various factors to consider in the design and production of OMV-based vaccines. OMVs may be isolated directly from the target pathogen, though this often requires them to be extracted using detergents that remove certain toxic components from the OMV surface such as lipopolysaccharide (LPS) found on Gram-negative bacteria (Gnopo et al. 2017; van de Waterbeemd et al. 2010). Alternatively, mutant

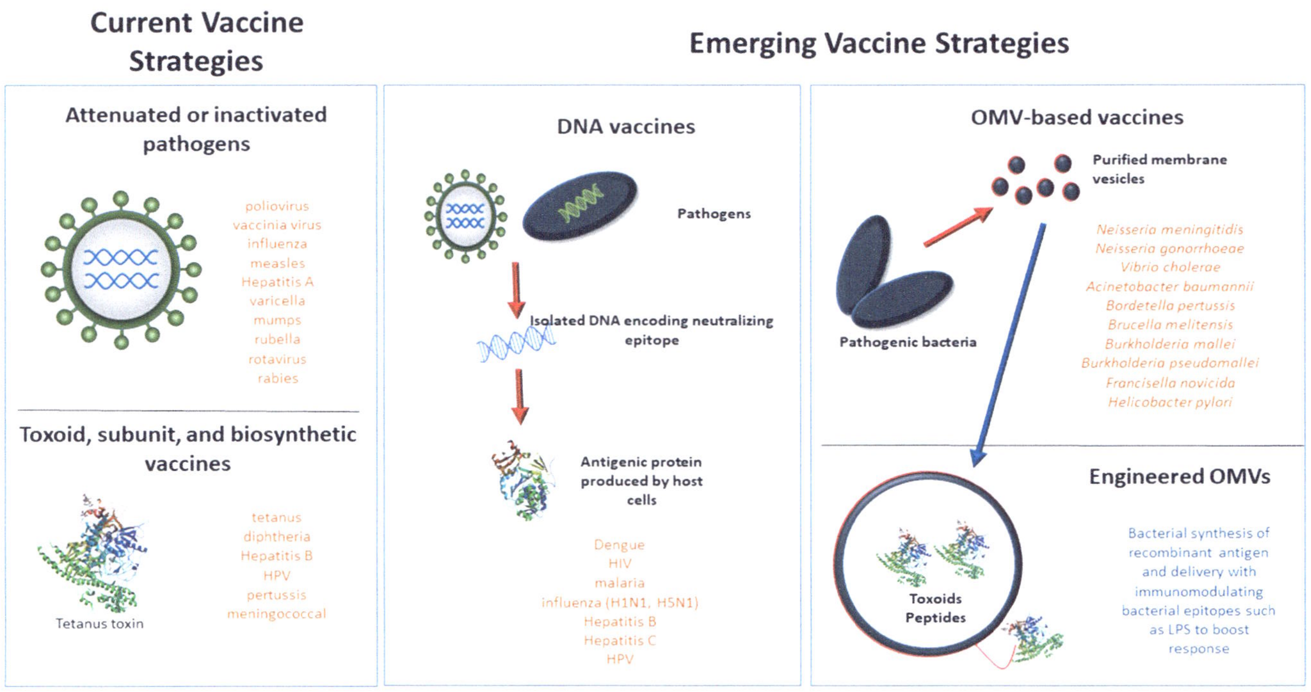


**Fig. 10.2** Current and emerging vaccine strategies. Current vaccines employ live and/or attenuated pathogens (upper left panel) or inactivated biomolecules such as proteins, carbohydrates, or peptides (lower left panel). DNA vaccines are based on isolated genetic sequences that encode an antigenic, neutralizing protein or peptide for a target pathogen. DNA is taken up by antigen-presenting cells, then transcribed and translated into a protein sequence that is presented on the cell surface to stimulate an immune response (middle panel). Membrane vesicles can be used in their natural form as they mimic the cell surface of pathogenic bacteria, thereby allowing the host immune system to respond to a wide range of antigenic features of the bacteria (upper right panel). Membrane vesicles can be engineered to contain recombinant proteins and peptides thus allowing for the development of vaccines for any number of bacterial or viral targets determined solely by the cargo within or on the surface of the vesicle itself (lower right panel)

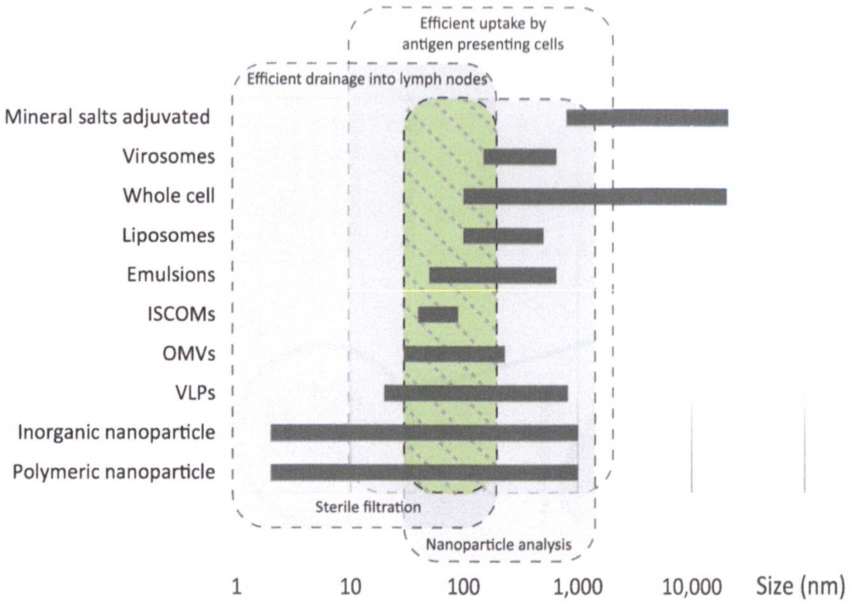


**Fig. 10.3** Comparison of current vaccine platforms. Vaccine platforms vary in their size, methods of purification, and their mode of interaction with host immune cells. For each system, these properties have to be weighed to determine their feasibility in administration and commercialization. For example, the possibility of sterile filtrations simplifies the production design while the analysis of nanoparticle size distribution can increase the difficulty of standardizing the final product. Within this figure, the green box highlights the overall preferred size window for vaccine production. Used with permission from Gerritzen et al. *Bioengineering bacterial outer membrane vesicles as vaccine platform*. Biotechnology Advances, 2017. **35**(5): p. 565–574

pathogen strains producing modified, less-toxic LPS can be used as the OMV source, or antigens can be heterologously expressed in engineered host species with modified LPS. In either case, this can be done in strains carrying mutations that increase vesicle formation to increase the OMV yield. Genetic engineering can also be used to introduce multiple antigens into a single OMV platform, thus increasing the strength or broadening the scope of the immune response to protect against additional strains of the pathogen. These considerations will be discussed in detail below in the context of the meningitis B vaccine as a primary case study, followed by several other examples of OMV-based vaccines against various diseases. This topic has also been extensively reviewed by other researchers and their work may be referred to for additional information (Gerritzen et al. 2017; Gnopo et al. 2017; Tan et al. 2018; van der Pol et al. 2015).

### 10.2.1 OMV Vaccines for Meningitis

To date, the development of OMV vaccines has seen the most attention for Meningitis type B (MenB), which remains the only disease for which they have been approved for human use. *Neisseria meningitidis* strains are classified into serogroups based on their capsular type, and effective vaccines containing a capsular polysaccharide coupled to a carrier protein have been developed against several serogroups. However, the serogroup B capsule bears structural similarity to a neural cell adhesion molecule in the human brain, which means that it is poorly immunogenic and also that its use in a vaccine has the potential to induce an autoimmune response (Finne et al. 1983; Rosenstein et al. 2001). Consequently, vaccine development for MenB has focused on other antigens, particularly the immunodominant outer membrane protein porin A (PorA) (Holst et al. 2009, 2013). As PorA is abundant in the outer membrane, it is naturally also present at high levels in OMVs isolated from *N. meningitidis* strains, making them an ideal candidate for vaccine development.

OMVs isolated from local serogroup B meningococcal strains proved effective as vaccines in Cuba, Norway, Chile, Brazil, and New Zealand, with the significant limitation that the immune response was strain-specific due to the high sequence variability of PorA (reviewed in Holst et al. 2009). To combat this limitation, an OMV vaccine was developed at the Netherlands Vaccine Institute that is produced from two genetically modified *N. meningitidis* strains that each express three PorA subtypes (Claassen et al. 1996; van der Ley et al. 1995). This hexavalent vaccine "HexaMen" was safe and effective in clinical trials and has been further improved to include a third trivalent OMV to provide coverage against the nine most frequently occurring subtypes of MenB in industrialized countries (de Kleijn et al. 2001; van den Dobbelsteen et al. 2007).

While OMV vaccines are safe and effective, their primary dependence on the PorA antigen necessitates that even the multivalent vaccines would still require periodic reformulation as the dominant subtypes of MenB change over time. Thus, work continued toward the identification of new, conserved antigens that would allow for the development of a "universal" vaccine for serogroup B meningococcal strains. The availability of whole genome sequencing technology gave rise to a new strategy of vaccine development termed "reverse vaccinology," in which candidate antigens identified in silico are individually expressed in *E. coli* and tested for immunogenicity (Rappuoli 2000). This approach allowed researchers to identify a number of novel antigens that are conserved across multiple *N. meningitidis* strains (Pizza et al. 2000). A new vaccine was then developed that includes five of these novel antigens in the form of three recombinant proteins: a fragment of NadA, plus two fusion proteins consisting of NHBA–GNA1030 and GNA2091-fHbp (Giuliani et al. 2006). NadA is an adhesin, NHBA is a heparin binding protein, and fHbp is a lipoprotein that binds to human complement factor H; the functions of GNA1030 and GNA2091 remain unknown (Comanducci et al. 2002; Madico et al. 2006; Serruto et al. 2010). The final formulation of this vaccine, registered as Bexsero by Novartis Vaccines, includes the five recombinant antigens plus OMVs prepared from the New Zealand epidemic strain of *N. meningitidis*, which contribute both the PorA antigen to increase strain coverage and additional adjuvant activity (Gorringe and Pajón 2012).

In addition to the choice of antigenic component(s), other important factors to be considered in the development of OMV-based vaccines include the source organism and method of isolation for the OMVs. In the case of MenB vaccines, OMVs have traditionally been isolated from wild-type or recombinant strains of *N. meningitidis*, which requires them to be prepared by detergent extraction due to the highly toxic nature of lipopolysaccharide (LPS) found on the OMV surface (Holst et al. 2009). LPS (also known as endotoxin) is a potent adjuvant that stimulates the innate immune system through Toll-like receptor 4 (TLR4) found on macrophages, however, high levels of LPS can lead to fever, inflammation, and septic shock (Copeland et al. 2005; Raetz and Whitfield 2002). While detergent extraction of OMVs removes most of the LPS and can also increase OMV yield, there are also drawbacks to this process. Detergent use can result in aggregation of OMVs due to removal of negatively charged LPS and phospholipid molecules, which can increase size variability and decrease vaccine stability and shelf life (Holst et al. 2009; van de Waterbeemd et al. 2010). It also alters the proteomic profile of OMVs, through both contamination of OMVs with cytoplasmic proteins as a result of bacterial cell lysis and removal of components that contribute to immunogenicity and adjuvanticity (van de Waterbeemd et al. 2013, 2010; Zariri et al. 2016a). For example, detergent treatment removes the fHbp antigen, a surface-exposed lipoprotein that is highly immunogenic against various *N. meningitidis* strains and is naturally present on native OMVs (Masignani et al. 2003). The absence of antigens and PAMPs such as lipoproteins and phospholipids can reduce the immunogenic response to OMVs, requiring the use of additional adjuvants (Gnopo et al. 2017; Zariri et al. 2016a).

Recent efforts toward a second-generation OMV vaccine for MenB have focused on genetic engineering of *N. meningitidis* strains to reduce LPS toxicity and eliminate the need for detergent extraction of OMVs. The *lpxL1* and *lpxL2* deletion mutants produce modified lipid A, which is the major component of LPS responsible for its toxicity (Fig. 10.4) (van der Ley et al. 2001; Zariri et al. 2016b). The resulting lipid A is penta-acylated rather than hexa-acylated and is no longer toxic as it shows little to no stimulation of human TLR4 (Steeghs et al. 2008; van der Ley et al. 2001). Thus, OMVs isolated from *lpxL1* and *lpxL2* mutant strains do not require detergent treatment and are still immunogenic, though they may require additional adjuvants in order to be effective vaccines in humans due to their lack of stimulation of innate immunity through TLR4 (Fisseha et al. 2005; Koeberling et al. 2008; van de Waterbeemd et al. 2010). Further bioengineering of LPS biosynthesis or modification through the use of the *lptA* or *lgtB* mutations and heterologous expression of the *pagL* gene encoding lipid A 3-*O*-deacylase from *Bordetella bronchiseptica* has shown that the potential exists to produce OMVs that display a broad range of TLR4 activation, which show promise to be used as effective stand-alone vaccines (Geurtsen et al. 2006; Pupo et al. 2014; Zariri et al. 2016b).

Finally, there are additional genetic mutations available that can increase overall OMV yield. A mutation in the RmpM protein, which links the outer membrane of

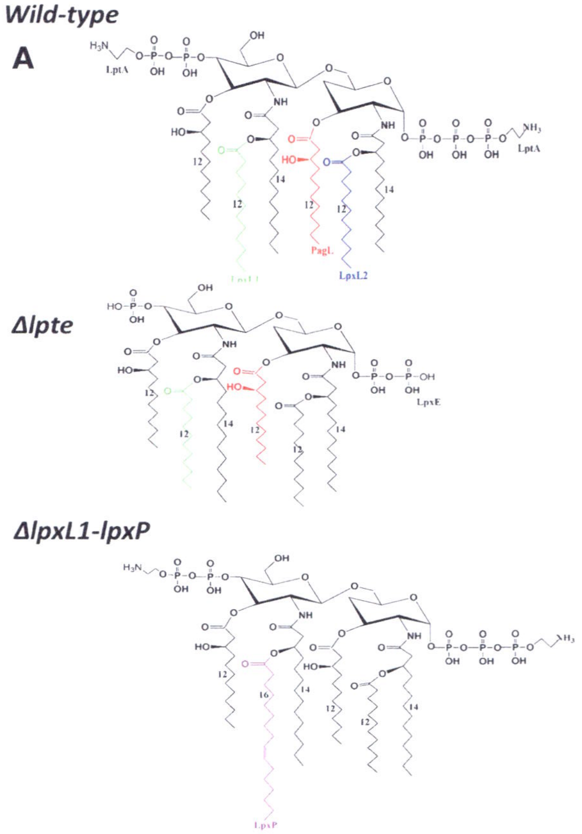


**B**

| Enzyme | Activity | Organism source |
|---|---|---|
| LpxL1 | Adds $C_{12}$ to the primary linked acyl chain at 2′-position. | *Neisseria meningitidis* |
| LpxL2 | Adds $C_{12}$ to the primary linked acyl chain at 2 position. | *Neisseria meningitidis* |
| LpxE | Removes 1 phosphate group. | *Bordetella bronchiseptica* |
| LpxP | Adds palmitoleate to the primary linked acyl-chain at 2′-position. | *Escherichia coli* |
| PagL | Removes acyl-chain from 3-position | *Bordetella bronchiseptica* |
| LptA | Adds phosphoethonalamine groups at 1 or/and 4′-position. | *Neisseria meningitidis* |

**Fig. 10.4** Structural changes in Lipid A. Lipid A can be enzymatically modified by a number of enzymes as shown in the color panel (**a**). LpxL1 (green), LpxL2 (blue), LpxP (pink) and LptA (brown) all add the corresponding group to the molecule, whereas PagL (red) and LpxE (orange) remove the group. The *ΔlptA* and *ΔlptL1-lpxP* mutants produce modified LPS that is less toxic than LPS from wild-type *E. coli*. The abbreviation of the enzymes, organism source, and activity are presented (**b**). Used with permission from Zariri et al. *Modulating endotoxin activity by combinatorial bioengineering of meningococcal lipopolysaccharide*. Scientific Reports, 2016. **6**: p. 36575

the cell to the peptidoglycan layer, leads to a more loosely attached outer membrane and increased OMV release, but does not affect bacterial growth or OMV immunogenicity (Arigita et al. 2004; Klugman et al. 1989; van de Waterbeemd et al. 2010).

Research to date on MenB vaccines highlights both the advantages and challenges of using OMVs as a vaccine platform. Recent developments in bioengineering have done much to expand the potential in this area, and it is likely that the next generation MenB vaccine will make use of the various options in antigen choice and display, LPS detoxification, and OMV yield to result in improved effectiveness, safety, and ease of production.

### *10.2.2 OMV Vaccines for Gonorrhea*

Gonorrhea is one of the most frequently reported communicable diseases in the USA, with a worldwide incidence estimated at 78 million new cases per year (Bolan et al. 2012; Newman et al. 2015). Despite more than a century of research, efforts to develop a vaccine against gonorrhea have been unsuccessful, while the need for such a vaccine has only increased as a number of antibiotic-resistant strains of gonorrhea have emerged (Bolan et al. 2012; Edwards et al. 2016). Interestingly, ecological data suggest a decline in gonorrhea during the time period immediately after the use of the OMV-based MenB vaccines in Cuba and Norway, indicating that the OMV vaccine may afford some protection against gonorrhea (Pérez et al. 2009; Whelan et al. 2016). Furthermore, a retrospective case-control study done in New Zealand estimated that the OMV vaccine used in that country, MeNZB, was 31% effective in preventing gonorrhea (Petousis-Harris et al. 2017). This landmark finding represents the first example of any vaccine being associated with protection against gonorrhea in humans.

Despite causing significantly different diseases, the causal pathogens of gonorrhea and meningitis, *Neisseria gonorrhoeae* and *N.meningitidis*, are closely related. The two are estimated to share 80–90% homology in DNA sequence, and several of the antigens found in the Bexsero MenB vaccine are also present in various *N. gonorrhoeae* strains at approximately 60–90% amino acid sequence identity to the *N. meningitidis* reference strain (Hadad et al. 2012; Semchenko et al. 2018; Tinsley and Nassif 1996). It was recently shown that both the OMV and recombinant protein antigen components of the Bexsero vaccine (which contains the same OMV component as the MeNZB vaccine) could elicit antibodies against *N. gonorrhoeae* in rabbits and humans (Semchenko et al. 2018). This provides some explanation of the cross-protection afforded by the MenB vaccines and suggests that development of a similar OMV-based strategy might be the key to an effective vaccine against gonorrhea.

The promising findings of Petousis-Harris et al. and Semchenko et al. with regard to vaccine cross-protection are very recent, and as such there have not yet been reports on OMV-based vaccines designed specifically against gonorrhea. However, some candidate antigens have been identified. For example, *N. gonorrhoeae* MetQ, a subunit of the methionine binding ABC transporter, was recently shown to be highly conserved, localize to the bacterial and OMV surface, and to play a role in adherence to cervical epithelial cells. Most importantly, antibodies against MetQ are bactericidal and can block adherence, indicating its potential as a candidate vaccine antigen (Semchenko et al. 2017). Additionally, it was recently demonstrated that a vaccine consisting of formalin-inactivated whole bacterial cells encapsulated in microparticles was effective in a mouse model (Gala et al. 2018), further supporting the hypothesis that a vaccine presenting surface-exposed antigens in their native state may be the most effective avenue for a successful gonorrhea vaccine.

### 10.2.3 OMV Vaccines for Influenza

While the above examples demonstrate the use of pathogen-derived OMVs as vaccines, other species of bacteria can be engineered to heterologously express and display antigens on their outer membrane to produce antibacterial or antiviral OMV vaccines. Despite the availability of seasonal vaccines, influenza infection remains an ongoing threat, particularly with the ability of influenza A viruses to form pandemic strains. Current influenza vaccines generally produce an immune response against the immunodominant glycoproteins hemagglutinin and neuraminidase, however, these proteins are extremely variable and thus strain-specific vaccines must be redeveloped annually (Sato et al. 2001; Treanor 2015). Variations in these surface epitopes and others also contribute to a low efficacy for influenza vaccinations (Osterholm et al. 2012). Therefore, there is significant interest in the development of a universal influenza vaccine that could reduce the need for annual redesign and revaccination and would provide protection should a new pandemic strain arise.

One of the most promising target antigens for a universal vaccine is M2e, an integral membrane protein of influenza A virus. Unlike hemagglutinin and neuraminidase, the M2e sequence is highly conserved across strains, however, it is not as immunogenic as the aforementioned antigens and requires adjuvants to be effective (Lamb et al. 1985). To produce a self-adjuvant, OMV-based vaccine, the probiotic *E. coli* strain Nissle 1917 was engineered to express a fusion protein consisting of M2e4xHet, a multimeric construct containing four M2e variants, as a C-terminal fusion to ClyA, an *E. coli* transmembrane protein that is known to be enriched in OMVs (Rappazzo et al. 2016). Mice vaccinated with the resulting recombinant OMVs showed a 100% survival rate after challenge with a lethal dose of a mouse-adapted H1N1 strain (Rappazzo et al. 2016). The OMV-based vaccine was further improved by production in ClearColi, an *E. coli* strain engineered to contain only the LPS precursor lipid IVa instead of full LPS, and these OMVs provided equal protection against influenza in mice and ferrets without LPS-based endotoxicity, which would otherwise hamper translation of this therapy to use in humans (Watkins et al. 2017).

### 10.2.4 OMV Vaccines for Cholera

Cholera, a secretory diarrheal disease caused by the Gram-negative bacterium *Vibrio cholerae*, is a major cause of mortality in developing countries, particularly for infants and young children. While cholera vaccines are currently available, they still suffer from drawbacks of high cost, short shelf life, and the need for cold storage, all of which limit their implementation in developing countries and highlight the need for new candidate vaccines (Bishop and Camilli 2011). *V. cholerae* OMVs could induce a specific, high-titer, and long-lasting immune response in mice, and immunization of female mice also resulted in protection of their neonatal offspring via the transfer of IgG and IgA antibodies in the mother's milk (Schild et al. 2009, 2008). As was the case for the *N. meningitidis* and *E. coli* OMVs, genetic modification of lipid A resulted in reduced endotoxicity without diminishing the immunogenic potential of the vaccine (Leitner et al. 2013).

It has already been shown that there is a protective effect of breastfeeding against cholera due to the presence of IgA antibodies directed against *V. cholerae* surface structures and cholera toxin (Clemens et al. 1990; Glass et al. 1983; Hanson et al. 1985; Qureshi et al. 2006). The major protective antigen of the OMV vaccine is the O-antigen, which is present in high amounts on the OMV surface, and cholera toxin is known to be packaged into OMVs (Chatterjee and Chaudhuri 2011; Leitner et al. 2013). Thus, an OMV vaccine may be the ideal candidate to deliver effective antigens that can induce immunity via high IgA titers, and immunization of adult women would hopefully lead to significant protection of newborns and young children, often the population most affected during cholera epidemics (Leitner et al. 2013).

In addition to its effectiveness, the OMV vaccine is also a promising candidate to overcome the cost, stability, and transport limitations of the other available vaccines. Large-scale production of OMVs at reasonable cost has already been demonstrated for the meningitis vaccines. OMVs purified from *V. cholerae* were stable after 1 month at 37 °C, and could be easily administered for immunizations without accessory buffer solutions, indicating that cold storage and trained medical professionals may not be required for vaccine distribution (Leitner et al. 2013; Schild et al. 2009).

### 10.2.5 OMV and EV Vaccines for Other Diseases

In addition to the above examples, OMVs from various other pathogenic species have been tested as a vaccine platform against their associated diseases. These include *Acinetobacter baumannii*, *Bordetella pertussis* and *B. parapertussis*, *Brucella melitensis*, *Burkholderia mallei* and *B. pseudomallei*, *Francisella novicida*, *Heliobacter pylori* and *H. felis*, *Klebsiella pneumoniae*, *Porphyromonas gingivalis*, *Salmonella enterica ssp. enterica ser. Typhimurium*, and *Shigella* spp. (Alaniz et al.

2007; Asensio et al. 2011; Avila-Calderón et al. 2012; Bottero et al. 2013; Keenan et al. 1998; Kesavalu et al. 1992; Lee et al. 2012; McConnell et al. 2011; Mitra et al. 2013; Nieves et al. 2014; Pierson et al. 2011; Roberts et al. 2008).

While the above examples all pertain to the use of OMVs from Gram-negative organisms, the relatively recent discovery that Gram-positive bacteria also produce membrane vesicles has led several groups to test whether Gram-positive MVs might also be effectively used as vaccines (Brown et al. 2015; Lee et al. 2009; Liu et al. 2018b) (See Chap. 2). Promising results from vaccinations with MVs have been demonstrated for *Bacillus anthracis, Clostridium perfringens, Mycobacterium tuberculosis, Staphylococcus aureus,* and *Streptococcus pneumoniae* (Choi et al. 2015; Jiang et al. 2014; Olaya-Abril et al. 2014; Prados-Rosales et al. 2014; Rivera et al. 2010). One potential advantage of using MVs from Gram-positive bacteria is that they do not contain LPS and thus are unlikely to require detoxification; for example, Choi et al. reported that unmodified MVs from *S. aureus* were self-adjuvanting and could stimulate effective immunity against *S. aureus* with no observable toxic effects (Choi et al. 2015).

In each of the abovementioned cases, the administered OMVs or MVs induced an immune response in cell lines or animal models, and usually promoted survival or prolonged the time to death after challenge with the associated pathogen. As antibiotic-resistant strains are a serious problem for some of these pathogens, the potential for the development of OMV/MV vaccines that protect against these diseases is highly attractive.

### *10.2.6 OMV Vaccines Based on Recombinant Antigens*

There are numerous reports of OMV vaccines based on heterologous expression of antigens. *E. coli* is often used as the host species for OMV production, as is the case for vaccine concepts against *A. baumannii, Francisella tularensis, S. pneumoniae, Campylobacter jejuni, Plasmodium sp., Chlamydia sp.,* and *Leishmania sp.* (Bartolini et al. 2013; Chen et al. 2016; Fantappiè et al. 2014; Huang et al. 2016; Pritsch et al. 2016; Schroeder and Aebischer 2009).

Other host species have also been used for recombinant OMV production, particularly those whose OMVs have already been studied as vaccine candidates including *N. meningitidis* expressing antigens against Lyme disease or genital herpes, *V. cholerae* expressing enterotoxigenic *E. coli* antigens and *S. enterica ssp. enterica ser. Typhimurium* expressing protective antigens to prevent pneumococcal disease, tuberculosis, and chlamydia infections (Daleke-Schermerhorn et al. 2014; Del Campo et al. 2010; Kuipers et al. 2015; Leitner et al. 2015; Muralinath et al. 2011; Salverda et al. 2016).

As in the influenza vaccine described above, the target protein-based antigens are generally expressed as fusions with native membrane-localized proteins such as ClyA or OmpA in *E. coli* or fHbp in *N. meningitidis*. The *E. coli* autotransporter Hemoglobin protease (Hbp) was also recently engineered as a platform that can be used for simultaneous display of multiple heterologous antigens on a single, stable scaffold that localizes to the OMV surface at high densities (Daleke-Schermerhorn et al. 2014; Kuipers et al. 2015). Similarly, a system has been designed in which multiple proteins can be produced separately from OMVs, then assembled onto a scaffold located on the OMV surface (see the below section on biomass conversion for more detail) (Park et al. 2014). These developments open up the exciting possibility of creating multivalent OMV vaccines that present several antigens from the same pathogen or even antigens from several different pathogens.

### *10.2.7 OMV Vaccines Based on Bacterial Glycans*

In some cases, the target vaccine antigens are not proteins but rather glycans. Many successful current vaccines, such as those for non-serogroup B meningitis, are based on the use of glycoconjugates that consist of glycans coupled to T cell-dependent protein antigens, as glycans alone usually elicit T cell-independent responses which are weaker and short-lived. However, a major drawback to this glycan-based vaccine strategy is that current production techniques are technically demanding, costly, and unreliable (Price et al. 2016). As an alternative strategy, *E. coli* can be engineered to produce and display pathogen-specific polysaccharides on OMVs. For example, Chen et al. introduced the gene cluster required for synthesis of *F. tularensis* O-Polysaccharide (O-PS), a subunit of LPS, into a hypervesiculating laboratory *E. coli* strain which is O-PS-deficient but still produces the lipid A core to which O-PS attaches (Chen et al. 2016). In this model, pathogen-specific O-PS is synthesized on the cytoplasmic face of the inner membrane by plasmid-encoded enzymes, to which endogenous *E. coli* proteins complete the translocation of O-PS to the outer membrane and attach it to the lipid A core. The result of this process is that *F. tularensis*—specific O-PS is displayed on the *E. coli* outer membrane and consequently the OMVs as well (Fig. 10.5). In parallel, the *E. coli* host strain was also engineered to produce the less inflammatory, penta-acylated lipid A to circumvent LPS-based toxicity. Vaccination of mice using the subsequent glycoengineered OMVs significantly delayed time to death after lethal *F. tularensis* challenge. This is a promising result for the development of a vaccine against *F. tularensis,* which is a class A bioterrorism agent for which no licensed vaccine currently exists (Oyston et al. 2004). A similar strategy was employed by Price et al., who engineered *E. coli* to produce OMVs displaying the *S. pneumoniae* serotype 14 capsular polysaccharide, which elicited an immune response comparable to a commercial pneumococcal vaccine in mice (Price et al. 2016). They also designed glycoengineered OMVs that display the *C. jejuni* N-glycan, which resulted in an unprecedented level of protection against *C. jejuni* in chickens (Price et al. 2016). Collectively, these studies highlight the potential of glycoengineering in vaccine development for pathogens that have to date proven incompatible with other methods.

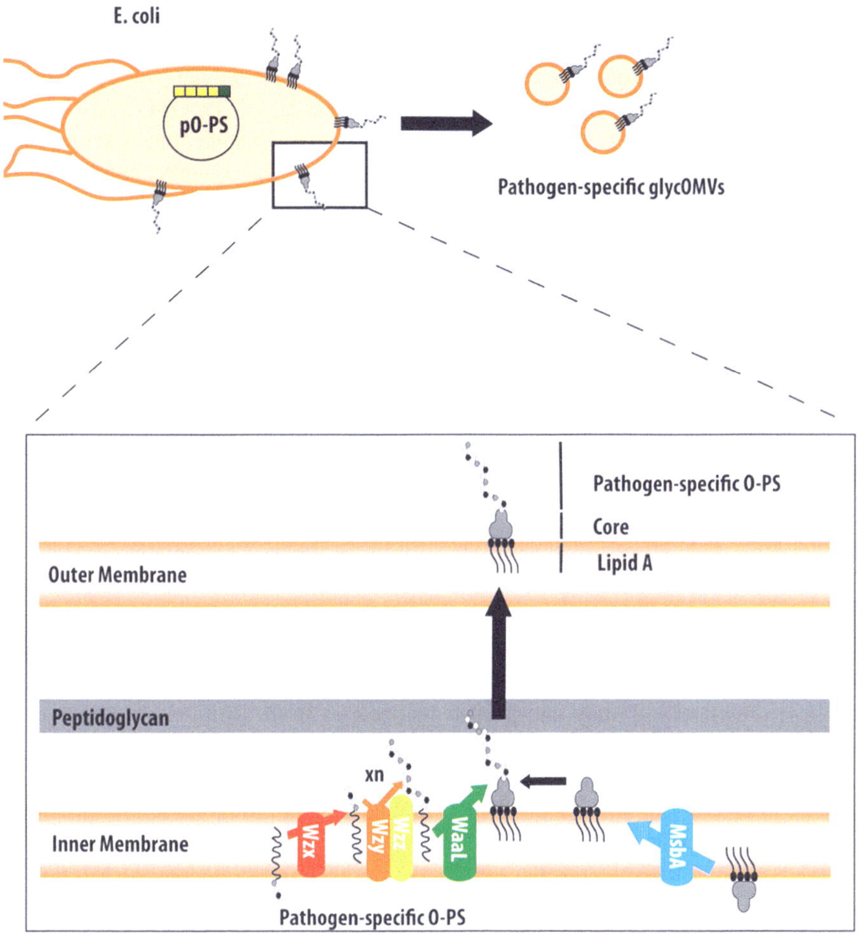


**Fig. 10.5** Schematic of the assembly of pathogen-specific antigens on the surface of OMVs. Cellular machinery is responsible for synthesis and transport to the outer membrane where flippase (Wzx) is responsible for membrane translocation and surface expression. Membrane-bound antigen is released in budding vesicles. Reproduced with permission from Chen et al. *Outer membrane vesicles displaying engineered glycotopes elicit protection antibodies*. PNAS, 2016. **113**(26): p. E3609-E3618

### *10.2.8 OMV Vaccines for Host Glycans*

Glycoengineering of OMVs also carries the exciting possibility for use in generating immune responses against clinically important host glycans such as those associated with certain types of cancer. For example, Valentine et al. engineered *E. coli* OMVs to display two clinically important human glycan structures, namely the tumor-specific carbohydrate antigens polysialic acid (PSA) and Thomsen-Friedenreich antigen (T-antigen) (Valentine et al. 2016). Both of these glycans are highly expressed in several different cancers but not in normal cells and antibodies recognizing these antigens could have clinical benefits, however, on their own, they have low intrinsic immunogenicity (Heimburg-Molinaro et al. 2011). OMVs displaying these antigens could elicit strong IgG antibody titers in mice, indicating that this may be an effective strategy for generating functional antibodies against clinically relevant carbohydrates. These results highlight the advantages of using OMVs to display glycan antigens: their production is less complicated and expensive than traditional glycoconjugate vaccines, the strategy can be easily tailored to various glycan antigens (provided the biosynthetic pathway for the target is known), and they provide the necessary adjuvanticity to activate long-lasting, T cell-dependent immunity.

## 10.3 OMV-Based Therapeutics

Biologically derived nanoparticles isolated from bacteria or mammalian cells have received substantial attention in recent years for their potential as biodegradable carriers that can specifically deliver cargo to targeted sites (Yoo et al. 2011). While many drugs face limitations in application due to toxicity, poor stability, and inability to cross cell membranes, engineered OMVs/MVs can circumvent many of these drawbacks as they can protect their cargo from degradation, target it to specific cells, and deliver it into those cells efficiently.

The capacity for engineered OMVs to target a specific cell population and deliver cargo was effectively demonstrated by Gujrati et al., who developed *E. coli* OMVs for use in cancer therapy (Gujrati et al. 2014). They engineered *E. coli* to express an affibody specific to HER2, a transmembrane receptor overexpressed in many cancers, as a fusion protein with the C-terminus of the native ClyA protein, which targets the resulting protein to the OMV surface (Fig. 10.6). Isolated OMVs were loaded via electroporation with a therapeutic siRNA that targets the kinesin spindle protein (KSP), which is overexpressed in rapidly proliferating cells such as those found in tumor tissue. Silencing of KSP blocks the formation of mitotic spindles, leading to cell-cycle arrest and apoptosis. When injected into mice, the resulting OMVs targeted HER2-overexpressing cells with high affinity and were rapidly internalized, leading to significant inhibition of tumor growth and reduction of tumor size that could be attributed to knockdown of KSP expression. As these OMVs were isolated from an *E. coli* strain with modified LPS, they have low endotoxicity and did not cause significant side effects, which is a major limiting factor of many drug therapies. This work highlights the potential of vesicle-based nanoparticles as cell-specific delivery vehicles that can overcome many of the drawbacks and limitations that hamper the development and release of new cancer therapies.

Various other health benefits have been observed for membrane vesicles, particularly those produced by probiotic bacteria. These include bacterial species that

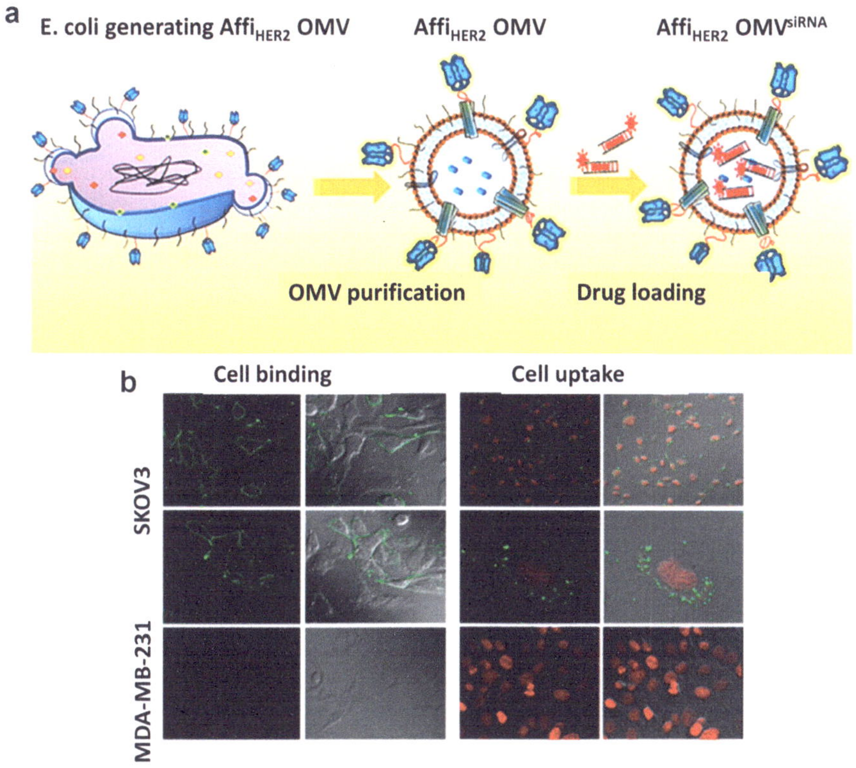


**Fig. 10.6** OMV as therapeutic agents for cancer. (**a**) *E. coli* OMVs were labeled with anti-HER2 affibodies then loaded with cytotoxic siRNA labeled with a fluorophore for visualization. (**b**) For cell binding and uptake studies, HER2-overexpressing SKOV3 cells and HER2-negative MDA-MB-231 cells were co-incubated with Affi$_{HER2}$ OMV and stained with an anti-affibody antibody (green). Receptor-specific cell binding and uptake were seen only with HER2-overexpressing SKOV3 cells. Reproduced with permission Gujrati et al. (2014) *Bioengineered Bacterial Outer Membrane Vesicles as Cell-Specific Drug-Delivery Vehicles for Cancer Therapy*. ACS Nano 8(2): pp. 1525–1537

colonize the gastrointestinal tract and confer benefits to the host through either direct communication with host cells or through interaction with other probiotic or pathogenic bacteria (Bron et al. 2011). Often, the beneficial effects result from the interaction of probiotic bacteria (and their associated membrane vesicles) with the host immune system. For example, recent studies have suggested that probiotic bacteria can suppress inflammatory and allergic responses through modulation of immune responses. MVs from *Bifidobacterium longum* could alleviate a food allergy response in a mouse model by penetrating through intestinal epithelial cells and selectively targeting and inducing apoptosis of mast cells (Kim et al. 2016). The commensal species *Bacteroides fragilis* delivers the immunomodulatory molecule Polysaccharide A to dendritic cells via OMVs, thereby suppressing immune responses that drive inflammation (Shen et al. 2012). A similar result was also shown for *Lactobacillus rhamnosus,* for which MVs could recapitulate the previously demonstrated immunoregulatory and neuronal effects of whole bacteria (Al-Nedawi et al. 2015). Finally, MVs of Kefir-derived *Lactobacillus* strains reduced inflammatory cytokine production and alleviated symptoms in a mouse model of inflammatory bowel disease (Seo et al. 2018).

The potential modulation of the host immune system by probiotic bacteria can also have a protective role against various pathogens. While it has not been definitely shown that these benefits are conveyed solely by membrane vesicles, researchers have performed studies using culture media and other preparations devoid of cells and seen similar results as discussed above. As specific examples of MV-mediated protection, Li et al. showed that MVs derived from *L. plantarum* provided protection to *C. elegans* against vancomycin-resistant *Enterococcus faecium* through upregulation of multiple host defense genes (Li et al. 2017). The authors also performed these experiments with the common human intestinal epithelial cell line Caco-2 and observed similar changes in gene expression. Further exploration into the mechanisms of this and other protective effects induced by probiotic MVs might allow for the development of new treatments for antibiotic-resistant pathogens.

Probiotic bacteria have also been suggested to have cancer prevention properties, particularly against colon cancer (Commane et al. 2005; dos Reis et al. 2017; Paolillo et al. 2009). While the exact mechanisms of this are unknown, it has been reported that many probiotic species exert this effect through the secretion of factors that induce apoptosis, and it is likely that these factors are delivered to host cells via bacterial membrane vesicles (Oelschlaeger 2010). Indeed, it was recently shown that purified MVs from *L. rhamnosus* have significant cytotoxic effects on hepatic cancer cells (Behzadi et al. 2017). As membrane vesicles derived from the microbiota in the gastrointestinal tract can travel to the liver and other nearby organs through the bloodstream, this provides further evidence for the potent anticancer properties of probiotic bacteria and their membrane vesicles (Salminen et al. 2004). Furthermore, as probiotic bacteria generally do not have deleterious effects on the host, exploitation of these properties may make possible the development of new cancer treatments that do not also cause the damaging side effects of traditional chemotherapeutic agents (Behzadi et al. 2017).

### 10.3.1 Emerging Therapeutic Applications

In addition to the potential anticancer and other uses discussed above, there are a number of other therapeutics that could benefit from delivery via OMVs/MVs. Compounds such as antimicrobial peptides which would be susceptible to degradation in their free form, or antibiotics that would otherwise not be able to cross the cell membrane, could be packaged into bacterial membrane vesicles for protection and delivery into target cells (Liu et al. 2018b). This may be particularly important for treatment of pathogens that can ordinarily resist antibiotics through mechanisms such as outer membranes with low permeability, or biofilms that delay penetration of the antibiotic or directly inactivate it via secreted enzymes such as β-lactamases (Messiaen et al. 2013). Unlike free molecules, vesicles may be transported through the biofilm and taken up into cells, bypassing these defense mechanisms. For example, packaging of the antibiotic tobramycin into artificial liposomes greatly increased its effectiveness against multiple pathogens including *Burkholderia cepacia, Pseudomonas aeruginosa*, and *S. aureus* (Beaulac et al. 1998). Several liposome-based drugs are currently in clinical trials; however OMV/MV-based vesicles may offer additional benefits related to ease of production or the ability to incorporate specific cell targeting motifs through genetic engineering.

Phage therapy has also received significant attention in recent years as an alternative treatment for antibiotic-resistant bacteria (Lin et al. 2017). One challenge to the implementation of phage therapies is that many phages have narrow host ranges, limiting their applicability. As membrane vesicles naturally play a role in broadening phage host ranges through the transfer of phage receptors between bacterial hosts, this capability could be harnessed for clinical use (Tzipilevich et al. 2017). For example, vesicles derived from phage-sensitive bacteria could be given to a patient prior to administration of the phage itself. If those vesicles are taken up by bacteria already infecting the patient, the phage receptor could be transferred and could potentially enhance targeting of the phage to the infectious bacteria (Liu et al. 2018b).

It has also been proposed that vesicles could be used for delivery of genome editing tools, such as the Cas9-guide RNA ribonucleoprotein complex required for CRISPR-based genome editing, to correct genetic disorders or to combat pathogens (Knott and Doudna 2018; Liu et al. 2019).

Membrane vesicles could also be used as vehicles for delivery of nutritional compounds to the host gastrointestinal tract. For example, vitamin K2 (menaquinone) is a cofactor required for the production of blood coagulation factors and osteocalcin (a bone-forming protein) and may also prevent osteoporosis, coronary heart disease, and liver cancer, but recent data indicate that subclinical vitamin K deficiency is not uncommon (DiNicolantonio et al. 2015). Vitamin K cannot be synthesized by humans, but rather is mainly produced by bacteria in the intestine such as *B. subtilis* and certain strains of lactic acid bacteria (Liu et al. 2018a). In particular, a strain of *B. subtilis* isolated from the traditional Japanese fermented soybean product natto could be engineered to produce very high amounts of vitamin K2 (Sato et al. 2001). As menaquinones are hydrophobic compounds that accumulate in the bacterial cell membrane, the potential exists for vitamin K-containing MVs isolated from *B. subtilis* or lactic acid bacteria to be administered as dietary supplements (Liu et al. 2018a). A similar strategy could be used for other important nutritional or medically relevant compounds that are not naturally produced in high amounts by these bacteria, if they could be genetically engineered to produce and package such compounds in vesicles.

### 10.3.2 Commercial Application of EVs

The protein composition of bacterial membrane vesicles is highly variable dependent upon growth conditions, culture age, and many other factors. Despite the variability, abundant membrane proteins such as porins, membrane channels, and others consistently appear in proteomic analysis of bacterial membrane vesicles and offer potential anchors for their functionalization (Dean et al. 2019; Kroniger et al. 2018; Kwon et al. 2009; Lee et al. 2007; Schwechheimer et al. 2013; Yun et al. 2017). With the targeted loading of bacterial membrane vesicles, researchers can begin to exploit the inherent natural advantages afforded by these vesicles, such as protection from environmental conditions, as well as others of specific design such as enzyme localization and assembly. This engineering of OMVs has the potential for designer probiotics, biological catalysts, and even development for new therapeutic platforms. In the subsequent section, we highlight some of the areas where engineered OMVs and MVs are already making inroads as new tools for a range of commercial, environmental, and health-related applications.

### 10.3.3 OMVs for Biomass Conversion

The successful engineering of *E. coli* and other bacteria to express target proteins and display them on the OMV surface makes possible their use not just in biomedical applications, but for other purposes in which nanoparticles are needed. Many biological processes, such as the Krebs TCA cycle in mitochondria or cellulose hydrolysis by cellulosomes on the surface of anaerobic bacteria, are carried out through complex multienzyme cascades that achieve specificity and efficiency through compartmentalization and precise spatial organization of individual proteins. In an effort to mimic this organization for biotechnological applications, several studies have demonstrated synthetic assembly of multiple enzymes onto liposomes or polymersomes; however, the process is complex, costly, and the resulting liposomes are often fragile, making this approach impractical for large-scale applications (Fischer et al. 2002; van Dongen et al. 2009; Vriezema et al. 2007). Conversely, OMVs could provide the ideal backbone for synthetic nanoreactors if some or all of the enzyme production and assembly could be driven by genetic engineering of the *E. coli* host strain.

This approach was successfully demonstrated by Park et al., who engineered *E. coli* OMVs to display a functional multienzyme complex similar to a cellulosome (Park et al. 2014). Natural cellulosomes are composed of a structural scaffold containing repeating cohesin domains that are bound individually to cellulases via corresponding dockerin domains (Fontes and Gilbert 2010). To mimic this structure on the surface of OMVs, Park and colleagues engineered *E. coli* to express a scaffold consisting of three different cohesin domains and a cellulose binding domain, attached to the outer membrane by the ice nucleation protein anchor (Fig. 10.7).

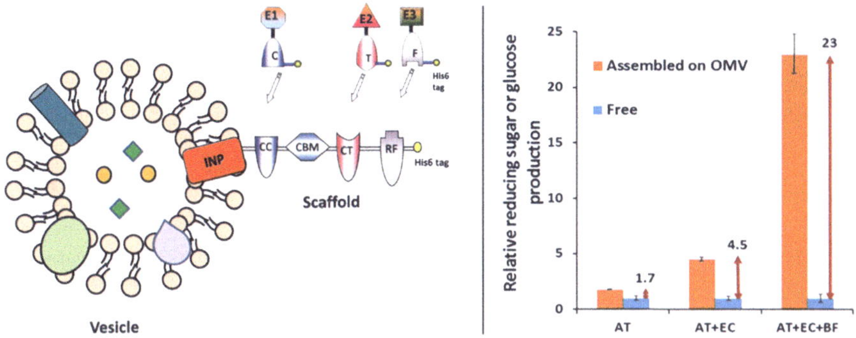


**Fig. 10.7** Multienzyme assembly on engineered OMVs. *Left panel*: A trivalent scaffold was developed that contained three orthogonal cohesion domains that enabled the assembly of cellulose enzymes tagged with complementary dockerin domains. *Right panel*: Assembly of the enzyme system on OMV surfaces significantly improved enzyme activity (orange bars) compared to the free enzyme controls (blue bars). Reproduced with permission from Park et al. (2014) *Positional Assembly of Enzymes on Bacterial Outer Membrane Vesicles for Cascade Reactions*. PLoS ONE 9(5): e97103

Three different cellulases, each possessing a dockerin domain corresponding to one of the cohesins, were produced separately in *E. coli* and subsequently incubated with the OMVs to allow cohesin–dockerin interactions to assemble the full multienzyme complex. The resulting OMVs showed 23-fold enhancement of cellulose hydrolysis as compared to free enzymes. This result is very promising for the potential of OMVs as nanobioreactors in biotechnology, with the added benefit that the same strategy could be used for various other enzymatic cascades by replacing the dockerin-bound cellulases with other enzymes engineered to contain the dockerin domain.

## *10.3.4 OMVS for Bioremediation*

Fortunately for mankind, microbial populations are able to exploit ancestral, bi-functional, or newly evolved cellular processes to degrade and consume many of the chemical contaminants we have produced or inadvertently released into the environment. Take for example, the rapid degradation of the oil plume released following the Deep Sea Horizon disaster (Atlas and Hazen 2011; Scoma et al. 2016) or the more recent identification of marine bacteria capable of degrading some of the plastics released into the ocean (Dash et al. 2013; Urbanek et al. 2018). Observations such as these and others stimulate the continued efforts of researchers and government agencies to develop biological tools for environmental remediation. Natural bioremediation, where an indigenous organism degrades the environmental contaminant is, of course, the ideal scenario. Unfortunately, this is often not achievable as spills and targeted release of toxic compounds can occur in locations where such

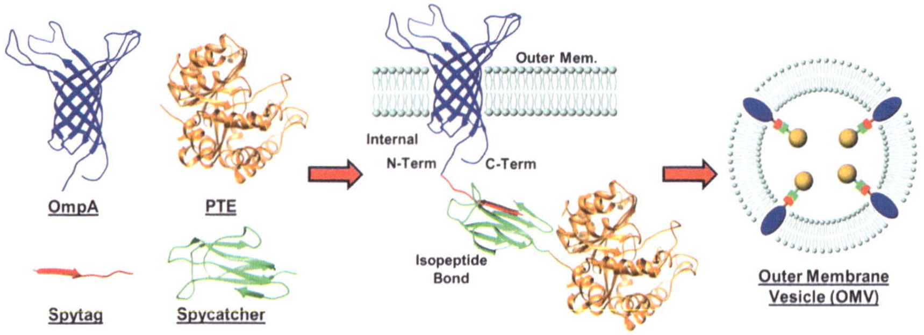


**Fig. 10.8** Directed packaging of the OMV lumen. A protein–protein ligation system (SpyCatcher/SpyTag) was used to anchor a recombinant phosphotriesterase (PTE) enzyme to the bacterial outer membrane enabling the directed loading of nascent OMVs. Reproduced with permission from Alves, N.J., et al., *Bacterial Nanobioreactors—Directing Enzyme Packaging int Bacterial Outer Membrane Vesicles*. ACS Appl Mater Interfaces, 2015. **7**(44): p. 24963–72

microbes are not native. As an alternative, the enzymatic systems from these organisms can be isolated, produced recombinantly, and deployed at the point of concern to facilitate decontamination. Enzymes, however, have their own limitations such as stability and cost of manufacture.

Similar to their function in the natural world, engineered bacterial membrane vesicles afford protection to encapsulated biomolecules from harsh environmental conditions. The controlled loading of enzymes into bacterial OMVs/MVs provides for a method of producing reagents that can easily be isolated from bacterial cultures, lyophilized for storage and distribution, then rehydrated to facilitate environmental cleanup. In a series of publications, Alves et al. showed that a protein–protein ligation system could be used to direct the packaging of a phosphotriesterase (PTE) enzyme capable of degrading an organophosphate compound into *E. coli* OMVs (Alves et al. 2015a, 2016). The authors employed a recombinant version of a native porin protein (ompA) presenting a small peptide (SpyTag) and a PTE fusion with its counterpart (SpyCatcher) (Fig. 10.8). As described by Zakeri and colleagues, the SpyCatcher/SpyTag system allows for the spontaneous formation of an isopeptide bond between the two components which in this system, facilitated the anchoring of PTE to the outer membrane and subsequent packaging of the enzyme into the OMVs (Zakeri et al. 2012). The authors demonstrated that not only did the enzyme maintain activity but also survived a number of storage and environmental conditions better than the free enzyme. In subsequent studies, Alves et al. also showed that these materials had relevance outside the laboratory by testing their materials in environmental water samples spiked with substrate and on a number of materials chosen to mimic military vehicle paint and surfaces (Alves et al. 2018). In these studies, the authors also examined enzyme activity under nonideal conditions such as variable pH, high salinity, and in environmental water samples with varying microbial populations and debris composition. Given the aversion of many societies to employ genetically modified organisms or release nonindigenous species into the wild, OMV-based reagents offer an alternative approach to the development of Green reagents for environmental remediation.

### 10.3.5 OMVs for Imaging and Biosensing

Today, synthetically manufactured liposomes are routinely used for the delivery of therapeutics and in many diagnostic assays (Akbarzadeh et al. 2013; Alavi et al. 2017; Xing et al. 2016). Despite their many successes, liposomes can be difficult to manufacture, load, and store for prolonged periods of time (Alves et al. 2015b). In contrast, engineered OMVs offer a potential path to a simple manufacture platform that allows for controlled packaging and a final product that exhibits biophysical properties making them ideal reagents for drug delivery and the building of biological sensors. As an example, Chen et al. demonstrated how the ability to modify both the exterior proteins and interior cargo could be used in combination to develop reagents that have both assay and imaging capabilities (Fig. 10.9) (Chen et al. 2017). Here the authors expanded upon a previous construct for OMV modification that presents a cohesion–dockerin scaffold on the OMV exterior, adding a terminal domain (Z-domain) that allows for attachment of an antibody (Chen et al. 2010; Park et al. 2014). Interior packaging is accomplished using a modified bacterial lipoprotein (SlyB) fusion which is known to localize to the outer membrane of *E. coli* (Tokuda and Matsuyama 2004). This system allows for versatility in assay and imaging development as any number of antibodies could be added to the exterior and used in combination with a wide array of reporter proteins, from fluorescent proteins to luminescent proteins.

### 10.3.6 Future Commercial Applications of Bacterial Membrane Vesicles

While therapeutic applications of OMVs/MVs are at the forefront of research efforts, the encapsulation of biomolecules within bacterial membrane vesicles has great potential for commercial applications. The ease of production and the protection afforded to encapsulated proteins make this a highly versatile system that could easily be adapted to benefit many applications that are not currently explored.

Recombinant production of proteins, enzymes, and other biomolecules is a critical component of both the commercial and medical industries (Adrio and Demain 2010; Demain and Vaishnav 2009; Ferrer-Miralles and Villaverde 2013; Sanchez-Garcia et al. 2016). While successes such as the *E. coli*-produced insulin serve as banners for this technology, there are numerous other biomolecules that are never able to come to market due to complications associated with biomanufacturing. Often, large-scale production of biomolecules is hindered by

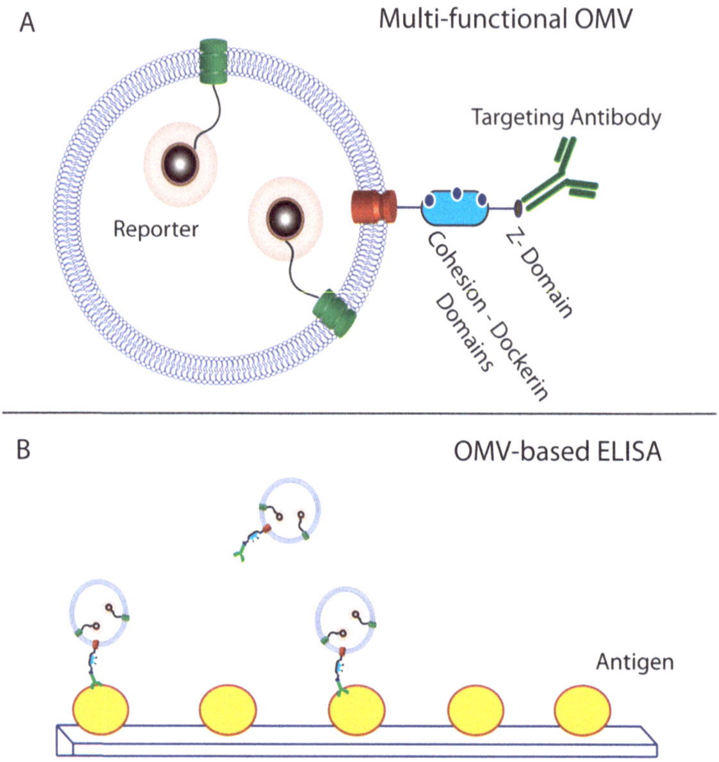


**Fig. 10.9** OMV-based reagents for imaging and biosensing. OMVs from Gram-negative bacteria offer numerous opportunities for modification by targeting both the exterior and interior domains. (**a**) Researchers have shown that exterior domains of OMVs can be modified to add targeting moieties while the interior cavity can be loaded with a range of reporter molecules using membrane anchors, fusions, and various other methods. (**b**) Multifunctional OMVs produced in this manner can be used as a one-pot material for bioassays such as the traditional ELISA or used in cell targeting and imaging

low yields that can be associated with toxicity, insolubility, or any number of other issues. Membrane vesicles could potentially serve as a method of "off-loading" recombinant products before they could accumulate in the engineered organism leading to toxicity. Enclosed within OMVs, these biomolecules could be isolated directly from batch cultures through engineered epitopes on the OMV surface as shown by Alves et al. (2017).

Researchers have shown that OMVs function in many microbial community interactions including interspecies communication and regulation of microbial populations. Additionally, antimicrobial peptides and other cytotoxic compounds are readily packaged within vesicles and purified OMVs alone can display bactericidal activity (Dean et al. 2019; Park 2018; Schulz et al. 2018). While this has obvious therapeutic benefits, these properties of OMVs and MVs could also be harnessed for nonmedical purposes. Microbial communities and the biofilms they form also plague many industrial, military, and environmental systems. In addition to obvious examples such as food production facilities (Glass et al. 1983; Marchand et al. 2012), damaging biofilms can also be found in fuel storage containers (Bücker et al. 2014), oil and gas pipelines (Tingyue Gu 2015), on the hulls of ships (Schultz et al. 2011), and countless other locations. The use of biological reagents that are easily resorbed into the environment to eliminate or reduce biofilms could prove beneficial in developing technologies that can readily be implemented without concern for additional harmful environmental effects.

## 10.4 Conclusion

Scientists continue to gain a greater understanding of the microbial world and the valuable tools it provides to society. We have long exploited them for food processing, as a pipeline for drug discovery, and as miniature factories for the production of target biomolecules; however, as shown here, bacteria can begin to aid the development of new vaccines or even as reagents for environmental cleanup. With the rapid emergence of antimicrobial resistance, OMV/MV derived vaccines may serve as mankind's next line of defense in the battle against pathogenic microbes, particularly those pathogens that can be weaponized such as *B. anthracis* or intracellular pathogens such as *F. tularensis* that are difficult to treat with conventional methods. Beyond vaccines, the observations and theories that support the presence of OMVs in the bloodstream and passing through the blood-brain barrier suggest that with advances in synthetic biology and a greater understanding of the human microbiome, scientists may be able to harness microbial communities to effect changes and treat disorders. Well beyond traditional probiotics, advances in these areas of research will dramatically alter the way we think about the foods we eat as well as how medical therapies are administered. Finally, by mimicking the bacteria themselves and loading enzyme systems into OMVs, we may be able to develop cell-free catalytic reagents that can be used to remediate environmental disasters, treat waste water, or even synthesize novel materials. Both natural and engineered OMVs and MVs have great potential in many areas of human society and will be an area of scientific investigation for years to come.

**Acknowledgments** Funding for this work was provided by Core funds of the Naval Research Laboratory and the Office of the Secretary of Defense through the Applied Research for the Advancement of S&T Priorities Synthetic Biology for Military-Relevant Environments program. The authors would also like to thank and acknowledge Dr. Claretta Sullivan (Air Force Research Laboratory) for previous efforts in imaging OMV/MV formation in *Lactobacillus acidophilus* which contributed to Fig. 10.1.

# References

Adrio J-L, Demain AL (2010) Recombinant organisms for production of industrial products. BIoeng Bugs 1:116–131

Akbarzadeh A et al (2013) Liposome: classification, preparation, and applications. Nanoscale Res Lett 8:102

Alaniz RC, Deatherage BL, Lara JC, Cookson BT (2007) Membrane vesicles are immunogenic facsimiles of *Salmonella typhimurium* that potently activate dendritic cells, prime B and T cell responses, and stimulate protective immunity in vivo. J Immunol 179:7692–7701

Alavi M, Karimi N, Safaei M (2017) Application of various types of liposomes in drug delivery systems. Adv Pharm Bull 7:3–9

Al-Nedawi K et al (2015) Gut commensal microvesicles reproduce parent bacterial signals to host immune and enteric nervous systems. FASEB J 29:684–695

Alves NJ, Turner KB, Daniele MA, Oh E, Medintz IL, Walper SA (2015a) Bacterial nanobioreactors—directing enzyme packaging into bacterial outer membrane vesicles. ACS Appl Mater Interfaces 7:24963–24972

Alves NJ, Turner KB, Medintz IL, Walper SA (2015b) Emerging therapeutic delivery capabilities and challenges utilizing enzyme/protein packaged bacterial vesicles. Ther Delv 6:873–887

Alves NJ, Turner KB, Medintz IL, Walper SA (2016) Protecting enzymatic function through directed packaging into bacterial outer membrane vesicles. Sci Rep 6:24866

Alves NJ, Turner KB, DiVito KA, Daniele MA, Walper SA (2017) Affinity purification of bacterial outer membrane vesicles (OMVs) utilizing a His-tag mutant. Res Microbiol 168:139–146

Alves NJ, Moore M, Johnson BJ, Dean SN, Turner KB, Medintz IL, Walper SA (2018) Environmental decontamination of a chemical warfare simulant utilizing a membrane vesicle-encapsulated phosphotriesterase. ACS Appl Mater Interfaces 10:15712–15719

Arigita C, Jiskoot W, Westdijk J, van Ingen C, Hennink WE, Crommelin DJA, Kersten GFA (2004) Stability of mono- and trivalent meningococcal outer membrane vesicle vaccines. Vaccine 22:629–642

Asensio CJA et al (2011) Outer membrane vesicles obtained from *Bordetella pertussis* Tohama expressing the lipid a deacylase PagL as a novel acellular vaccine candidate. Vaccine 29:1649–1656

Atlas RM, Hazen TC (2011) Oil biodegradation and bioremediation: a tale of the two worst spills in U.S. history. Environ Sci Technol 45:6709–6715

Avila-Calderón ED et al (2012) Characterization of outer membrane vesicles from *Brucella melitensis* and protection induced in mice. Clin Dev Immunol 2012:352493–352493

Barak I, Muchova K (2013) The role of lipid domains in bacterial cell processes. Int J Mol Sci 14:4050–4065

Bartolini E et al (2013) Recombinant outer membrane vesicles carrying *Chlamydia muridarum* HtrA induce antibodies that neutralize chlamydial infection in vitro. J Extracell Vesicles 2:20181

Beaulac C, Sachetelli S, Lagace J (1998) In-vitro bactericidal efficacy of sub-MIC concentrations of liposome-encapsulated antibiotic against gram-negative and gram-positive bacteria. J Antimicrob Chemother 41:35–41

Behzadi E, Mahmoodzadeh Hosseini H, Imani Fooladi AA (2017) The inhibitory impacts of *Lactobacillus rhamnosus* GG-derived extracellular vesicles on the growth of hepatic cancer cells. Microb Pathog 110:1–6

Berleman JE et al (2014) The lethal cargo of *Myxococcus xanthus* outer membrane vesicles. Front Microbiol 5:474

Bishop AL, Camilli A (2011) *Vibrio cholerae*: lessons for mucosal vaccine design. Expert Rev Vaccines 10:79–94

Bolan GA, Sparling PF, Wasserheit JN (2012) The emerging threat of untreatable gonococcal infection. New Engl J Med 366:485–487

Bottero D et al (2013) Outer membrane vesicles derived from *Bordetella parapertussis* as an acellular vaccine against *Bordetella parapertussis* and *Bordetella pertussis* infection. Vaccine 31:5262–5268

Bron PA, van Baarlen P, Kleerebezem M (2011) Emerging molecular insights into the interaction between probiotics and the host intestinal mucosa. Nat Rev Microbiol 10:66

Brown L, Wolf JM, Prados-Rosales R, Casadevall A (2015) Through the wall: extracellular vesicles in gram-positive bacteria, mycobacteria and fungi. Nat Rev Microbiol 13:620

Bücker F et al (2014) Fuel biodegradation and molecular characterization of microbial biofilms in stored diesel/biodiesel blend B10 and the effect of biocide. Int Biodeterior Biodegradation 95:346–355

Chatterjee D, Chaudhuri K (2011) Association of cholera toxin with *Vibrio cholerae* outer membrane vesicles which are internalized by human intestinal epithelial cells. FEBS Lett 585:1357–1362

Chen DJ, Osterrieder N, Metzger SM, Buckles E, Doody AM, DeLisa MP, Putnam D (2010) Delivery of foreign antigens by engineered outer membrane vesicle vaccines. Proc Natl Acad Sci U S A 107:3099–3104

Chen L et al (2016) Outer membrane vesicles displaying engineered glycotopes elicit protective antibodies. Proc Natl Acad Sci U S A 113:E3609–E3618

Chen Q, Rozovsky S, Chen W (2017) Engineering multi-functional bacterial outer membrane vesicles as modular nanodevices for biosensing and bioimaging. Chem Commun (Camb) 53:7569–7572

Choi SJ et al (2015) Active immunization with extracellular vesicles derived from *Staphylococcus aureus* effectively protects against staphylococcal lung infections, mainly via Th1 cell-mediated immunity. PLoS One 10:e0136021–e0136021. https://doi.org/10.1371/journal.pone.0136021

Claassen I et al (1996) Production, characterization and control of a *Neisseria meningitidis* hexavalent class 1 outer membrane protein containing vesicle vaccine. Vaccine 14:1001–1008. d

Clemens JD et al (1990) Breast feeding and the risk of severe cholera in rural Bangladeshi children. Am J Epidemiol 131:400–411

Coffman RL, Sher A, Seder RA (2010) Vaccine adjuvants: putting innate immunity to work. Immunity 33:492–503

Comanducci M et al (2002) NadA, a novel vaccine candidate of *Neisseria meningitidis*. J Exp Med 195:1445

Commane D, Hughes R, Shortt C, Rowland I (2005) The potential mechanisms involved in the anti-carcinogenic action of probiotics. Mutat Res 591:276–289

Copeland S, Warren HS, Lowry SF, Calvano SE, Remick D (2005) Acute inflammatory response to endotoxin in mice and humans. Clin Diagn Lab Immunol 12:60–67. https://doi.org/10.1128/cdli.12.1.60-67.2005

Daleke-Schermerhorn MH et al (2014) Decoration of outer membrane vesicles with multiple antigens by using an autotransporter approach. Appl Environ Microbiol 80:5854–5865

Dash HR, Mangwani N, Chakraborty J, Kumari S, Das S (2013) Marine bacteria: potential candidates for enhanced bioremediation. Appl Microbiol Biotechnol 97:561–571

de Kleijn E et al (2001) Serum bactericidal activity and isotype distribution of antibodies in toddlers and schoolchildren after vaccination with RIVM hexavalent PorA vesicle vaccine. Vaccine 20:352–358

Dean SN, Leary DH, Sullivan CJ, Oh E, Walper SA (2019) Isolation and characterization of *lactobacillus*-derived membrane vesicles. Sci Rep 9:877. https://doi.org/10.1038/s41598-018-37120-6

Deatherage BL, Lara JC, Bergsbaken T, Rassoulian Barrett SL, Lara S, Cookson BT (2009) Biogenesis of bacterial membrane vesicles. Mol Microbiol 72:1395–1407

Del Campo J et al (2010) Intranasal immunization with a proteoliposome-derived cochleate containing recombinant gD protein confers protective immunity against genital herpes in mice. Vaccine 28:1193–1200

Demain AL, Vaishnav P (2009) Production of recombinant proteins by microbes and higher organisms. Biotechnol Adv 27:297–306
DiNicolantonio JJ, Bhutani J, O'Keefe JH (2015) The health benefits of vitamin K. Open Heart 2: e000300–e000300. https://doi.org/10.1136/openhrt-2015-000300
dos Reis SA, da Conceição LL, Siqueira NP, Rosa DD, da Silva LL, Peluzio MCG (2017) Review of the mechanisms of probiotic actions in the prevention of colorectal cancer. Nutr Res 37:1–19
Edwards JL, Jennings MP, Apicella MA, Seib KL (2016) Is gonococcal disease preventable? The importance of understanding immunity and pathogenesis in vaccine development. Crit Rev Microbiol 42:928–941
Elhenawy W, Bording-Jorgensen M, Valguarnera E, Haurat MF, Wine E, Feldman MF (2016) LPS remodeling triggers formation of outer membrane vesicles in *Salmonella*. MBio 7:e00940–e00916. https://doi.org/10.1128/mBio.00940-16
Ellis TN, Kuehn MJ (2010) Virulence and immunomodulatory roles of bacterial outer membrane vesicles. Microbiol Mol Biol Rev 74:81–94 d
Fantappiè L et al (2014) Antibody-mediated immunity induced by engineered *Escherichia coli* OMVs carrying heterologous antigens in their lumen. J Extracell Vesicles 3:24015. https://doi.org/10.3402/jev.v3.24015
Ferrer-Miralles N, Villaverde A (2013) Bacterial cell factories for recombinant protein prodcution; expanding the catalogue. Microb Cell Factories 12:113
Finne J, Leinonen M, Mäkelä PH (1983) Antigenic similarities between brain components and bacteria causing meningitis: implications for vaccine development and pathogenesis. Lancet 322:355–357
Fischer A, Frnaco A, Oberholzer T (2002) Giant vesicles as microreactors for enzymatic mRNA synthesis. Chembiochem 3:409–417
Fisseha M, Chen P, Brandt B, Kijek T, Moran E, Zollinger W (2005) Characterization of native outer membrane vesicles from mutant strains of *Neisseria meningitidis* for use in parenteral vaccination. Infect Immun 73:4070–4080
Fontes CM, Gilbert HJ (2010) Cellulosomes: highly efficient nanomachines designed to deconstruct plant cell wall omplex carbohydrates. Annu Rev Biochem 79:655–681
Gala RP, Zaman RU, D'Souza MJ, Zughaier SM (2018) Novel whole-cell inactivated *Neisseria gonorrhoeae* microparticles as vaccine formulation in microneedle-based transdermal immunization. Vaccine 6:60
Gerritzen MJH, Martens DE, Wijffels RH, van der Pol L, Stork M (2017) Bioengineering bacterial outer membrane vesicles as vaccine platform. Biotechnol Adv 35:565–574
Geurtsen J et al (2006) Expression of the lipopolysaccharide-modifying enzymes pagP and pagL modulates the endotoxic activity of *Bordetella pertussis*. Infect Immun 74:5574–5585
Giuliani MM et al (2006) A universal vaccine for serogroup B meningococcus. Proc Natl Acad Sci U S A 103:10834
Glass RI, Svennerholm A-M, Stoll BJ, Khan MR, Hossain KMB, Hug MI, Holmgren J (1983) Protection against Cholera in breast-fed children by antibodies in breast milk. New Engl J Med 308:1389–1392
Gnopo YMD, Watkins HC, Stevenson TC, DeLisa MP, Putnam D (2017) Designer outer membrane vesicles as immunomodulatory systems – reprogramming bacteria for vaccine delivery. Adv Drug Del Rev 114:132–142
Gorringe AR, Pajón R (2012) Bexsero. Hum Vaccin Immunother 8:174–183
Grande R et al (2017) Detection and physicochemical characterization of membrane vesicles (MVs) of *Lactobacillus reuteri* DSM 17938. Front Microbiol 8:1040. https://doi.org/10.3389/fmicb.2017.01040
Gujrati V, Kim S, Kin S-H, Min JJ, Choy HE, Kim SC, Jon S (2014) Bioengineered bacterial outer membrane vesicles as cell-specific drug-delivery vehicles for cancer therapy. ACS Nano 8:1525–1537

Hadad R, Jacobsson S, Pizza M, Rappuoli R, Fredlund H, Olcén P, Unemo M (2012) Novel meningococcal 4CMenB vaccine antigens – prevalence and polymorphisms of the encoding genes in Neisseria gonorrhoeae. APMIS 120:750–760

Hanson LA, Hofvander Y, Lindquist B, Zetterstrom R (1985) Breast-feeding as a protection against gastroenteritis and other infections. Acta Paediatr Scand 74:641–642

Heimburg-Molinaro J, Lum M, Vijay G, Jain M, Almogren A, Rittenhouse-Olson K (2011) Cancer vaccines and carbohydrate epitopes. Vaccine 29:8802–8826

Holst J, Martin D, Arnold R, Huergo CC, Oster P, O'Hallahan J, Rosenqvist E (2009) Properties and clinical performance of vaccines containing outer membrane vesicles from *Neisseria meningitidis*. Vaccine 27:B3–B12

Holst J et al (2013) Vaccines against meningococcal serogroup B disease containing outer membrane vesicles (OMV): lessons from past programs and implications for the future. Hum Vaccin Immunother 9:1241–1253. d

Huang W et al (2016) Employing *Escherichia coli*-derived outer membrane vesicles as an antigen delivery platform elicits protective immunity against *Acinetobacter baumannii* infection. Sci Rep 6:37242

Jan AT (2017) Outer membrane vesicles (OMVs) of Gram-negative bacteria: a perspective update. Front Microbiol 8:1053. https://doi.org/10.3389/fmicb.2017.01053

Jiang Y, Kong Q, Roland KL, Curtiss R (2014) Membrane vesicles of *Clostridium perfringens* type A strains induce innate and adaptive immunity. Int J Med Microbiol 304:431–443

Kaparakis-Liaskos M, Ferrero RL (2015) Immune modulation by bacterial outer membrane vesicles. Nat Rev Immunol 15:375

Keenan JI, Allardyce RA, Bagshaw PF (1998) Lack of protection following immunisation with *H. pylori* outer membrane vesicles highlights antigenic differences between *H. felis* and *H. pylori*. FEMS Microbiol Lett 161:21–27

Kesavalu L, Ebersole JL, Machen RL, Holt SC (1992) *Porphyromonas gingivalis* virulence in mice: induction of immunity to bacterial components. Infect Immun 60:1455–1464

Kim J-H et al (2016) Extracellular vesicle–derived protein from *Bifidobacterium longum* alleviates food allergy through mast cell suppression. J Allegy Clin Immunol 137:507–516.e508

Klugman KP, Gotschlich EC, Blake MS (1989) Sequence of the structural gene (rmpM) for the class 4 outer membrane protein of *Neisseria meningitidis*, homology of the protein to gonococcal protein III and *Escherichia coli* OmpA, and construction of meningococcal strains that lack class 4 protein. Infect Immun 57:2066–2071

Knott GJ, Doudna JA (2018) CRISPR-Cas guides the future of genetic engineering. Science 361:866–869

Knox KW, Cullen J, Work E (1967) An extracellular lipopolysaccharide-phospholipid-protein complex produced by *Escherichia coli* grown under lysine-limiting conditions. Biochem J 103:192–200

Koeberling O, Seubert A, Granoff DM (2008) Bactericidal antibody responses elicited by a meningococcal outer membrane vesicle vaccine with overexpressed factor H-binding protein and genetically attenuated endotoxin. J Infect Dis 198:262–270

Kroniger T, Otto A, Becher D (2018) Proteomic analysis of bacterial (outer) membrane vesicles: progress and clinical potential. Expert Rev Proteomics 15:623–626

Kuipers K et al (2015) *Salmonella* outer membrane vesicles displaying high densities of pneumococcal antigen at the surface offer protection against colonization. Vaccine 33:2022–2029

Kulp A, Kuehn MJ (2010) Biological functions and biogenesis of secreted bacterial outer membrane vesicles. Annu Rev Microbiol 64:163–184

Kwon SO, Gho YS, Lee JC, Kim SI (2009) Proteome analysis of outer membrane vesicles from a clinical *Acinetobacter baumannii* isolate. FEMS Microbiol Lett 297:150–156

Lamb RA, Zebedee SL, Richardson CD (1985) Influenza virus M2 protein is an integral membrane protein expressed on the infected-cell surface. Cell 40:627–633

Lee EY et al (2007) Global proteomic profiling of native outer membrane vesicles derived from *Escherichia coli*. Proteomics 7:3143–3153

Lee EY, Choi DS, Kim KP, Gho YS (2008) Proteomics in gram-negative bacterial outer membrane vesicles. Mass Spectrom Rev 27:535–555
Lee E-Y et al (2009) Gram-positive bacteria produce membrane vesicles: proteomics-based characterization of *Staphylococcus aureus*-derived membrane vesicles. Proteomics 9:5425–5436
Lee JC et al (2012) *Klebsiella pneumoniae* secretes outer membrane vesicles that induce the innate immune response. FEMS Microbiol Lett 331:17–24
Leitner DR, Feichter S, Schild-Prüfert K, Rechberger GN, Reidl J, Schild S (2013) Lipopolysaccharide modifications of a cholera vaccine candidate based on outer membrane vesicles reduce endotoxicity and reveal the major protective antigen. Infect Immun 81:2379–2393
Leitner D et al (2015) A combined vaccine approach against *Vibrio cholerae* and ETEC based on outer membrane vesicles. Front Microbiol 6:823. https://doi.org/10.3389/fmicb.2015.00823
Li M, Lee K, Hsu M, Nau G, Mylonakis E, Ramratnam B (2017) *Lactobacillus*-derived extracellular vesicles enhance host immune responses against vancomycin-resistant enterococci. BMC Microbiol 17:66–66
Lin DM, Koskella B, Lin HC (2017) Phage therapy: an alternative to antibiotics in the age of multidrug resistance. World J Gastrointest Pharmacol Ther 8:162–173
Liu Y, Alexeeva S, Defourny KA, Smid EJ, Abee T (2018a) Tiny but mighty: bacterial membrane vesicles in food biotechnological applications. Curr Opin Biotechnol 49:179–184
Liu Y, Defourny KAY, Smid EJ, Abee T (2018b) Gram-positive bacterial extracellular vesicles and their impact on health and disease. Front Microbiol 9:1502–1502 d
Liu Y, Smid EJ, Abee T, Notebaart RA (2019) Delivery of genome editing tools by bacterial extracellular vesicles. Microb Biotechnol 12:71–73
Madico G et al (2006) The meningococcal vaccine candidate GNA1870 binds the complement regulatory protein factor H and enhances serum resistance. J Immunol 177:501–510
Manolova V, Flace A, Bauer M, Schwarz K, Saudan P, Bachmann MF (2008) Nanoparticles target distinct dendritic cell populations according to their size. Eur J Immunol 38:1404–1413. d
Marchand S, De Block J, De Jonghe V, Coorevits A, Heyndrickx M, Herman L (2012) Biofilm formation in milk production and processing environments; influence on milk quality and safety. Compr Rev Food Sci Food Saf 11:133–147
Masignani V et al (2003) Vaccination against *Neisseria meningitidis* using three variants of the lipoprotein GNA1870. J Exp Med 197:789–799
McConnell MJ, Rumbo C, Bou G, Pachón J (2011) Outer membrane vesicles as an acellular vaccine against *Acinetobacter baumannii*. Vaccine 29:5705–5710
Messiaen A-S, Forier K, Nelis H, Braeckmans K, Coenye T (2013) Transport of nanoparticles and tobramycin-loaded liposomes in *Burkholderia cepacia* complex biofilms. PLoS One 8:e79220–e79220. https://doi.org/10.1371/journal.pone.0079220
Mitra S, Chakrabarti MK, Koley H (2013) Multi-serotype outer membrane vesicles of *Shigellae* confer passive protection to the neonatal mice against shigellosis. Vaccine 31:3163–3173
Muralinath M, Kuehn MJ, Roland KL, Curtiss R 3rd (2011) Immunization with *Salmonella enterica* serovar Typhimurium-derived outer membrane vesicles delivering the pneumococcal protein PspA confers protection against challenge with *Streptococcus pneumoniae*. Infect Immun 79:887–894
Newman L et al (2015) Global estimates of the prevalence and incidence of four curable sexually transmitted infections in 2012 based on systematic review and global reporting. PLoS One 10: e0143304. https://doi.org/10.1371/journal.pone.0143304
Nieves W et al (2014) A *Burkholderia pseudomallei* outer membrane vesicle vaccine provides protection against lethal sepsis. Clin Vaccine Immunol 21:747–754
Oelschlaeger TA (2010) Mechanisms of probiotic actions – a review. Int J Med Microbiol 300:57–62
Olaya-Abril A et al (2014) Characterization of protective extracellular membrane-derived vesicles produced by *Streptococcus pneumoniae*. J Proteome 106:46–60
Osterholm MT, Kelley NS, Sommer A, Belongia EA (2012) Efficacy and effectiveness of influenza vaccines: a systematic review and meta-analysis. Lancet Infect Dis 12:36–44

Oyston PCF, Sjöstedt A, Titball RW (2004) Tularaemia: bioterrorism defence renews interest in *Francisella tularensis*. Nat Rev Microbiol 2:967

Paolillo R, Romano Carratelli C, Sorrentino S, Mazzola N, Rizzo A (2009) Immunomodulatory effects of *lactobacillus plantarum* on human colon cancer cells. Int Immunopharmacol 9:1265–1271

Park K (2018) Inherent antimicrobial activity by bacteria-derived vesicles. J Control Release 290:180

Park M, Sun Q, Liu F, DeLisa MP, Chen W (2014) Positional assembly of enzymes on bacterial outer membrane vesicles for cascade reactions. PLoS One 9:e97103. https://doi.org/10.1371/journal.pone.0097103

Pérez O et al (2009) Mucosal approaches in *Neisseria* vaccinology. Vaccimonitor 18:55–57

Petousis-Harris H, Paynter J, Morgan J, Saxton P, McArdle B, Goodyear-Smith F, Black S (2017) Effectiveness of a group B outer membrane vesicle meningococcal vaccine against gonorrhoea in New Zealand: a retrospective case-control study. Lancet 390:1603–1610

Pierson T, Matrakas D, Taylor YU, Manyam G, Morozov VN, Zhou W, van Hoek ML (2011) Proteomic characterization and functional analysis of outer membrane vesicles of *Francisella novicida* suggests possible role in virulence and use as a vaccine. J Proteome Res 10:954–967

Pizza M et al (2000) Identification of vaccine candidates against Serogroup B Meningococcus by whole-genome sequencing. Science 287:1816

Prados-Rosales R et al (2014) Mycobacterial membrane vesicles administered systemically in mice induce a protective immune response to surface compartments of *Mycobacterium tuberculosis*. MBio 5:e01921

Price NL, Goyette-Desjardins G, Nothaft H, Valguarnera E, Szymanski CM, Segura M, Feldman MF (2016) Glycoengineered outer membrane vesicles: a novel platform for bacterial vaccines. Sci Rep 6:24931. https://doi.org/10.1038/srep24931

Pritsch M et al (2016) Comparison of intranasal outer membrane vesicles with cholera toxin and injected MF59C.1 as adjuvants for malaria transmission blocking antigens AnAPN1 and Pfs48/45. J Immunol Res 2016:11

Pupo E, Hamstra H-J, Meiring H, van der Ley P (2014) Lipopolysaccharide engineering in *Neisseria meningitidis*: Structural analysis of different pentaacyl lipid A mutants and comparison of their modified agonist properties. J Biol Chem 289:8668–8680

Qureshi K et al (2006) Breast milk reduces the risk of illness in children of mothers with cholera: observations from an epidemic of cholera in Guinea-Bissau. Pediatr Infect Dis J 25:1163–1166

Raetz CRH, Whitfield C (2002) Lipopolysaccharide endotoxins. Annu Rev Biochem 71:635–700

Raposo G, Stoorvogel W (2013) Extracellular vesicles: exosomes, microvesicles, and friends. J Cell Biol 200:373–383

Rappazzo CG et al (2016) Recombinant M2e outer membrane vesicle vaccines protect against lethal influenza A challenge in BALB/c mice. Vaccine 34:1252–1258

Rappuoli R (2000) Reverse vaccinology. Curr Opin Microbiol 3:445–450

Rivera J, Cordero RJB, Nakouzi AS, Frases S, Nicola A, Casadevall A (2010) *Bacillus anthracis* produces membrane-derived vesicles containing biologically active toxins. Proc Natl Acad Sci U S A 107:19002–19007

Roberts R et al (2008) Outer membrane vesicles as acellular vaccine against pertussis. Vaccine 26:4639–4646

Roier S et al (2016) A novel mechanism for the biogenesis of outer membrane vesicles in Gram-negative bacteria. Nat Commun 7:10515. https://doi.org/10.1038/ncomms10515

Rosenstein NE, Perkins BA, Stephens DS, Popovic T, Hughes JM (2001) Meningococcal disease. N Engl J Med 344:1378–1388. d

Salminen MK, Rautelin H, Tynkkynen S, Poussa T, Saxelin M, Valtonen V, Järvinen A (2004) *Lactobacillus* bacteremia, clinical significance, and patient outcome, with special focus on probiotic L. Rhamnosus GG. Clin Infect Dis 38:62–69

Salverda MLM et al (2016) Surface display of a borrelial lipoprotein on meningococcal outer membrane vesicles. Vaccine 34:1025–1033

Sanchez-Garcia L, Martin L, Mangues R, Ferrer-Miralles N, Vazquez E, Villaverde A (2016) Recombinant pharmaceuticals from microbial cells: a 2015 update. Microb Cell Factories 15:33. https://doi.org/10.1186/s12934-016-0437-3

Sato T, Yamada Y, Ohtani Y, Mitsui N, Murasawa H, Araki S (2001) Efficient production of menaquinone (vitamin K2) by a menadione-resistant mutant of *Bacillus subtilis*. J Ind Microbiol Biotechnol 26:115–120

Schild S, Nelson EJ, Camilli A (2008) Immunization with *Vibrio cholerae* outer membrane vesicles induces protective immunity in mice. Infect Immun 76:4554–4563

Schild S, Nelson EJ, Bishop AL, Camilli A (2009) Characterization of *Vibrio cholerae* outer membrane vesicles as a candidate vaccine for cholera. Infect Immun 77:472–484

Schroeder J, Aebischer T (2009) Recombinant outer membrane vesicles to augment antigen-specific live vaccine responses. Vaccine 27:6748–6754

Schultz MP, Bendick JA, Holm ER, Hertel WM (2011) Economic impact of biofouling on a naval surface ship. Biofouling 27:87–98

Schulz E et al (2018) Biocompatible bacteria-derived vesicles show inherent antimicrobial activity. J Control Release 290:46–55

Schwechheimer C, Sullivan CJ, Kuehn MJ (2013) Envelope control of outer membrane vesicle production in gram-negative bacteria. Biochemistry 52:3031–3040

Scoma A, Yakimov MM, Boon N (2016) Challenging oil bioremediation at deep-sea hydrostatic pressure. Front Microbiol 7:1203

Semchenko EA, Day CJ, Seib KL (2017) MetQ of *Neisseria gonorrhoeae* is a surface-expressed antigen that elicits bactericidal and functional blocking antibodies. Infect Immun 85:e00898–e00816

Semchenko EA, Tan A, Borrow R, Seib KL (2018) The serogroup B meningococcal vaccine Bexsero elicits antibodies to *Neisseria gonorrhoeae*. Clin Infect Dis 69:1101–1111. https://doi.org/10.1093/cid/ciy1061

Seo MK, Park EJ, Ko SY, Choi EW, Kim S (2018) Therapeutic effects of kefir grain lactobacillus-derived extracellular vesicles in mice with 2,4,6-trinitrobenzene sulfonic acid-induced inflammatory bowel disease. J Dairy Sci 101:8662–8671

Serruto D et al (2010) *Neisseria meningitidis* GNA2132, a heparin-binding protein that induces protective immunity in humans. Proc Natl Acad Sci 107:3770–3775. https://doi.org/10.1073/pnas.0915162107

Shen Y, Giardino Torchia ML, Lawson GW, Karp CL, Ashwell JD, Mazmanian SK (2012) Outer membrane vesicles of a human commensal mediate immune regulation and disease protection. Cell Host Microbe 12:509–520

Steeghs L et al (2008) Differential activation of human and mouse toll-like receptor 4 by the adjuvant candidate LpxL1 of *Neisseria meningitidis*. Infect Immun 76:3801–3807. https://doi.org/10.1128/iai.00005-08

Tan K, Li R, Huang X, Liu Q (2018) Outer membrane vesicles: current status and future direction of these novel vaccine adjuvants. Front Microbiol 9:783–783

Tingyue Gu DX (2015) The war against problematic biofilms in the oil and gas industry. J Microb Biochem Technol 07:1000e124. https://doi.org/10.4172/1948-5948.1000e124

Tinsley CR, Nassif X (1996) Analysis of genetic difference between *Neisseria meningitidis* and *Neisseria gonorrhoeae*: two closely related bacteria expressing two different pathogenicities. Proc Natl Acad Sci U S A 93:11109–11114

Tokuda H, Matsuyama S (2004) Sorting of lipoproteins to the outer membrane in *E. coli*. Biochim Biophys Acta 1693:5–13. https://doi.org/10.1016/j.bbamcr.2004.02.005

Treanor JJ (2015) Prospects for broadly protective influenza vaccines. Vaccine 33:D39–D45

Tricarico C, Clancy J, D'Souza-Schorey C (2017) Biology and biogenesis of shed microvesicles. Small GTPases 8:220–232

Tzipilevich E, Habusha M, Ben-Yehuda S (2017) Acquisition of phage sensitivity by bacteria through exchange of phage receptors. Cell 168:186–199.e112

Urbanek AK, Rymowicz W, Mironczuk AM (2018) Degradation of plastics and plastic-degrading bacteria in cold marine habitats. Appl Microbiol Biotechnol 102:7669–7678

Valentine Jenny L et al (2016) Immunization with outer membrane vesicles displaying designer glycotopes yields class-switched, glycan-specific antibodies. Cell Chem Biol 23:655–665

van de Waterbeemd B et al (2010) Improved OMV vaccine against *Neisseria meningitidis* using genetically engineered strains and a detergent-free purification process. Vaccine 28:4810–4816

van de Waterbeemd B, Mommen GPM, Pennings JLA, Eppink MH, Wijffels RH, van der Pol LA, de Jong APJM (2013) Quantitative proteomics reveals distinct differences in the protein content of outer membrane vesicle vaccines. J Proteome Res 12:1898–1908

van den Dobbelsteen GPJM, van Dijken HH, Pillai S, van Alphen L (2007) Immunogenicity of a combination vaccine containing pneumococcal conjugates and meningococcal PorA OMVs. Vaccine 25:2491–2496

van der Ley P, van der Biezen J, Poolman JT (1995) Construction of *Neisseria meningitidis* strains carrying multiple chromosomal copies of the porA gene for use in the production of a multivalent outer membrane vesicle vaccine. Vaccine 13:401–407. d

van der Ley P, Steeghs L, Hamstra HJ, ten Hove J, Zomer B, van Alphen L (2001) Modification of lipid a biosynthesis in *Neisseria meningitidis lpxL* mutants: influence on lipopolysaccharide structure, toxicity, and adjuvant activity. Infect Immun 69:5981–5990. https://doi.org/10.1128/iai.69.10.5981-5990.2001

van der Pol L, Stork M, van der Ley P (2015) Outer membrane vesicles as platform vaccine technology. Biotechnol J 10:1689–1706

van Dongen SFM, Nallani M, Cornelissen JJLM, Nolte RJM, van Hest JCM (2009) A three-enzyme cascade reaction through positional assembly of enzymes in a Polymersome Nanoreactor. Chem Eur J 15:1107–1114

Vriezema DM et al (2007) Positional assembly of enzymes in polymersome nanoreactors for cascade reactions. Angew Chem 119:7522–7526

Watkins HC et al (2017) Safe recombinant outer membrane vesicles that display M2e elicit heterologous influenza protection. Mol Ther 25:989–1002

Whelan J, Kløvstad H, Haugen IL, Holle MR-DRB, Storsaeter J (2016) Ecologic study of meningococcal B vaccine and *Neisseria gonorrhoeae* infection, Norway. Emerg Infect Dis 22:1137–1139. https://doi.org/10.3201/eid2206.151093

Xiang SD, Scholzen A, Minigo G, David C, Apostolopoulos V, Mottram PL, Plebanski M (2006) Pathogen recognition and development of particulate vaccines: does size matter? Methods 40:1–9

Xing H, Hwang K, Lu Y (2016) Recent developments of liposomes as nanocarriers for theranostic applications. Theranostics 6:1336–1352

Yoo J-W, Irvine DJ, Discher DE, Mitragotri S (2011) Bio-inspired, bioengineered and biomimetic drug delivery carriers. Nat Rev Drug Discov 10:521

Yun SH et al (2017) Proteomic characterization of the outer membrane vesicle of the halophilic marine bacterium *Novosphingobium pentaromativorans* US6-1. J Microbiol 55:56–62

Zakeri B, Fierer JO, Celik E, Chittock EC, Schwarz-Linek U, Moy VT, Howarth M (2012) Peptide tag forming a rapid covalent bond to a protein, through engineering a bacterial adhesin. Proc Natl Acad Sci U S A 109:E690–E697

Zariri A et al (2016a) Meningococcal outer membrane vesicle composition-dependent activation of the innate immune response. Infect Immun 84:3024–3033

Zariri A, Pupo E, van Riet E, van Putten JPM, van der Ley P (2016b) Modulating endotoxin activity by combinatorial bioengineering of meningococcal lipopolysaccharide. Sci Rep 6:36575. https://doi.org/10.1038/srep36575

Zepp F (2010) Principles of vaccine design—lessons from nature. Vaccine 28:C14–C24

Zomer A, Vendrig T, Hopmans ES, van Eijndhoven M, Middeldorp JM, Pegtel DM (2010) Exosomes: Fit to deliver small RNA. Commun Integr Biol 3:447–450

The manufacturer's authorised representative in the EU is Springer Nature Customer Service Centre GmbH, Europaplatz 3, 69115 Heidelberg, Germany. If you have any concerns regarding our products, please contact ProductSafety@springernature.com

Printed and bound by CPI Group (UK) Ltd, Croydon, CR0 4YY
30/09/2026
02225507-0009